Y0-AAF-229

PROCEEDINGS OF THE INTERNATIONAL CONFERENCE ON STOCHASTIC ANALYSIS AND APPLICATIONS

Proceedings of the International Conference on Stochastic Analysis and Applications

Hammamet, 2001

Edited by

Sergio Albeverio
University of Bonn,
Bonn, Germany

Anne Boutet de Monvel
University Paris 7,
Paris, France

and

Habib Ouerdiane
Tunis University,
Tunis, Tunisia

KLUWER ACADEMIC PUBLISHERS
DORDRECHT / BOSTON / LONDON

A C.I.P. Catalogue record for this book is available from the Library of Congress.

ISBN 1-4020-2467-3 (HB)
ISBN 1-4020-2468-1 (e-book)

Published by Kluwer Academic Publishers,
P.O. Box 17, 3300 AA Dordrecht, The Netherlands.

Sold and distributed in North, Central and South America
by Kluwer Academic Publishers,
101 Philip Drive, Norwell, MA 02061, U.S.A.

In all other countries, sold and distributed
by Kluwer Academic Publishers,
P.O. Box 322, 3300 AH Dordrecht, The Netherlands.

Printed on acid-free paper

Printed in the Netherlands.

This book is dedicated to Paul Krée on the occasion of his 65th birthday.

Contents

Preface

This book, dedicated to Professor Paul Krée on the occasion of his 65th birthday, contains 18 papers in the areas of stochastic and infinite dimensional analysis and their applications.

Some articles are surveys, others contain new results, methods or applications.

The International Conference on Stochastic Analysis and Applications was hed October 22-27, 2001 in the hotel Abou Nawas, Hammamet, Tunisia.

The conference was sponsored by grant from Tunisian Ministry for Higher Education, Research and Technology, the University of Tunis El Manar, the Tunisian Mathematical Society and other Tunisian Research Institutes.

This effort included 50 one-hour lectures.

The Opening ceremony included welcoming remarks by Professor Youssef Alouane, President of the Tunis El Manar University. They were followed by some introductions words from Professor Sergio Albeverio from the University of Bonn in Germany.

We are grateful to Professor Jean-Jacques Sansuc for his great help for the editorial procedure.

A large number of colleagues have anonymously and graciously contributed to this volume as referees, special thanks are due to them.

Finally we hope that this conference has contributed not only to the promotion scientific activities but also to mutual international understanding.

Sergio Albeverio, Anne Boutet de Monvel, Habib Ouerdiane

MATHEMATICAL ASPECTS OF DECOHERENCE INDUCED CLASSICALITY IN QUANTUM SYSTEMS

Philippe Blanchard

Physics Faculty and BiBoS, University of Bielefeld, 33615 Bielefeld, Germany

blanchard@physik.uni-bielefeld.de

Robert Olkiewicz

Physics Faculty and BiBoS, University of Bielefeld, 33615 Bielefeld, Germany, Institute of Theoretical Physics, University of Wrocław, 50-204 Wrocław, Poland

rolek@ift.uni.wroc.pl

Abstract Framework for a general discussion of environmentally induced classical properties, like superselection rules, privileged basis and classical behavior, in quantum systems with both finite and infinite number of degrees of freedom is proposed.

Keywords: Decoherence

1. Introduction

The origin of deterministic laws that govern the classical domain of our everyday experience has attracted much attention in recent years. For example, the question in which asymptotic regime non-relativistic quantum mechanics reduces to its ancestor, i.e. Hamiltonian mechanics, was addressed in [12, 13]. It was shown there that for very many bosons with weak two-body interactions there is a class of states for which time evolution of expectation values of certain operators in these states is approximately described by a non-linear Hartree equation. The problem under what circumstances such an equation reduces to the Newtonian mechanics of point particles was also discussed. A different point of view was taken in a seminal paper by Gell-Mann and Hartle [14]. They gave a thorough analysis of the role of decoherence in the derivation of phenomenological classical equations of motion. Various forms of

S. Albeverio et al. (eds.),
Proceedings of the International Conference on Stochastic Analysis and Applications, 1–15.

decoherence (weak, strong) and realistic mechanisms for the emergence of various degrees of classicality were also presented. Since quantum interferences are damped in the presence of an environment, so one may hope that the classical $\hbar \to 0$ limit for quantum dissipative dynamics may exists for arbitrary large time. Such a problem was discussed in [16]. In this work we adopt a different point of view and follow the idea of environmentally induced decoherence whose potential impact on behavior of quantum open systems was briefly described by Zeh: 'All quasi-classical phenomena, even those representing reversible mechanics, are based on de facto irreversible decoherence' [24]. The main objective of the present paper is to provide an algebraic framework which will enable a general discussion of the environmentally induced decoherence and, as a consequence, the appearance of classical properties in quantum systems with both finite and infinite number of particles. A number of examples showing that classical concepts do not have to be presumed as an independent fundamental ingredient are also discussed. This paper is a shortened version of a more comprehensive work [4], where proofs of the results presented here will appear.

2. Mathematical description of quantum and classical systems

Suppose $\mathcal{H}_S$ is the Hilbert space associated with a quantum system and let $\mathcal{M}$ be a von Neumann algebra representing bounded observables of the system. Since we intend to generate classical properties in the system we assume that the center of $\mathcal{M}$ is trivial, i.e. that $\mathcal{M}$ is a factor. Physical observables are Hermitian operators from the algebra $\mathcal{M}$, or, more generally, self-adjoint operators affiliated to $\mathcal{M}$. Generalizing the notion of a density matrix representing mixture of states we say that statistical states of the system are represented by positive normal and normalized functionals on $\mathcal{M}$. The set of statistical states we denote by D. Hence $\phi \in D$ iff $\phi(A) \geq 0$ whenever $A \geq 0$, $\phi(\mathbf{1}) = 1$, where $\mathbf{1}$ is the identity operator, and ϕ is continuous in the σ-weak topology on $\mathcal{M}$ (see, for example, [7] for definition of these terms). The linear space generated by D is called the predual space of $\mathcal{M}$ and denoted by $\mathcal{M}_*$.

Let us now consider the dynamics of a quantum system. If a system is closed (conservative), then the time development of any observable is given by a continuous symmetry transformation, i.e. $A \to A(t) = \alpha_t(A)$, where α_t is a σ-weakly continuous one parameter group of *-automorphisms of $\mathcal{M}$. If there exists an energy observable H for the system, then automorphisms α_t are inner, given by $\alpha_t(A) = \mathrm{e}^{\frac{\mathrm{i}}{\hbar}tH} A \mathrm{e}^{-\frac{\mathrm{i}}{\hbar}tH}$. However, if a system interacts with an environment, then its evolution becomes irreversible. In fact, although the whole system evolves unitarily according to the total Hamiltonian $H = H_S + H_E + H_I$, where the three parts represent respectively the system, environment and interaction Hamiltonians, the evolution of a system observable A is given by

$$T_t(A) = P_E(\mathrm{e}^{\frac{\mathrm{i}}{\hbar}tH}(A \otimes \mathbf{1}_E)\mathrm{e}^{-\frac{\mathrm{i}}{\hbar}tH}), \tag{2.1}$$

where $\mathbf{1}_E$ is the identity operator in the environment and P_E denotes the conditional expectation onto the algebra $\mathcal{M}$ with respect to a reference state ϕ_E of the environment. Equivalently, we may define T_t as the adjoint map to the operator $T_{t*}\colon \mathcal{M}_* \to \mathcal{M}_*$ given by

$$T_{t*}(\phi) = \mathrm{Tr}_E(\mathrm{e}^{-\frac{\mathrm{i}}{\hbar}tH}(\phi \otimes \phi_E)\mathrm{e}^{\frac{\mathrm{i}}{\hbar}tH}), \tag{2.2}$$

where Tr_E denotes the partial trace with respect to the environmental variables. T_t being the composition of *-automorphisms and a conditional expectation is a family of maps (superoperators) which in general satisfies a complicated integro-differential equation describing an irreversible dynamics. However, in the Markovian approximation the dynamics of the quantum system is represented by a dynamical semigroups $T_t\colon \mathcal{M} \to \mathcal{M}$ such that:

a) For any observable $A \in \mathcal{M}$ the function $t \to T_t(A)$ is σ-weakly continuous.

b) For all $t \geq 0$ the superoperators T_t are completely positive, normal and unital. Moreover, T_t are contractive in the operator norm, i.e. $\|T_t A\|_\infty \leq \|A\|_\infty$.

c) There is a faithful, normal and semifinite weight ω_0 on $\mathcal{M}$ such that $\omega_0 \circ T_t = \omega_0$ for all $t \geq 0$.

Having described the framework for quantum systems we now turn to the Hilbert space description of classical dynamical systems. Everybody agrees that concepts of classical and quantum physics are opposite in many aspects. Therefore, in order to demonstrate how quanta become classical, it is necessary to express them in one mathematical framework. Since a natural language for quantum systems is that of von Neumann algebras we reformulate now the concept of classical dynamical systems in a similar way.

Suppose that M is a configuration space of a classical system. We assume that M is a locally compact metric space. A continuous evolution of the system is given by a (continuous) flow on M, i.e. a continuous mapping $g\colon \mathbb{R} \times M \to M$ such that $g_t\colon M \to M$ is a homeomorphism for all $t \in \mathbb{R}$, and $t \to g_t$ is a group homomorphism. The map $t \to g_t(x)$ is called a trajectory of a point $x \in M$. From the very definition, all trajectories are continuous. We assume also that there exists a σ-finite Borel measure μ_0 on M, finite on compact sets, and such that $\mu_0(g_t^{-1}(B)) = \mu_0(B)$ for all $t \in \mathbb{R}$ and all μ_0-finite Borel subsets $B \subset M$. In addition, we assume that $\int f d\mu_0 > 0$ whenever $f \geq 0$ and $f \neq 0$. The triple (M, g_t, μ_0) is called a (classical) topological dynamical system. The following result is clear.

Proposition 2.1. *Suppose that g_t is a flow on M. Then $\gamma_t : C_0(M) \to C_0(M)$, $\gamma_t(f)(x) = f(g_t x)$ is a strongly continuous one parameter group of *-automorphisms of $C_0(M)$, where $C_0(M)$ is the C^*-algebra of continuous functions on M vanishing at infinity.*

It follows that a dynamical system may be equivalently described by the triple $(C_0(M), \gamma_t, \phi_0)$, where ϕ_0 is a γ_t-invariant weight on $C_0(M)$ determined by the measure μ_0. If M is compact and therefore μ_0 is finite, we always assume that μ_0 is a probability measure, what implies that ϕ_0 is a state on $C(M)$. So far we have made a half-way. What we really need is a Hilbert space representation of the system. Suppose $\mathcal{H} = L^2(M, \mu_0)$. There is a natural representation of $C_0(M)$ in $\mathcal{H}$ given by $\pi(f)\psi(x) = f(x)\psi(x)$. Let us define $\mathcal{A} = \pi(C_0(M))''$. Then $\mathcal{A}$ is the von Neumann algebra $L^\infty(M, \mu_0)$ of essentially bounded functions on M, acting in the Hilbert space $\mathcal{H}$. Moreover, γ_t extends uniquely to a σ-weakly continuous group of *-automorphisms of $\mathcal{A}$, and μ_0 determines a γ_t-invariant, faithful, normal and semifinite weight ϕ_0 on $\mathcal{A}$. We call the triple $(\mathcal{A}, \gamma_t, \phi_0)$ a Hilbert space representation of the dynamical system (M, g_t, μ_0).

3. Decoherence in action

In recent years decoherence has been widely discussed and accepted as the mechanism responsible for the appearance of classicality in quantum measurements and the absence, in the real world, of Schrödinger-cat-like states [3, 15, 17, 23, 25, 26]. The basic idea behind it is that classicality is an emergent property induced in quantum systems by unavoidable and practically irreversible interaction with their environment. It is marked by the dynamical suppression of quantum interferences and so the transition of the vast majority of pure states of the system to statistical mixtures. A loss of phase coherence as the consequence of the coupling with an environment has been established both in the Markovian regime [21, 22] and for a system with a non-Markovian evolution [9].

In spite of the progress in the theoretical and experimental understanding of decoherence, the models studied so far do not answer the question concerning its nature satisfactorily. Dynamical diagonalization of pure states with respect to a preferred basis explains essentially the measurements results but it is only an example of possible scenarios. Other possibilities include: Environmentally induced superselection rules of discrete and continuous types, and completely classical behavior of the quantum system. Let us now discuss this issue in a detailed way.

Definition 3.1. Environmentally induced decoherence is said to take place in the system, if there exists at least one projection $P \in \mathcal{M}$ such that $T_t(P)$ is not a projection for some instant $t > 0$.

The above definition excludes only automorphic evolutions. For the discussion on emergence of classical properties we find it more useful to strengthen it in the following way.

Definition 3.2. We say that environmentally induced decoherence takes place in the system, if there are two Banach *-invariant subspaces $\mathcal{M}_1$ and $\mathcal{M}_2$ in $\mathcal{M}$ such that:

(i) $\mathcal{M} = \mathcal{M}_1 \oplus \mathcal{M}_2$ with $\mathcal{M}_2 \neq 0$. Moreover, both $\mathcal{M}_1$ and $\mathcal{M}_2$ are T_t-invariant.

(ii) $\mathcal{M}_1$ represents a decoherence free part of the system. It is a von Neumann algebra (the image of a conditional expectation of $\mathcal{M}$) generated by all projections P in $\mathcal{M}$ such that $T_t(P)$ remains a projection for all $t > 0$. We additionally assume that for any projection $P \in \mathcal{M}_1$ and any $t > 0$ there exists a projection $Q \in \mathcal{M}_1$ such that $T_t(Q) = P$.

(iii) $\mathcal{M}_2$ represents those observables of the system which, after some time, are not detectable by measurements, i.e. all their expectation values vanish with time. More precisely

$$\lim_{t \to \infty} \phi(T_t B) = 0 \tag{3.1}$$

for all $\phi \in D$ and any $B = B^* \in \mathcal{M}_2$.

If the process of decoherence is efficient, and usually it is, then Hermitian operators from $\mathcal{M}_1$ are those which can be detected in practice. Hence, we call $\mathcal{M}_1$ the algebra of effective observables. Let us observe that the evolution restricted to this subalgebra has a nice automorphic property.

Theorem 3.3. *For any $t \geq 0$, $T_t|_{\mathcal{M}_1}$ is a *-automorphism.*

Definition 3.4. If $\mathcal{M}_1$ is noncommutative with $Z(\mathcal{M}_1) \neq \mathbb{C} \cdot \mathbf{1}$, where $Z(\mathcal{M}_1)$ denotes the center of $\mathcal{M}_1$, then we speak of environmentally induced superselection rules in the system.

In such a case the system dynamically loses its pure quantum character and behaves like a conservative one, however, with a non-trivial superselection operator.

Definition 3.5. We say that environment induces a classical structure in the system, if $\mathcal{M}_1$ is a commutative algebra greater than $\mathbb{C} \cdot \mathbf{1}$. If $\mathcal{M}_1 = \mathbb{C} \cdot \mathbf{1}$, then we say that the system is ergodic.

Definition 3.6. Classical structure is said to represent a classical dynamical system, if $(\mathcal{M}_1, T_t, \omega_0)$ is isomorphic with $(L^\infty(M), \hat{T}_t, \mu_0)$, where M is a locally compact space, $\hat{T}_t$ is a one parameter group of *-automorphisms on $L^\infty(M)$ induced by a continuous flow g_t on M, and μ_0 is a $\hat{T}_t$-invariant σ-finite Borel measure on M.

4. Examples

We start with the following general theorem.

Theorem 4.1. *Suppose $\mathcal{M}$ is a type I factor, i.e. $\mathcal{M} = B(\mathcal{H}_S)$, where $\mathcal{H}_S$ is a separable (finite or infinite dimensional) Hilbert space. Let the evolution of the system be given by a family of maps $\{T_t\}_{t\geq 0}$ which fulfils the conditions* a)–c) *from Section 2 with $\omega_0 = \mathrm{Tr}$, the standard trace. If $\{T_t\}$ satisfies the semigroup property $T_t \circ T_s = T_{t+s}$, and if there exists a faithful density matrix ρ_0 subinvariant with respect to T_t, i.e. $\mathrm{Tr}\,\rho_0 T_t(A) \leq \mathrm{Tr}\,\rho_0 A$ for all $A \geq 0$, then the decomposition $\mathcal{M} = \mathcal{M}_1 \oplus \mathcal{M}_2$ from Definition 3.2 always exists. Moreover, the effective part of any observable $A \in \mathcal{M}$ is given by a* Tr*-compatible conditional expectation from $\mathcal{M}$ onto $\mathcal{M}_1$, the automorphic evolution of the algebra $\mathcal{M}_1$ is a Hamiltonian one, and the limit in equation (3.1) is uniform on bounded sets of $\mathcal{M}_2$.*

Example 4.2 (The Araki-Zurek model). We follow a mathematical description of the model given by Kupsch [18]. Suppose the total Hamiltonian

$$H = H_S \otimes \mathbf{1}_E + \mathbf{1}_S \otimes H_E + A \otimes B, \tag{4.1}$$

defined on a Hilbert space $\mathcal{H}_S \otimes \mathcal{H}_E$, where $\mathcal{H}_E = L^2(\mathbb{R}, \mathrm{d}a)$ satisfies the following assumptions:

- $[H_E, B] = 0$,
- $B = \hat{p}$, the momentum operator on $\mathcal{H}_E$,

- $A = \sum_{n=1}^{\infty} \lambda_n P_n$, $\lambda_n \in \mathbb{R}$, $\lambda_n \neq \lambda_m$ if $n \neq m$, and P_n are mutually orthogonal one-dimensional projections summing up to the identity operator,
- $[H_S, A] = 0$, i.e. $H_S = \sum_n \gamma_n P_n$, $\gamma_n \in \mathbb{R}$,
- $\omega_E = |\psi_E >< \psi_E|$, where $\psi_E(a) = \frac{1}{\sqrt{2\pi}} \int \frac{\mathrm{e}^{\mathrm{i}ap}}{\sqrt{\pi(1+p^2)}} \mathrm{d}p$.

Then the evolution of an observable $X \in B(\mathcal{H}_S)$ is given by

$$T_t(X) = \sum_{n,m=1}^{\infty} \chi_{n,m}(t) \mathrm{e}^{\mathrm{i}t(\gamma_n - \gamma_m)} P_n X P_m, \tag{4.2}$$

where

$$\chi_{n,m}(t) = \frac{1}{\pi} \int \mathrm{e}^{\mathrm{i}t(\lambda_n - \lambda_m)p} \frac{\mathrm{d}p}{1+p^2} = \mathrm{e}^{-|\lambda_n - \lambda_m| t}.$$

Because now $\chi_{n,m}(t)\chi_{n,m}(s) = \chi_{n,m}(t+s)$ so T_t is a quantum Markov semigroup. It is also clear that it preserves Tr. Let $\{b_n\}$ be a sequence of positive numbers which sum up to 1. Then $\rho_0 = \sum b_n P_n$ is a faithful T_t-invariant density matrix. Hence all assumptions of Theorem 4.1 are satisfied and we may conclude that the decomposition from Definition 3.2 holds true with

$$\mathcal{M}_1 = \sum_{n=1}^{\infty} P_n B(\mathcal{H}_S) P_n \equiv l^{\infty}(\mathbb{N}).$$

The evolution T_t restricted to $\mathcal{M}_1$ is trivial. Let us point out that due to Theorem 4.1, all expectation values of observables belonging to $\mathcal{M}_2$ tend to zero, and the limit is uniform on bounded sets of $\mathcal{M}_2$. It is worth noting that this result has been obtained without assuming that there is a minimal gap between distinct eigenvalues λ_n of the operator A.

Example 4.3 (Quantum stochastic process). Quantum stochastic processes were introduced by Davies [10] to describe rigorously certain continuous measurement processes. They can be constructed from two infinitesimal generators. The first is the generator Z of a strongly continuous semigroup on a Hilbert space $\mathcal{H}_S$, and the second is a stochastic kernel J, describing how the measuring apparatus interacts with the system. Let us recall that a stochastic kernel is a measure defined on the σ-algebra of Borel sets in some locally compact space and with values in the space of bounded positive linear operators on tr, the Banach space of trace class operators on $\mathcal{H}_S$. In this example we take the Poincaré disc $D_1 = \{\zeta \in \mathbb{C} : |\zeta| < 1\}$ as the underlying topological space, and define $Z = iH_S - \frac{\kappa}{2}\mathbf{1}_S$ where H_S is the Hamiltonian of the system, $\kappa > 0$ is the coupling constant. For $E \subset D_1$ and $\rho \in$ tr the stochastic kernel is defined by

$$\mathrm{Tr}[J(E, \rho)A] = \kappa \int_E \mathrm{d}\mu(\zeta)\,\mathrm{Tr}(e_\zeta \rho e_\zeta A),$$

where $A \in B(\mathcal{H}_S)$, $e_\zeta = |\zeta><\zeta|$ with $|\zeta>$ being a SU(1,1) coherent state, i.e. a holomorphic function on D_1 [20]

$$|\zeta>(z) = (1 - |\zeta|^2)(1 - z\zeta)^{-2}$$

and

$$d\mu(\zeta) = \frac{1}{\pi}\frac{d\zeta d\bar{\zeta}}{(1 - |\zeta|^2)^2}$$

is a SU(1,1) invariant measure on D_1. It should be pointed out that $|\zeta>$ are coherent states in Hilbert space $\mathcal{H}_S = L^2(D_1, \frac{dzd\bar{z}}{\pi})$, which is the space of a unitary irreducible representation π of the group SU(1,1) given by

$$\pi(g)f(z) = (\beta z + \bar{\alpha})f(\frac{\alpha z + \bar{\beta}}{\beta z + \bar{\alpha}}),$$

where

$$g = \begin{pmatrix} \alpha & \beta \\ \bar{\beta} & \bar{\alpha} \end{pmatrix}$$

with $|\alpha|^2 - |\beta|^2 = 1$. In order to define a quantum stochastic process Z and J have to satisfy the following relation

$$\mathrm{tr}[J(D_1, e_\psi)] = -2\,\mathrm{Re}\,\langle\psi, Z\psi\rangle$$

$e_\psi = |\psi\rangle\,\langle\psi|$, for all normalized vectors $\psi \in D(Z)$, the domain of Z. It is straightforward to check that

$$\mathrm{Tr}[J(D_1, e_\psi)] = \kappa \int_{D_1} \mathrm{d}\mu(\zeta)\,\mathrm{Tr}(e_\zeta e_\psi e_\zeta) = \kappa = -2\mathrm{Re}\,\langle\psi, Z\psi\rangle\,.$$

As was shown in [5], the generator of the semigroup T_t associated with the process is given by

$$L(X) = \mathrm{i}[H_S, X] + \kappa \int_{D_1} \mathrm{d}\mu(\zeta) e_\zeta X e_\zeta - \kappa X \tag{4.3}$$

From equation (?? it is clear that T_t satisfies all but the last assumption of Theorem 4.1. However, although the decomposition $B(\mathcal{H}_S) = \mathcal{M}_1 \oplus \mathcal{M}_2$ does not hold in this case, T_t describes a very efficient decoherence in the quantum system in the spirit of Definition 3.1. In fact, if H_S is the operator closure of $(d\pi(h), D_G)$, where $h \in su(1,1)$ — the Lie algebra of group SU(1, 1), and D_G is the Gårding domain, then

$$\lim_{t\to\infty} \|T_t A\|_\infty = 0,$$

for all $A \in K(\mathcal{H}_S)$, the space of compact operators on $\mathcal{H}_S$ (see Theorem 3.10 in [6]). It follows that $K(\mathcal{H}_S) \in \mathcal{M}_2$, and so all pure states of the system instantaneously deteriorate to statistical states. Let us notice that the pre-adjoint semigroup T_{t*} is asymptotically stable, i.e.

$$\lim_{t\to\infty} \|T_{t*}\rho_1 \; - \; T_{t*}\rho_2\|_1 = 0$$

for all density matrices ρ_1 and ρ_2.

So far we have restricted the discussion on the emergence of classical properties in quantum open systems to type I factors. A generic feature of such factors is that they possess minimal projections. Hence, the only possible classical structure induced in such factors is a discrete one, and so the dynamics restricted to it must be trivial. Now we turn to infinite quantum spin systems whose GNS representation with respect to a normalized trace tr is known to be a hyperfinite factor of type II_1 [11]. We start with the following general theorem.

Theorem 4.4. *Suppose $\mathcal{M}$ is a type* II_1 *factor. Let its evolution be given by a family of maps $\{T_t\}$ satisfying the conditions* a)–c) *from Section 2 with $\omega_0 = \mathrm{tr}$, the normalized trace on $\mathcal{M}$. If $\{T_t\}$ is a semigroup, then the decomposition $\mathcal{M} = \mathcal{M}_1 \oplus \mathcal{M}_2$ always exists. Moreover, the effective part of any observable in $\mathcal{M}$ is given by a* tr*-compatible conditional expectation from $\mathcal{M}$ onto $\mathcal{M}_1$.*

Example 4.5 (Apparatus with continuous readings). Suppose an apparatus, a semi-infinite linear array of spin-$\frac{1}{2}$ particles, fixed at positions $k = 1, 2, 3, \ldots$, interacts with a quantum particle moving along the x-axis. Then, the algebra $\mathcal{M}$ of the measuring device is a hyperfinite factor of type II_1, and the algebra of the system is $B(\mathcal{H}_S)$, where $\mathcal{H}_S = L^2(\mathbb{R}, dx)$. More precisely, $\mathcal{M} = \pi(\otimes_1^\infty M_{2\times 2})''$, where π is the GNS representation with respect to the normalized trace on the Glimm algebra $\otimes_1^\infty M_{2\times 2}$. The evolution of the joint system is determined by a Hamiltonian

$$H = H_A \otimes \mathbf{1}_S + \mathbf{1}_A \otimes H_S + A \otimes B, \tag{4.4}$$

where H_A, H_S, A and B are assumed to satisfy the following conditions:

- $[H_A, A] = 0$,
- $A = \pi(\sum_{n=1}^{\infty} (\frac{1}{2})^n \sigma_n^3)$, where σ_n^3 is the third Pauli matrix located at position n, and so $A \in \mathcal{M}$,
- $H_S = -\frac{1}{2m}\triangle$, the kinetic energy operator,
- $B = \hat{p}$, the momentum operator,
- $\omega_S = |\psi\rangle\langle\psi|$, where $\psi(x) = \frac{1}{\sqrt{2\pi}} \int \frac{e^{ixp}}{\sqrt{\pi(1+p^2)}} dp$.

Let P_S denote the conditional expectation from $\mathcal{M} \otimes B(\mathcal{H}_S)$ onto $\mathcal{M}$ with respect to the state ω_S. Then, for any $X \in \mathcal{M}$,

$$T_t(X) = e^{itH_A} P_S[e^{itA\otimes B} X \otimes \mathbf{1}_S e^{-itA\otimes B}] e^{-itH_A}. \tag{4.5}$$

By Theorem 12 in [19], T_t satisfies all assumptions of Theorem 4.7. Hence, $\mathcal{M} = \mathcal{M}_1 \oplus \mathcal{M}_2$ with $\mathcal{M}_1$ being the von Neumann algebra generated by spectral projections of all σ_n^3, $n \in \mathbb{N}$. It is easy to check that $\mathcal{M}_1 = L^\infty(\mathcal{C}, \mu)$, where $\mathcal{C}$ is the Cantor set and μ is a continuous, regular, Borel, probability measure on $\mathcal{C}$, see [19] for more details. It is worth pointing out that $\mathcal{M}_1$ is unitarily isomorphic with $L^\infty([0, 1], dx)$, and the trace state tr corresponds to the Lebesgue integral on $[0, 1]$. Because H_A commutes with all spectral projections of the operator A so the evolution restricted to $\mathcal{M}_1$ is trivial. Hence, this case may be considered as a continuous analog of the selection of pointer states from Example 4.2.

Example 4.6 (Classical dynamical system). Suppose that a quantum system is a semi-infinite linear array of spin-$\frac{1}{2}$ particles fixed at positions $k \in \mathbb{N}$. The algebra $\mathcal{M}$ of the system is a hyperfinite factor of type II_1 defined as $\mathcal{M} = \pi(\otimes_1^\infty M_{2\times 2})'' \subset B(\mathcal{H}_S)$, where π is the GNS representation with respect to the normalized trace on the Glimm algebra. The free evolution of the system is given by a σ-weakly continuous one parameter group of automorphisms $\alpha_t \colon \mathcal{M} \to \mathcal{M}$ constructed in the following way. Suppose $U(\frac{k}{2^n})$, $k = 0, 1, \ldots, 2^n - 1$, is a representation of a cyclic group $\{\frac{k}{2^n}\}$, with addition modulo 1, in the space $\mathbb{C}^{2^n}$, such that

$$U(\frac{1}{2^n})(z_1, \ldots, z_{2^n}) = (z_{2^n}, z_1, \ldots, z_{2^n-1}).$$

Since it is a restriction of the standard unitary representation of the permutation group S_{2^n}, the $U(\frac{k}{2^n})$ are unitary matrices in $M_{2^n\times 2^n}$. Because there is an embedding of $M_{2^n\times 2^n}$ into $\otimes_1^\infty M_{2\times 2}$, so they may be considered as operators in the Glimm algebra. Hence, they induce a discrete group of unitary automorphisms of $\mathcal{M}$ by the formula

$$\alpha_{\frac{k}{2^n}}(X) = \pi(U(\frac{k}{2^n}))X\pi(U(\frac{k}{2^n}))^*.$$

Because n was arbitrary so we obtain in this way a homomorphism $d \to \alpha_d$, where d is a dyadic number, i.e. $d = \frac{k}{2^n}$ for some $n \in \mathbb{N}$ and some $k = 0, 1, \dots, 2^n - 1$. By Theorem 13 in [19], this homomorphism extends to the whole set of real numbers yielding a group of unitary (but not inner) automorphisms $\alpha_t(X) = \mathrm{e}^{\mathrm{i}tH} X \mathrm{e}^{-\mathrm{i}tH}$, where H is a self-adjoint operator on $\mathcal{H}_S$. It is clear from the construction that $\alpha_m = \mathrm{id}$, for any integer m.

The reservoir is chosen to consists of phonons of an infinitely extended harmonic crystal. The Hilbert space $\mathcal{H}$ representing pure states of a single phonon is given by $\mathcal{H} = L^2(\mathbb{R}^3, \mathrm{d}\mathbf{k})$. It follows that the Hilbert space of the environment is the symmetric Fock space $\mathcal{F}$ over $\mathcal{H}$, and its algebra $\mathcal{M}_E$ is a von Neumann algebra generated by Weyl operators $W(f) = \mathrm{e}^{\mathrm{i}\phi(f)}$, $\phi(f) = \frac{1}{\sqrt{2}}(a^*(f) + a(f))$, where $a^*(f)$ denotes the creation operator of one particle state $f \in \mathcal{H}$, and $a(f) = (a^*(f))^*$ [8]. Because the Fock representation is irreducible so $\mathcal{M}_E = B(\mathcal{F})$. The reference state of the reservoir ω_E is taken to be a pure state $\omega_E = |\tilde{f} >< \tilde{f}|$, where $|\tilde{f} >$ is a coherent state in the Fock space, i.e.

$$|\tilde{f} >= \mathrm{e}^{-\frac{1}{2}\|f\|^2} \sum_{n=0}^{\infty} \frac{[a^*(f)]^n}{n!} \Omega,$$

where $f \in \mathcal{H}$, and Ω is the vacuum state. Such a state represents a state of phonons associated with a classical acoustic wave. The free evolution of the reservoir is determined by the Hamiltonian

$$H_E = \int d(\mathbf{k})\omega(\mathbf{k})a^*(\mathbf{k})a(\mathbf{k}),$$

where $\omega(\mathbf{k})$ is the dispersion function.

Suppose now that these two systems interact. The coupling between the matter and the boson field is given by an interaction Hamiltonian H_I, a self-adjoint operator on $\mathcal{H}_S \otimes \mathcal{F}$. To derive an explicit form of H_I we use the formula (I.20) in [2], in which we put $G(\mathbf{k}) = A\frac{g(\mathbf{k})}{\sqrt{2}}$, $g \neq 0 \in \mathcal{H}$, and A is the same as in Example 4.5. Hence, $H_I = A \otimes \phi(g)$. For simplicity we put the coupling constant equal to one. It should be pointed out that, due to the form of A, H_I is a straightforward generalization of the interacting term for the spin-boson model. Because H_I commutes neither with H nor with H_E, so in order to determine the reduced dynamics of the system we do not follow a general strategy as in the previous cases. Instead we use a simplified procedure: First we calculate the reduced dynamics of the H_I only, and next add to it the automorphic evolution α_t. Hence, for any $X \in \mathcal{M}$,

$$T_t(X) = \alpha_t(P_E[\mathrm{e}^{\mathrm{i}tH_I} X \otimes \mathbf{1}_E \mathrm{e}^{-\mathrm{i}tH_I}]), \tag{4.6}$$

where P_E is the conditional expectation onto the algebra $\mathcal{M}$ with respect to the state ω_E. In order to calculate the explicit form of the superoperators T_t we suppose that $X \in \pi(M_{2^n \times 2^n})$. Then

$$\mathrm{e}^{\mathrm{i}tH_I} X \otimes \mathbf{1}_E \mathrm{e}^{-\mathrm{i}tH_I} = \sum_{\substack{i_1,\ldots,i_n \\ j_1,\ldots,j_n}} P_{i_1\ldots i_n} X P_{j_1\ldots j_n} \otimes \mathrm{e}^{\mathrm{i}t(a_{i_1\ldots i_n} - a_{j_1\ldots j_n})\phi(g)},$$

where $P_{i_1\ldots i_n} = P_{i_1} \otimes \cdots \otimes P_{i_n}$, $i_k \in \{0, 1\}$, and P_{i_k} are spectral projections of $\pi(\sigma_k^3)$. Parameters $a_{i_1\ldots i_n}$ are given by

$$a_{i_1\ldots i_n} = \sum_{k=1}^{n} (-1)^{i_k} \frac{1}{2^k}.$$

Hence,

$$\begin{aligned} &P_E[\mathrm{e}^{\mathrm{i}tH_I} X \otimes \mathbf{1}_E \mathrm{e}^{-\mathrm{i}tH_I}] \\ &\quad = \sum_{\substack{i_1,\ldots,i_n \\ j_1,\ldots,j_n}} P_{i_1\ldots i_n} X P_{j_1\ldots j_n} \langle \tilde{f}, W(t(a_{i_1\ldots i_n} - a_{j_1\ldots j_n})g)\tilde{f}\rangle, \end{aligned}$$

and so

$$\begin{aligned} T_t(X) = \alpha_t\Big[&\sum_{\substack{i_1,\ldots,i_n \\ j_1,\ldots,j_n}} P_{i_1\ldots i_n} X P_{j_1\ldots j_n} \times \\ &\times \mathrm{e}^{-\frac{1}{4}t^2(a_{i_1\ldots i_n} - a_{j_1\ldots j_n})^2\|g\|^2 + \mathrm{i}t\sqrt{2}(a_{i_1\ldots i_n} - a_{j_1\ldots j_n})\mathrm{Re}\langle f, g\rangle}\Big] \end{aligned} \tag{4.7}$$

Because $a_{i_1\ldots i_n} \neq a_{j_1\ldots j_n}$ if $i_k \neq j_k$ for at least one k, so the off-diagonal terms vanish when $t \to \infty$. Let us point out that in this case T_t is not a semigroup. However, the subalgebra of effective observables may be determined in the same way as in Example 4.5, yielding the same result, i.e. $\mathcal{M}_1 = L^\infty(\mathcal{C}, \mu)$, where $\mathcal{C}$ is the Cantor set. If $X \in \mathcal{M}_1$, then $T_t(X) = \alpha_t(X)$. In this way we have obtained an abstract commutative dynamical system $(\mathcal{M}_1, \alpha_t, \mathrm{tr})$.

Theorem 4.7. *$(\mathcal{M}_1, \alpha_t, \mathrm{tr})$ is isomorphic with $(L^\infty(S^1), g_t, d\mu)$, where $d\mu$ is the normalized Lebesgue measure on a circle S^1 and $g_t(\mathrm{e}^{\mathrm{i}a}) = \mathrm{e}^{\mathrm{i}(a+2\pi t)}$ is a flow on S^1.*

Example 4.8 (Ergodic spin system). Suppose again that a quantum system is a semi-infinite linear array of spin-$\frac{1}{2}$ particles fixed at positions $k \in \mathbb{N}$. The quasi-local algebra $\mathcal{A}$ is the norm closure of the algebra $\mathcal{A}_0 = \bigcup \mathcal{A}_n$ of local observables. Here, by $\mathcal{A}_n$ we denote the local algebra associated with the set $\Lambda_n = \{1, 2, \ldots, n\}$. It is clear that $\mathcal{A}_n = \otimes_{k=1}^{n} \mathcal{A}_{(k)}$, where $\mathcal{A}_{(k)}$ is isomorphic with the algebra of 2×2 matrices. $\mathcal{M} = \pi(\mathcal{A})'' \subset B(\mathcal{H}_S)$, as in the previous example. Suppose that the system interacts with its environment represented by the algebra $\mathrm{B}(\mathcal{H}_E)$. The evolution of the joint system is determined by a Hamiltonian

$$H = H_S \otimes \mathbf{1}_E + \mathbf{1}_S \otimes H_E + A \otimes B, \tag{4.8}$$

where H_S and A are given by

- $H_S = \pi(\prod\limits_{k=1}^{\infty} (\mathbf{1} + b_k \sigma_k^1))$, $\sigma_k^1 \in \mathcal{A}_{(k)}$ is the first Pauli matrix, $b_k > 0$, and $\sum\limits_{k=1}^{\infty} b_k < \infty$,
- $A = \pi(\sum\limits_{k=1}^{\infty} \frac{1}{2^k} \sigma_k^3)$, as in the Example 4.5.

Because $\|H_S\|_\infty = \prod_{k=1}^{\infty}(1 + b_k) < \infty$ so both H_S and A are bounded and belong to $\pi(\mathcal{A})$. We do not specify the form of the operators H_E and B. Instead, we assume that the so called singular coupling limit [1] may be applied for derivation of the reduced dynamics of the system. Hence, the Markovian master equation for $x \in \mathcal{M}$ reads

$$\dot{x} = L(x) = \mathrm{i}[H, x] + L_D(x),$$

where $H = H_S + \alpha A^2$ and

$$L_D(x) = \gamma A x A - \frac{\gamma}{2}\{x, A^2\}.$$

The coefficients $\alpha \in \mathbb{R}$ and $\gamma > 0$ are given by the formula

$$\int_0^\infty \mathrm{Tr}(\rho_E \mathrm{e}^{\mathrm{i}tH_E} B \mathrm{e}^{-\mathrm{i}tH_E} B) dt = \frac{\gamma}{2} + \mathrm{i}\alpha,$$

where ρ_E is a density matrix of the environment. It is clear that the semigroup $T_t = \mathrm{e}^{tL}$ on $\mathcal{M}$ preserves the trace tr and so it satisfies the assumptions of Theorem 4.4. Hence, the decomposition $\mathcal{M} = \mathcal{M}_1 \oplus \mathcal{M}_2$ holds true.

Theorem 4.9. *The system* $(\mathcal{M}, T_t, \mathrm{tr})$ *is ergodic, i.e.* $\mathcal{M}_1 = \mathbb{C} \cdot \mathbf{1}$.

Acknowledgments

One of the authors (R.O.) would like to thank A. von Humboldt Foundation for the financial support.

References

[1] Alicki R., Lendi K.: *Quantum Dynamical Semigroups and Applications*, Lecture Notes in Physics 286, Springer, Berlin, 1987.

[2] Bach V., Fröhlich J., Sigal I. M.: Return to equilibrium, *J. Math. Phys.* **41** (2000) 3985–4060.

[3] Blanchard Ph. et al. (Eds.): *Decoherence: Theoretical, Experimental and Conceptual Problems*, Lecture Notes in Physics 538, Springer, Berlin, 2000.

[4] Blanchard Ph., Olkiewicz R.: Decoherence induced transition from quantum to classical dynamics, *Rev. Math. Phys.* **15**(3) (2003) 217–243.

[5] Blanchard Ph., Olkiewicz R.: Effectively classical quantum states for open systems, *Phys. Lett. A* **273**(4) (2000) 223–231.

[6] Blanchard Ph., Olkiewicz R.: Interacting quantum and classical continuous systems. II. Asymptotic behavior of the quantum subsystem, *J. Stat. Phys.* **94**(5-6) (1999) 933–953.

[7] Bratteli O., Robinson D. W.: *Operator Algebras and Quantum Statistical Mechanics I*, Springer, New York, 1979.

[8] Bratteli O., Robinson D. W.: *Operator Algebras and Quantum Statistical Mechanics II*, Springer, New York, 1981.

[9] Breuer H.-P., Petruccione F.: Destruction of quantum coherence through emission of bremsstrahlung, *Phys. Rev. A* **63** (2001) 032102.

[10] Davies E. B.: Quantum stochastic processes, *Comm. Math. Phys.* **15** (1969) 277–304.

[11] Evans D.E., Kawahigashi Y.: *Quantum Symmetries on Operator Algebras*, Clarendon Press, Oxford, 1998.

[12] Fröhlich J., Tsai T.-P., Yau H.-T.: On a classical limit of quantum theory and the non-linear Hartree equation, *Geom. Funct. Anal.* **2000**, Special Volume GAFA 2000 (Tel Aviv, 1999), Part I, 57–78.

[13] Fröhlich J., Tsai T.-P., Yau H.-T.: On the point-particle (Newtonian) limit of the non-linear Hartree equation, *Comm. Math. Phys.* **225**(2) (2002) 223–274.

[14] Gell-Mann M., Hartle J. B.: Classical equations for quantum systems, *Phys. Rev. D* **47**(8) (1993) 3345–3382.

[15] Giulini D., Joos E., Kiefer C., Kupsch J., Stamatescu I.-O., Zeh H. D.: *Decoherence and The Appearance of a Classical World in Quantum Theory*, Springer, Berlin, 1996, 2nd ed., 2003.

[16] Haba Z.: Classical limit of quantum dissipative systems, *Lett. Math. Phys.* **44**(2) (1998) 121–130.

[17] Joos E., Zeh H.D.: The Emergence of Classical Properties through Interaction with the Environment, *Z. Phys. B* **59** (1985) 223–243.

[18] Kupsch, J.: Mathematical Aspects of Decoherence, In *Decoherence: Theoretical, Experimental and Conceptual Problems*, Lecture Notes in Physics 538, Springer, Berlin, 2000, 125–136.

[19] Ługiewicz P., Olkiewicz R.: Decoherence in infinite quantum systems, *J. Phys. A* **35** (2002) 6695–6712.

[20] Perelomov A. M.: Coherent states for arbitrary Lie group, *Comm. Math. Phys.* **26** (1972) 222–236.

[21] Twamley J.: Phase-space decoherence: A comparison between consistent histories and environment-induced superselection, *Phys. Rev. D* **48** (1993) 5730–5745.

[22] Unruh W. G., Zurek W. H.: Reduction of a wave packet in quantum Brownian motion, *Phys. Rev. D* **40**(4) (1989) 1071–1094.

[23] Zeh H. D.: On the Interpretation of Measurement in Quantum Theory, *Found. Phys.* **1** (1970) 69–76.

[24] Zeh H. D.: *The Physical Basis of The Direction of Time*, 4th ed., Springer, Berlin, 2001.

[25] Zurek W. H.: Environment-induced superselection rules, *Phys. Rev. D* **26**(8) (1982) 1862–1880.

[26] Zurek W. H.: Preferred States, Predictability, Classicality and the Environment-Induced Decoherence, *Progr. Theor. Phys.* **89**(2) (1993) 281–312.

HANKEL OPERATORS ON SEGAL-BARGMANN SPACES

Thomas Deck

Fakultät für Mathematik und Informatik, Universität Mannheim, D–68131 Mannheim, Germany

Abstract Hankel operators H_b on Segal–Bargmann spaces, with respect to finitely and infinitely many variables, are investigated. A regularity condition on the symbol b which guarantees boundedness of H_b is provided, the Hilbert–Schmidtness of H_b is characterized, and an integral representation for Hankel operators of Hilbert Schmidt type is given. The proofs partially employ the hypercontractivity of the Ornstein-Uhlenbeck semigroup. The case with infinitely many variables is treated via approximations with finitely many variables.

Keywords: Small Hankel operators, holomorphic Wiener functionals, complex hypercontractivity.
Primary: 47B35, 46E20. Secondary: 47D03, 47D07

1. Introduction

This work connects two fields of mathematics which until recently had no overlap: *Hankel operators* and *Wiener spaces*. Both subjects are of notable size: One can find more than 2000 references containing the key word "Hankel operator" in the AMS database, and about 1900 references containing "Wiener space". But there is none containing both.

This article is based on joint work with Leonard Gross, cf. [4, 3]. The first work concerns Hankel operators over complex, finite dimensional manifolds, and requires the theory of holomorphic Dirichlet forms as developed in [8]. The second one concerns Hankel operators over a complex Wiener space. Most results for the Wiener space context are based on [4] for the following special case: The complex manifold $\mathbb{C}^d$ is equipped with a Gaussian measure

$$d\gamma_c(z) = \frac{1}{(2\pi c)^d}\, e^{-\frac{|z|^2}{2c}}\, dz, \quad z \in \mathbb{C}^d.$$

S. Albeverio et al. (eds.),
Proceedings of the International Conference on Stochastic Analysis and Applications, 17–36.

In that case the background on holomorphic function spaces and Dirichlet forms, as given in [4], is not required. Instead one can give more direct proofs of those results from [4] relevant to understand the Wiener case [3]. Thus, after some preparations in §2 I provide these results with streamlined proofs in §3, so that a reader interested in Hankel operators over Wiener space need not go through all of [4]. Some background on holomorphic Wiener functionals is then given in §4. The main results for Hankel operators over Wiener space are finally discussed (partly with sketches of proofs) in §5. It turns out that some interesting deviations form the $\mathbb{C}^d$–case arise in infinite dimensions.

There are two main conclusions: Firstly, the results show that a theory of Hankel operators over an infinite dimensional base space can be developed quite analogously to the finite dimensional situation. Secondly, the results show that typical questions from the field of Hankel operators can serve as a source for new questions about holomorphic Wiener functionals. This latter field has not found many connections to other fields of mathematics as yet.

The following notations will be used subsequently: $\mathcal{H}(\mathbb{C}^d)$ denotes the space of holomorphic functions on $\mathbb{C}^d$, $\mathcal{L}^p(\mathbb{C}^d, \gamma_c)$ are the p-th integrable functions w.r.t. Gauss measure γ_c and $\mathcal{H}L^p(\mathbb{C}^d, \gamma_c) := \mathcal{H}(\mathbb{C}^d) \cap \mathcal{L}^p(\mathbb{C}^d, \gamma_c)$. For $p > 0$ this space of functions (not function classes) equipped with the $\mathcal{L}^p$–metric is a complete metric space. The space of polynomials in the variables $z_1, \ldots, z_d$, denoted $\mathcal{P}(\mathbb{C}^d)$, is dense in $\mathcal{H}L^p(\mathbb{C}^d, \gamma_c)$. A *Hankel form* on the Hilbert space $\mathcal{H}L^2(\mathbb{C}^d, \gamma_c)$ is a jointly continuous bilinear form $\Gamma\colon \mathcal{H}L^2(\mathbb{C}^d, \gamma_c) \times \mathcal{H}L^2(\mathbb{C}^d, \gamma_c) \to \mathbb{C}$ which satisfies

$$\Gamma(f, g) = \Gamma(fg, 1), \quad \forall\, f, g \in \mathcal{P}(\mathbb{C}^d). \tag{1.1}$$

Since $f \mapsto \Gamma(f, 1)$ is a continuous linear functional on $\mathcal{H}L^2(\mathbb{C}^d, \gamma_c)$ there is a unique element $b \in \mathcal{H}L^2(\mathbb{C}^d, \gamma_c)$ such that $\Gamma(f, 1) = \int_{\mathbb{C}^d} \bar{b} f \, \mathrm{d}\gamma_c$. Then (1.1) implies

$$\Gamma(f, g) = \int_{\mathbb{C}^d} \bar{b} f g \, \mathrm{d}\gamma_c, \quad \forall\, f, g \in \mathcal{P}(\mathbb{C}^d). \tag{1.2}$$

By continuity this equation determines the given Hankel form Γ uniquely on $\mathcal{H}L^2(\mathbb{C}^d, \gamma_c)$. b will be called the *symbol of* Γ, and Γ will be denoted Γ_b. With Γ_b is associated a unique continuous *Hankel operator with symbol b*, $H_b\colon \mathcal{H}L^2(\mathbb{C}^d, \gamma_c) \to \mathcal{H}L^2(\mathbb{C}^d, \gamma_c)$, defined by

$$\Gamma(f, g) = \int_{\mathbb{C}^d} f \overline{H_b g} \, \mathrm{d}\gamma_c, \quad \forall f, g \in \mathcal{H}L^2(\mathbb{C}^d, \gamma_c). \tag{1.3}$$

Remarks. 1. Notice that H_b is an anti-linear operator, so that $H_b(\lambda f) = \bar{\lambda} H_b f$, for all $\lambda \in \mathbb{C}$. If one wishes to work with a linear operator one can add a complex conjugation, i.e., one defines $\overline{H}_b\colon f \mapsto \overline{H_b f}$. This operator maps into $\mathcal{H}L^2(\mathbb{C}^d,\gamma_c)^\perp$, the orthogonal complement of $\mathcal{H}L^2(\mathbb{C}^d,\gamma_c)$ in $L^2(\mathbb{C}^d,\gamma_c)$.

2. Let $b \in \mathcal{H}L^2(\mathbb{C}^d,\gamma_c)$ be the symbol of a Hankel operator and denote by $P\colon L^2(\mathbb{C}^d,\gamma_c) \to \mathcal{H}L^2(\mathbb{C}^d,\gamma_c)$ the orthogonal projection. The equality $\int \bar{b} fg \,\mathrm{d}\gamma_c = \int \overline{H_b \bar{f}} g \,\mathrm{d}\gamma_c$, which holds for all $f, g \in \mathcal{P}(\mathbb{C}^d)$, implies $H_b f = P(b\bar{f})$, provided $b\bar{f} \in L^2(\gamma_c)$. So if M_b denotes multiplication by b and C the complex conjugation we obtain (modulo domain questions) $H_b = P \circ M_b \circ C$. This is a more conventional form for a Hankel operator.

Equations (1.1) to (1.3) show that the following objects allow a definition of a Hankel operator: A *finite measure* μ, a Hilbert space of *holomorphic* $L^2(\mu)$*–functions*, and a dense subspace for which $\Gamma(f,g) = \Gamma(fg,1)$ holds. These notions can also be defined in an infinite dimensional setting: We are going to replace $(\mathbb{C}^d,\gamma_c)$ by a complex Wiener space $(W_{\mathbb{C}},\mu_c)$, $\mathcal{H}L^2(\mathbb{C}^d,\gamma_c)$ by $\mathcal{H}L^2(W_{\mathbb{C}},\mu_c)$, and $\mathcal{P}(\mathbb{C}^d)$ by $\mathcal{P}(Z)$ (a subspace of polynomials in $\mathcal{H}L^2(W_{\mathbb{C}},\mu_c)$). Details will be given in §4.

Among the classical questions concerning bilinear forms such as (1.2) is the relation between properties of b and continuity properties of Γ_b as well as trace ideal properties of the associated operator H_b. In this paper continuity, Hilbert-Schmidtness and an integral representation for Hankel operators are treated. These properties can successfully be investigated both in the finite and in the infinite dimensional context.

Remark. $\mathcal{H}L^2(\mathbb{C}^d,\gamma_c)$ was called Fock-space in [20]. This notation has then been used in several other works. Since in stochastic analysis (and in physics) a Fock-space usually means a closure of a (symmetric) tensor algebra [12, 15, 38] the notation Segal-Bargmann space, for $\mathcal{H}L^2(\mathbb{C}^d,\gamma_c)$ and $\mathcal{H}L^2(W_{\mathbb{C}},\mu_c)$, is used subsequently. Both names have their justification since these spaces are naturally isomorphic, cf. e.g. [9].

Hankel operators do not belong to the standard background in stochastic analysis. Therefore the following orientation may be helpful for readers from that field:

Historical sketch. The German mathematician Hermann Hankel (1839-1873), in his Ph.D. thesis [10] in 1861, studied (sub)determinants of complex matrices of type

$$b_{nm} := b_{n+m}, \quad \forall n,m \in \mathbb{N}, \tag{1.4}$$

for given sequences $b = (b_n)_{n\in\mathbb{N}}$. In 1881 Kronecker proved, that the matrix (b_{n+m}) has finite rank iff the function $b(z) := \sum_{n=1}^{\infty} b_n z^n$ is a rational function of $z \in \mathbb{C}$ [22]. In 1906 Hilbert showed that the operator $H\colon l^2(\mathbb{N}) \to l^2(\mathbb{N})$ defined by $(Hc)_m := \sum_{n=1}^{\infty} \frac{1}{n+m} c_n$, $m \in \mathbb{N}$, is continuous. Notice that the matrix elements of H w.r.t. the standard orthonormal basis (ONB) in $l^2(\mathbb{N})$ read $\frac{1}{n+m}$. In this way Hankel matrices became Hankel operators, i.e. operators H on a separable Hilbert space having matrix elements

$$\langle e_n, He_m \rangle = b_{n+m}, \tag{1.5}$$

in a suitable ONB $\{e_n, n \in \mathbb{N}\}$. Until 1957 only few progress was made. Then Nehari [25] characterized continuous H_b in terms of the function $\varphi \mapsto \sum_{n=1}^{\infty} b_n \mathrm{e}^{-\mathrm{i}n\varphi}$. In his proof Nehari used an approximation theorem for holomorphic functions on the unit disc. In view of this (and Kronecker's theorem) it is not quite unexpected that a Hilbert space of holomorphic functions on the unit disc provides a "natural" space to represent Hankel operators. This Hilbert space is the Hardy space $H(T)$, i.e. the space of holomorphic functions on the open unit disc which have "L^2–boundary values" on $T = \{\mathrm{e}^{\mathrm{i}\varphi}, \varphi \in [0, 2\pi]\}$. To be precise, $b(z) = \sum_{n=0}^{\infty} b_n z^n \in H(T)$ iff $b(\mathrm{e}^{-\mathrm{i}\varphi}) = \sum_{n=0}^{\infty} b_n \mathrm{e}^{-\mathrm{i}n\varphi} \in L^2([0, 2\pi], \mathrm{d}\varphi)$ i.e. iff $\sum_{n=0}^{\infty} |b_n|^2 < \infty$. The representation of Hankel operators on the Hardy space allows to prove results and to formulate theorems in a flexible way, cf. [26, 33, 49]. For example, Nehari's proof makes use of holomorphic functions, and the compactness of a Hankel operator is characterized in terms of the boundary function $b(\mathrm{e}^{-\mathrm{i}\varphi})$, by the theorem of Hartmann [11] given in 1958.

After these ground breaking results the theory of Hankel operators developed, first gradually, and then in the beginning of the seventies rapidly. This is partly due to the fact that Hankel operators are connected to various branches of mathematics and its applications (cf. the review article [32]). Connections exist with Toeplitz operators, interpolation and approximation theory, the Hamburger moment problem, stationary Gaussian processes (e.g. [30, 31, 35]), linear system theory, control theory, and others, cf. [26, 33, 49] and references given there.

The traditional concept of Hankel operators has soon been generalized in various directions [27, 34, 36, 20, 18, 14]. The notion of (small) Hankel operators considered in this paper coincides with the one in [20], where the investigation of Hankel operators on $\mathcal{H}L^p(\mathbb{C}^d, \gamma_c)$ was initiated. This notion connects to the traditional matrices (1.4) as follows: Assume $b(z) = \sum_{n=0}^{\infty} b_n z^n$ is the symbol of $H_b \colon \mathcal{H}L^2(\mathbb{C}, \gamma_c) \to \mathcal{H}L^2(\mathbb{C}, \gamma_c)$. With respect to the orthogonal basis (OGB) $\{z^n, n \in \mathbb{N}_0\}$ one obtains the Hankel matrix:

$$\langle H_b z^n, z^m \rangle = b_{n+m} \|z^{n+m}\|^2. \tag{1.6}$$

But in contrast to the Hardy space (where z^n has norm 1) we have $\|z^n\|^2_{L^2(\gamma_c)} = (2c)^n n!$, so the matrix element w.r.t. normalized z^n do not constitute a classical Hankel matrix.

Hankel operators on Fock type spaces were further investigated in several works, e.g. in [14, 19, 21, 28, 29, 37, 40, 47]. For the distinction between small and big Hankel operators, and for the choice of linear or anti-linear Hankel operators see e.g. [27, 20, 49].

2. Finite dimensional preparations

Recall that $\mathcal{H}L^p(\mathbb{C}^d, \gamma_c)$ is the set of holomorphic $\mathcal{L}^p$–functions on $\mathbb{C}^d$ w.r.t. Gauss measure γ_c. It is simple to check that when $p = 2$ this space of functions, equipped with the sesquilinear form

$$\langle f, g\rangle = \int_{\mathbb{C}^d} f\bar{g}\, \mathrm{d}\gamma_c,$$

is a genuine inner product space (i.e. the $\mathcal{L}^2$-seminorm $\|\cdot\|$ is really a norm, not just a seminorm). For multi–indices $\alpha = (\alpha_1, \ldots, \alpha_d) \in \mathbb{N}_0^d$ we abbreviate $|\alpha| := \alpha_1 + \cdots + \alpha_d$, and $\alpha! := \alpha_1! \cdots \alpha_d!$.

Theorem 2.1. *Let $c > 0$ and $0 < p < \infty$. Then*

(a) *The space $\mathcal{H}L^p(\mathbb{C}^d, \gamma_c)$ equipped with the $\mathcal{L}^p$–metric is a complete metric space.*

(b) *The monomials $z^\alpha = z_1^{\alpha_1} \cdots z_d^{\alpha_d}$, $\alpha \in \mathbb{N}_0^d$, form an OGB in $\mathcal{H}L^2(\mathbb{C}^d, \gamma_c)$ with*

$$\|z^\alpha\|^2_{L^2(\gamma_c)} = (2c)^{|\alpha|} \alpha!. \tag{2.1}$$

(c) *For an arbitrary holomorphic function $f(z) = \sum_{\alpha\in\mathbb{N}_0^d} a_\alpha z^\alpha$ the following equality (including the case with $\infty = \infty$) holds:*

$$\int_{\mathbb{C}^d} |f(z)|^2\, \mathrm{d}\gamma_c(z) = \sum_{\alpha\in\mathbb{N}_0^d} (2c)^{|\alpha|} \alpha!\, |a_\alpha|^2. \tag{2.2}$$

(d) *A formal power series $\sum_\alpha a_\alpha z^\alpha$ defines (pointwise) an*

$$f \in \mathcal{H}L^2(\mathbb{C}^d, \gamma_c)$$

iff the r.h.s. in (2.2) is finite. Moreover, if (2.2) is finite then $f(z) := \sum_\alpha a_\alpha z^\alpha$ is the orthogonal expansion of $f \in \mathcal{H}L^2(\mathbb{C}^d, \gamma_c)$, and the norm of f (squared) is given by (2.2).

(e) (Bargmann's pointwise bound) *For all $f \in \mathcal{H}L^2(\mathbb{C}^d, \gamma_c)$ it holds:*

$$|f(z)|^2 \leq \|f\|^2_{L^2(\gamma_c)}\, \mathrm{e}^{|z|^2/2c} \quad \forall z \in \mathbb{C}^d. \tag{2.3}$$

For a proof of this theorem see [1]. A readable presentation, which also connects to infinite dimensions, is given in [9]. Notice that (2.3) simply follows with (2.2):

$$|f(z)|^2 \le \Big(\sum_{\alpha\in\mathbb{N}_0^d} |a_\alpha z^\alpha|\Big)^2 = \Big(\sum \sqrt{(2c)^{|\alpha|}\alpha!}|a_\alpha|\frac{|z^\alpha|}{\sqrt{(2c)^{|\alpha|}\alpha!}}\Big)^2$$
$$\le \Big(\sum (2c)^{|\alpha|}\alpha!|a_\alpha|^2\Big)\Big(\sum \frac{|z^\alpha|^2}{(2c)^{|\alpha|}\alpha!}\Big) = \|f\|^2_{L^2(\gamma_c)}\, \mathrm{e}^{|z|^2/2c}.$$

Remark. For readers less familiar with holomorphic function spaces Theorem 2.1 is remarkable in two respects:

Firstly it says that $\mathcal{H}L^p(\mathbb{C}^d,\gamma_c)$, for $p \ge 1$, is a Banach space, i.e. $\mathcal{L}^p$-Cauchy sequences of holomorphic functions have pointwise well–defined holomorphic limits. (In case $p = 2$ this follows right away from (2.3).)

Secondly the *c–independent* monomials z^α define, for any $c > 0$, an OGB in $\mathcal{H}L^2(\mathbb{C}^d,\gamma_c)$.

Definition. For $f \in \mathcal{H}(\mathbb{C}^d)$ define the *Ornstein-Uhlenbeck group* e^{-tN} (with parameter $c > 0$) by

$$\mathrm{e}^{-tN} f(z) := f(\mathrm{e}^{-t/c} z), \qquad t \in \mathbb{R}. \tag{2.4}$$

Remarks. 1. When $f \in \mathcal{H}(\mathbb{C}^d)$ is represented by $\sum_{\alpha\in\mathbb{N}_0^d} a_\alpha z^\alpha$ one has

$$\mathrm{e}^{-tN} f(z) = f(\mathrm{e}^{-t/c} z) = \sum_{\alpha\in\mathbb{N}_0^d} \mathrm{e}^{-t|\alpha|/c} a_\alpha z^\alpha. \tag{2.5}$$

2. The "full" OU–semigroup on $L^2(\mathbb{C}^d,\gamma_c) = L^2(\mathbb{R}^{2d},\gamma_c)$ can be defined as follows. Let N be the second order differential operator defined by

$$\int_{\mathbb{R}^{2d}} \sum_{k=1}^{2d} \frac{\partial f(x)}{\partial x_k}\frac{\partial \bar{g}(x)}{\partial x_k}\, \mathrm{d}\gamma_c(x) = \int_{\mathbb{R}^{2d}} f(x) N\bar{g}(x)\, \mathrm{d}\gamma_c(x), \quad \forall f, g \in C_c^\infty(\mathbb{R}^{2d}).$$

The closure of N in $L^2(\mathbb{R}^{2d},\gamma_c)$ gives rise to the OU–semigroup e^{-tN} in $L^2(\mathbb{R}^{2d},\gamma_c)$. The Laplacian part of N annihilates $f \in \mathcal{H}L^2(\mathbb{C}^d,\gamma_c)$, and one obtains

$$Nf = \frac{1}{c}\sum_{k=1}^{d} z_k \frac{\partial f}{\partial z_k}.$$

Thus N acts as a vector field Z on $f \in \mathcal{H}L^2(\mathbb{C}^d,\gamma_c)$. If one decomposes Z into real and imaginary parts, $Z = \frac{1}{2}(X - \mathrm{i}Y)$, then $z \mapsto \mathrm{e}^{t/c} z$ is the flow of X. Since this flow exists for all times the OU–semigroup on $\mathcal{H}L^2(\mathbb{C}^d, \gamma_c)$ naturally extends to the group (2.4).

Theorem 2.2. (a) *Suppose* $p > 0$, $c > 0$, $t \in \mathbb{R}$, *and* $f \in \mathcal{H}(\mathbb{C}^d)$. *Then*

$$\mathrm{e}^{-tN} f \in \mathcal{H}L^p(\mathbb{C}^d, \gamma_c) \iff f \in \mathcal{H}L^p(\gamma_{c\mathrm{e}^{-2t/c}}). \tag{2.6}$$

Moreover,

$$\|\mathrm{e}^{-tN} f\|_{L^p(\gamma_c)} = \|f\|_{L^p(\gamma_{c\mathrm{e}^{-2t/c}})}. \tag{2.7}$$

(b) *Suppose* $c > 0$ *and* $0 < p \leq q < \infty$. *Then the hypercontractivity estimate*

$$\|\mathrm{e}^{-tN} f\|_{L^q(\gamma_c)} \leq \|f\|_{L^p(\gamma_c)} \tag{2.8}$$

holds for all $f \in \mathcal{H}L^p(\mathbb{C}^d, \gamma_c)$, *and for all* $t \geq t_J(p,q) := \frac{c}{2}\log(\frac{q}{p})$.

Proof. (a) A simple calculation shows that the map $\psi \colon z \mapsto \mathrm{e}^{-t/c} z$ induces the measure $\psi_* \gamma_c = \gamma_{c\mathrm{e}^{-2t/c}}$ on $\mathbb{C}^d$. The claim follows with (2.4) by the transformation theorem for integrals. The proof of (b) is nontrivial. In fact there are several different proofs in the literature. The first one is due to Janson [16], others are given in [2, 48, 17, 8]. □

Remarks. 1. For $p = q$ the estimate (2.8) shows that the OU–group in $\mathcal{H}(\mathbb{C}^d)$ restricts to a contraction semigroup $(\mathrm{e}^{-tN})_{t \geq 0}$ on $\mathcal{H}L^p(\mathbb{C}^d, \gamma_c)$.

2. The equivalence (2.6) shows that $\varphi \in D((\mathrm{e}^{tN} | \mathcal{H}L^p(\mathbb{C}^d, \gamma_c))$ is simply characterized by the size requirement $\varphi \in \mathcal{H}L^p(\gamma_{c\mathrm{e}^{2t/c}})$, a fact which is false in the absence of holomorphy.

3. It is easy to see that for $t > \frac{c}{2} \ln \frac{p}{2}$ the map

$$\mathrm{e}^{-tN} \colon \mathcal{H}L^2(\mathbb{C}^d, \gamma_c) \to \mathcal{H}L^p(\mathbb{C}^d, \gamma_c)$$

is continuous: In view of (2.3) and (2.7) we have

$$\begin{aligned} \|\mathrm{e}^{-tN} f\|^p_{L^p(\gamma_c)} &= \int_{\mathbb{C}^d} |f(z)|^p \, \mathrm{d}\gamma_{c\mathrm{e}^{-2t/c}}(z) \\ &\leq \|f\|^p_{L^2(\gamma_c)} (2\pi c\mathrm{e}^{-2t/c})^{-d} \int_{\mathbb{C}^d} \mathrm{e}^{\frac{p}{2}\frac{|z|^2}{2c}} \mathrm{e}^{-\mathrm{e}^{2t/c}\frac{|z|^2}{2c}} \, \mathrm{d}z. \end{aligned}$$

The right side of this estimate is finite iff $\mathrm{e}^{2t/c} > \frac{p}{2}$, i.e. iff $t > \frac{c}{2} \ln \frac{p}{2}$.

The special form of the OU–semigroup (2.4) on $\mathcal{H}(\mathbb{C}^d)$ together with estimate (2.8) are the main ingredients for the proof of the continuity result, Theorem 3.1, in Section 3.

3. Hankel operators over $\mathbb{C}^d$

The first result concerns the continuity of Hankel operators. For its proof let me recall the notion of a reproducing kernel: Observe that estimate (2.3) implies that for each $z \in \mathbb{C}^d$ the point evaluation $\delta_z : f \mapsto f(z)$ is a continuous linear functional on $\mathcal{H}L^2(\mathbb{C}^d, \gamma_c)$. By the Riesz Lemma there is a unique element $K_z \in \mathcal{H}L^2(\mathbb{C}^d, \gamma_c)$, such that $\delta_z = \langle \cdot, K_z \rangle$, i.e. the reproduction property $f(z) = \langle f, K_z \rangle$ holds for all $f \in \mathcal{H}L^2(\mathbb{C}^d, \gamma_c)$. The reproducing kernel $K_z(w)$ can easily be computed [20], p. 98. It turns out to be

$$K_z(w) = \mathrm{e}^{\bar{z}w/2c}, \quad \forall z, w \in \mathbb{C}^d. \tag{3.1}$$

Theorem 3.1. (i) *Assume $c > 0$, $p \geq 2$, and put $t_J := \frac{c}{2} \ln p'$, with $\frac{1}{p} + \frac{1}{p'} = 1$. For $\varphi \in \mathcal{H}L^p(\mathbb{C}^d, \gamma_c)$ define $b := \mathrm{e}^{-tN}\varphi$, with $t \geq t_J$. Then*

$$\Big| \int_{\mathbb{C}^d} \bar{b} f g \, \mathrm{d}\gamma_c \Big| \leq \|\varphi\|_{L^p(\gamma_c)} \|f\|_{L^2(\gamma_c)} \|g\|_{L^2(\gamma_c)}, \quad \forall f, g \in \mathcal{P}(\mathbb{C}^d). \tag{3.2}$$

(ii) *Let $p = 2$, put $t_J = \frac{c}{2} \ln 2$ and $\operatorname{Ran} \mathrm{e}^{-tN} := \mathrm{e}^{-tN}(\mathcal{H}L^2(\mathbb{C}^d, \gamma_c))$. Assume the symbol $b \in \mathcal{H}L^2(\mathbb{C}^d, \gamma_c)$ is such that Γ_b is continuous. Then*

$$b \in \bigcap_{t < t_J} \operatorname{Ran} \mathrm{e}^{-tN}. \tag{3.3}$$

Remarks. 1. The estimate (3.2) implies that Γ_b is a bounded bilinear form on $\mathcal{H}L^2(\mathbb{C}^d, \gamma_c)$ because $\mathcal{P}(\mathbb{C}^d)$ is dense in $\mathcal{H}L^2(\mathbb{C}^d, \gamma_c)$.

2. The definition $b := \mathrm{e}^{-tN}\varphi$ in (i) may be viewed as a "regularization" of the function φ. In view of (2.2) and (2.5) this means that if the coefficients in $b(z) = \sum a_\alpha z^\alpha$ decay sufficiently fast then Γ_b is continuous. In particular every polynomial b defines a continuous Γ_b. (In fact a finite rank operator, as one easily verifies.) Notice that the larger p in the condition $\varphi \in \mathcal{H}L^p(\mathbb{C}^d, \gamma_c)$ is the less φ needs to be regularized (i.e. we need a smaller shortest time t_J).

3. (ii) shows that for $p = 2$ the result in (i) is sharp in the sense that for each $t < t_J$ the symbol b can be written as $b = \mathrm{e}^{-tN}\varphi_t$ with a suitable $\varphi_t \in \mathcal{H}L^2(\mathbb{C}^d, \gamma_c)$.

4. The proof of (i) given below only requires the hypercontractivity of $\mathrm{e}^{-tN} : \mathcal{H}L^2(\mathbb{C}^d, \gamma_c) \to \mathcal{H}L^{2p'}(\mathbb{C}^d, \gamma_c)$. For $t > t_J$ we observed (in the remark following Theorem 2.2) the continuity of this operator by simple arguments. Thus, under the slightly stronger condition $t > t_J$ the continuity of Γ_b follows by fairly elementary arguments.

Proof of Theorem 3.1. (i) For $f, g \in \mathcal{P}(\mathbb{C}^d) \subset \mathcal{H}L^2(\mathbb{C}^d, \gamma_c)$ we have:

$$|\Gamma_b(f, g)| = |\langle b, fg \rangle| = |\langle \mathrm{e}^{-tN}\varphi, fg \rangle| = |\langle \varphi, \mathrm{e}^{-tN}(fg) \rangle|.$$

Definition (2.4) immediately implies the factorization rule

$$\mathrm{e}^{-tN}(fg)(z) = \mathrm{e}^{-tN}f(z)\mathrm{e}^{-tN}g(z). \tag{3.4}$$

Combined with Hölder's inequality for indices $p^{-1}+(2p')^{-1}+(2p')^{-1}=1$ this yields

$$\begin{aligned}|\Gamma_b(f,g)| &= |\langle\varphi, (\mathrm{e}^{-tN}f)(\mathrm{e}^{-tN}g)\rangle| \\ &\le \|\varphi\|_{L^p(\gamma_c)}\|\mathrm{e}^{-tN}f\|_{L^{2p'}(\gamma_c)}\|\mathrm{e}^{-tN}g\|_{L^{2p'}(\gamma_c)}.\end{aligned}$$

The hypercontractivity estimate (2.8), for q replaced by $2p'$ and $p=2$, now implies (3.2).

(ii) For each $z\in\mathbb{C}^d$ we have $H_bK_z\in\mathcal{H}L^2(\mathbb{C}^d,\gamma_c)$. Explicitly this function is given by

$$H_bK_z(w) = \langle H_bK_z, K_w\rangle = \langle b, K_zK_w\rangle = \langle b, K_{z+w}\rangle = b(z+w). \tag{3.5}$$

Now $\|b(z+\cdot)\|_{L^2(\gamma_c)} = \|H_bK_z\|_{L^2(\gamma_c)} \le \|H_b\|\cdot\|K_z\|_{L^2(\gamma_c)}$, for all $z\in\mathbb{C}^d$. Taking the square and integrating over the probability measure γ_s on $\mathbb{C}^d$ yields

$$\|H_b\|^2 \ge \int_{\mathbb{C}^d}\|b(z+\cdot)\|^2_{L^2(\gamma_c)}\cdot\|K_z\|^{-2}_{L^2(\gamma_c)}\,\mathrm{d}\gamma_s(z).$$

Define u by $u^{-1}=c^{-1}+s^{-1}$. Since $\|K_z\|^2_{L^2(\gamma_c)} = K_z(z) = \mathrm{e}^{|z|^2/2c}$, we have

$$\begin{aligned}\|H_b\|^2 &\ge \int_{\mathbb{C}^d}\Big(\int_{\mathbb{C}^d}|b(z+w)|^2\,\mathrm{d}\gamma_c(w)\Big)\mathrm{e}^{-|z|^2/2c}\,\mathrm{d}\gamma_s(z)\\ &= \Big(\frac{u}{s}\Big)^d\int_{\mathbb{C}^d}\int_{\mathbb{C}^d}|b(z+w)|^2\,\mathrm{d}\gamma_c(w)\,\mathrm{d}\gamma_u(z)\\ &= \Big(\frac{u}{s}\Big)^d\cdot\|b\|^2_{L^2(\gamma_{c+u})}.\end{aligned} \tag{3.6}$$

So if $\|H_b\|<\infty$ then $\|b\|_{L^2(\gamma_{c+u})}<\infty$ whenever $0<s<\infty$, i.e. whenever $u<c$. Thus $b\in\mathcal{H}L^2(\mathbb{C}^d,\gamma_a)$ for all $a<2c$. If we write $a=c\mathrm{e}^{2t/c}$ this is equivalent to $\mathrm{e}^{2t/c}<2$, i.e. equivalent to $t<\frac{c}{2}\ln 2$. In view of (2.6) we find $\mathrm{e}^{tN}b\in\mathcal{H}L^2(\mathbb{C}^d,\gamma_c)$ for all $t<\frac{c}{2}\ln 2$. $D(\mathrm{e}^{tN}) = \mathrm{Ran}\;\mathrm{e}^{-tN}$ concludes the proof. □

Remark. It seems that one cannot apply hypercontractivity in a similar way for the study of Toeplitz operators over $\mathbb{C}^d$. At least the proof of Theorem 3.1 (i) breaks down at an essential point. To see this first recall that a Toeplitz operator T_b with symbol b can be defined by $T_b = P\circ M_b$, where M_b is multiplication by b and $P\colon L^2(\mathbb{C}^d,\gamma_c) \to \mathcal{H}L^2(\mathbb{C}^d,\gamma_c)$ is the orthogonal projection. Such an operator gives rise to the sesquilinear form $\int \bar{b}\bar{f}g \, \mathrm{d}\gamma_c$. The proof of Theorem 3.1 (i) breaks down when we want to factorize $\mathrm{e}^{-tN}(\bar{f}g)$ into $(\mathrm{e}^{-tN}\bar{f})(\mathrm{e}^{-tN}g)$: Equation (3.4) is generally false for non–holomorphic functions.

Next, let me recall the integral representation for Hankel operators already noticed in [20]. For later comparison with the infinite dimensional case the short proof is included.

Theorem 3.2. *Let $b \in \mathcal{H}L^2(\mathbb{C}^d,\gamma_c)$ be the symbol of a continuous Hankel operator. Then*

$$b(z+\cdot\,) \in \mathcal{H}L^2(\mathbb{C}^d,\gamma_c), \quad \forall z \in \mathbb{C}^d. \tag{3.7}$$

Moreover, for all $f \in \mathcal{H}L^2(\mathbb{C}^d,\gamma_c)$ the following integral representation holds:

$$H_b f(z) = \int_{\mathbb{C}^d} b(z+w)\overline{f(w)} \, \mathrm{d}\gamma_c(w). \tag{3.8}$$

Proof. (3.7) is an immediate consequence of (3.5). To prove (3.8) first notice that $V := \operatorname{span}\{K_z, z \in \mathbb{C}^d\}$ is a dense subspace in $\mathcal{H}L^2(\mathbb{C}^d,\gamma_c)$: $f \perp V$ implies $f(z) = \langle f, K_z\rangle = 0$ for all $z \in \mathbb{C}^d$, i.e. $f = 0$. Next proceed as follows: (3.5) gives

$$H_b K_u(z) = b(z+u) = \langle b(z+\cdot\,), K_u\rangle = \int_{\mathbb{C}^d} b(z+w)\overline{K_u(w)} \, \mathrm{d}\gamma_c(w).$$

By anti-linearity this implies (3.8) for all $f \in V$. By continuity and density this extends to all $f \in \mathcal{H}L^2(\mathbb{C}^d,\gamma_c)$. □

The following characterization of HS–symbols is also due to [20]. However, the proof given below is much more elementary than the original one, and it clearly "explains" the doubling of the variance parameter c.

Corollary 3.3. *Let $b \in \mathcal{H}L^2(\mathbb{C}^d,\gamma_c)$. Then*

$$H_b \text{ is HS on } \mathcal{H}L^2(\mathbb{C}^d,\gamma_c) \iff b \in \mathcal{H}L^2(\mathbb{C}^d,\gamma_{2c}). \tag{3.9}$$

Moreover, in case H_b is HS we have $\|H_b\|^2_{\mathrm{HS}(\gamma_c)} = \|b\|^2_{L^2(\gamma_{2c})}$.

Proof. First notice that when H_b is HS the representation (3.8) holds. Conversely, $b \in \mathcal{H}L^2(\mathbb{C}^d,\gamma_{2c})$ implies $\mathrm{e}^{t_J N} b \in \mathcal{H}L^2(\mathbb{C}^d,\gamma_c)$ by (2.6), so H_b is continuous by Theorem 3.1. Therefore each side in (3.9) implies that

H_b is given by (3.5). By the HS–characterization for integral operators ([38],Theorem VI.23) we thus obtain:

$$\begin{aligned} H_b \text{ is HS } &\iff \int_{\mathbb{C}^d \times \mathbb{C}^d} |b(z+w)|^2 \, \mathrm{d}\gamma_c(z) \, \mathrm{d}\gamma_c(w) < \infty \\ &\iff \int_{\mathbb{C}^d} |b(u)|^2 \, \mathrm{d}(\gamma_c * \gamma_c)(u) < \infty \\ &\iff \int_{\mathbb{C}^d} |b(u)|^2 \, \mathrm{d}\gamma_{2c}(u) < \infty. \end{aligned} \tag{3.10}$$

4. Infinite dimensional preparations

This section provides some background on holomorphic Wiener functionals so that readers from the field of Hankel operators can follow the discussion.

The complex Wiener space $(W_{\mathbb{C}}, \mathcal{B}, \mu_c)$ over $[0,T]$ consist of the following objects:

$$W_{\mathbb{C}} := \{\omega \in C([0,T], \mathbb{C}) \mid \omega(0) = 0\}.$$

$\mathcal{B}$ is the Borel σ–algebra on $W_{\mathbb{C}}$ induced by the $\|\cdot\|_\infty$–topology on $W_{\mathbb{C}}$.

$$Z_t(\omega) := X_t(\omega) + iY_t(\omega) := \omega(t), \quad \omega \in W_{\mathbb{C}}, \tag{4.1}$$

is the canonical process, and μ_c is the Wiener measure with variance parameter $c > 0$, i.e., μ_c is the unique probability measure on $\mathcal{B}$ which is such that:

(Z1) X and Y defined in (4.1) are independent, real processes.

(Z2) X and Y have independent, centered Gaussian increments.

(Z3) $\mathbb{E}[(X_t - X_s)^2] = \mathbb{E}[(Y_t - Y_s)^2] = c(t-s)$ for all $0 \leq s < t \leq T$.

Remark. The measures μ_c and $\mu_{\tilde{c}}$ with $c < \tilde{c}$ are singular against each other, $\mu_c \perp \mu_{\tilde{c}}$. So there is no natural relation, such as $L^p(\mathbb{C}^d, \gamma_{\tilde{c}}) \subset L^p(\mathbb{C}^d, \gamma_c)$, between $L^p(W_{\mathbb{C}}, \mu_{\tilde{c}})$ and $L^p(W_{\mathbb{C}}, \mu_c)$. This requires some care when one extends results for Hankel operators from finite to infinite dimensions. Therefore I will sometimes distinguish an equivalence class of functions $[f]_c \in L^2(\mu_c)$ from a pointwise defined representative $f \in [f]_c$.

Denote by $\mathcal{P}(Z)$ the *algebra of holomorphic polynomials* generated by the complex Gaussian variables Z_t, $t \in [0,T]$. Since the point evaluation $Z_t(\omega) = \omega(t)$ is a continuous linear functional on the complex Banach space $(W_{\mathbb{C}}, \|\cdot\|_\infty)$ the space $\mathcal{P}(Z)$ consists of holomorphic functions on $(W_{\mathbb{C}}, \|\cdot\|_\infty)$. (For background on holomorphic functions in Banach spaces see e.g. [13].) One can uniquely identify $Q \in \mathcal{P}(Z)$ with its equivalence class $[Q]_c \in L^p(\mu_c)$ because $[Q]_c$ contains exactly one representative in $\mathcal{P}(Z)$. (The simple proof is left to the reader.) Subsequently we identify a polynomial Q with its class $[Q]_c$, and thereby $\mathcal{P}(Z)$ with a subspace in $L^p(\mu_c)$. Notice that this allows to identify the classes $[Q]_c$ and $[Q]_{\tilde{c}}$, which a priori are not comparable.

For $p \in (1, \infty)$ define the space of *holomorphic L^p–Wiener functionals* to be

$$\mathcal{H}L^p(W_{\mathbb{C}}, \mu_c) := \mathcal{H}L^p(\mu_c) := L^p(\mu_c)\text{-closure of } \mathcal{P}(Z). \tag{4.2}$$

Remarks. 1. Holomorphic L^p–Wiener functionals can be introduced and characterized in several ways, and they have been studied to some extend in the literature [5, 6, 7, 23, 24, 39, 41, 42, 43, 45, 46]. The definition (4.2) requires the least amount of terminology.

2. An arbitrary class $[f]_c \in \mathcal{H}L^2(\mu_c)$ can in general not be identified in a unique way with a pointwise defined representative f (as for polynomials, or in the finite dimensional case). In particular $[f]_c$ need not contain a continuous representative f on $(W_{\mathbb{C}}, \|\cdot\|_\infty)$, see [44]. So in the strict sense $\mathcal{H}L^p(\mu_c)$ is not a space of holomorphic functions on $(W_{\mathbb{C}}, \|\cdot\|_\infty)$. (However, $\mathcal{H}L^p(\mu_c)$ is naturally isomorphic to a space of genuine holomorphic functions on the complex Cameron Martin subspace of $W_{\mathbb{C}}$, for details see [9, 42].) We will see that this "deficiency" does not seriously affect the basic framework for Hankel operators.

Let me now construct an orthogonal basis in $\mathcal{H}L^2(W_{\mathbb{C}}, \mu_c)$ so that every $\varphi \in \mathcal{H}L^2(\mu_c)$ can be represented by a power series w.r.t. complex "variables" $Z_1, Z_2, \ldots$, quite analogous to the $\mathbb{C}^d$-case. This representation allows to restrict every such φ to finitely many variables $Z_1, \ldots, Z_d$ and thereby to obtain a natural link to the spaces $\mathcal{H}L^2(\mathbb{C}^d, \gamma_c)$, as explained by Lemma 5.1. The basic variables Z_k are complex Gaussian random variables constructed by the elementary Wiener integral as follows: Let f be a step function on $[0, T]$, i.e. there are time points $0 = t_0 < t_1 < \cdots < t_n = T$ and constants $f_i \in \mathbb{C}$ such that $f = \sum_{i=1}^n f_i \, 1_{[t_{i-1}, t_i)}$. Denote by $S[0, T]$ the vector space of such step functions. Then the stochastic integral of $f \in S[0, T]$ w.r.t. complex Brownian motion is defined as

$$\int_0^T f(t)\, \mathrm{d}Z_t(\omega) := \sum_{i=1}^n f_i (Z_{t_i}(\omega) - Z_{t_{i-1}}(\omega)), \quad \forall \omega \in \Omega.$$

The properties (Z1), (Z2), (Z3) satisfied by the Brownian motion Z easily imply

$$\left\| \int_0^T f(t)\,\mathrm{d}Z_t \right\|^2_{L^2(\mu_c)} = 2c\|f\|^2_{L^2(\,\mathrm{d}x)}, \tag{4.3}$$

so the linear map $f \mapsto \int_0^T f(t)\,\mathrm{d}Z_t$ is isometric from $(S[0,T], \|\cdot\|_{L^2(\,\mathrm{d}x)})$ to $L^2(\mu_c)$, up to the normalization constant $2c$.

For general $f \in L^2([0,T], \mathrm{d}x)$ the Wiener integral $\int_0^T f(t)\,\mathrm{d}Z_t$ is simply obtained by continuous extension of this map, and it defines a complex Gaussian random variable. Clearly the Itô-isometry (4.3) remains valid under this extension, and by definition (4.2) we have $\int_0^T f(t)\,\mathrm{d}Z_t \in \mathcal{H}L^2(\mu_c)$.

The following fact is an immediate consequence of the Segal-Bargmann isomorphism [9] applied to the well–known Hermite–basis over the real Wiener space $W_{\mathbb{R}}$, and it generalizes parts of Theorem 2.1 in a natural way:

Theorem 4.1. *Let $\{e_1, e_2, \dots\}$ be an orthonormal basis in $L^2([0,T], \mathrm{d}x)$. Define*

$$Z_k := \int_0^T e_k(t)\,\mathrm{d}Z_t, \quad k \in \mathbb{N}. \tag{4.4}$$

Put $\mathbb{N}_c^\infty := \{(\alpha_1, \alpha_2, \dots) \mid$ only finitely many $\alpha_i \in \mathbb{N}_0$ are non-zero$\}$, $|\alpha| := \alpha_1 + \dots + \alpha_d$ and $\alpha! := \alpha_1! \cdots \alpha_d!$, where d is the largest index such that $\alpha_d \neq 0$. Define

$$Z^0 := 1, \quad Z^\alpha := Z_1^{\alpha_1} \cdots Z_d^{\alpha_d}.$$

Then $\{Z^\alpha, \alpha \in \mathbb{N}_c^\infty\}$ is an OGB in $\mathcal{H}L^2(\mu_c)$ with normalization

$$\|Z^\alpha\|^2_{L^2(\mu_c)} = (2c)^{|\alpha|}\alpha!.$$

By the previous theorem any $\varphi \in \mathcal{H}L^2(\mu_c)$ admits an orthogonal expansion

$$\varphi = \sum_{n=0}^\infty \varphi^{(n)} = \sum_{n=0}^\infty \sum_{|\alpha|=n} a_\alpha Z^\alpha = \sum_{\alpha \in \mathbb{N}_c^\infty} a_\alpha Z^\alpha, \tag{4.5}$$

with

$$\|\varphi\|^2_{L^2(\mu_c)} = \sum_{\alpha \in \mathbb{N}_c^\infty} (2c)^{|\alpha|}\alpha!\,|a_\alpha|^2 < \infty. \tag{4.6}$$

Remarks. 1. Although the functions Z_k given in (4.4) depend on the choice of basis $\{e_1, e_2, \dots\}$, the components $\varphi^{(n)}$ given in (4.5) do not. This follows from the corresponding property of the Hermite decomposition over $W_{\mathbb{R}}$ together with the Segal–Bargmann transformation.

2. Observe that the function $\tilde{Z}_k(z_1, \dots, z_d) := z_k$ defines a complex Gaussian random variable on $(\mathbb{C}^d, \gamma_c)$, so the expansion $f(z) = \sum a_\alpha z^\alpha$ for $f \in \mathcal{H}L^2(\mathbb{C}^d, \gamma_c)$ can also be viewed as a series of Gaussian random variables.

3. The representation (4.5) is called the *complex chaos expansion* of φ, and the $\varphi^{(n)}$ are called the *chaos components*.

From now on we fix an arbitrary ONB $\{e_1, e_2, \dots\}$ in $L^2([0,T], \mathrm{d}x)$, and thereby also the corresponding Z_k given in (4.4). None of the results in this paper depends on the choice of such a basis. The OU–semigroup on $\mathcal{H}L^2(\mathbb{C}^d, \gamma_c)$ now generalizes in the obvious (and basis independent) way to $\mathcal{H}L^2(\mu_c)$: For $\varphi = \sum_{n=0}^\infty \varphi^{(n)} \in \mathcal{H}L^2(\mu_c)$ we denote by

$$\mathrm{e}^{-tN} \sum_{n=0}^{\infty} \varphi^{(n)} := \sum_{n=0}^{\infty} \mathrm{e}^{-tn/c} \varphi^{(n)}. \tag{4.7}$$

the *Ornstein–Uhlenbeck semigroup* $(\mathrm{e}^{-tN})_{t \geq 0}$ on $\mathcal{H}L^2(W_{\mathbb{C}}, \mu_c)$. To distinguish the finite dimensional case we will write $\mathrm{e}^{-t\tilde{N}}$ for the OU–semigroup on $\mathcal{H}L^2(\mathbb{C}^d, \gamma_c)$. We don't need to recall the hypercontractivity of the semigroup e^{-tN} on $\mathcal{H}L^2(\mu_c)$ here, because subsequently we only use the hypercontractivity on $\mathcal{H}L^2(\mathbb{C}^d, \gamma_c)$.

5. Hankel operators on holomorphic Wiener functionals

Subsequently we study continuous Hankel forms Γ_b on $\mathcal{H}L^2(\mu_c)$ given by

$$\Gamma_b(f, g) = \langle fg, b \rangle = \int_{W_{\mathbb{C}}} \bar{b} f g \, \mathrm{d}\mu_c, \qquad \forall f, g \in \mathcal{P}(Z), \tag{5.1}$$

with suitable symbol functions $b \in \mathcal{H}L^2(\mu_c)$. Again the associated Hankel operator H_b on $\mathcal{H}L^2(\mu_c)$ is defined by $\Gamma_b(f, g) = \langle f, H_b g \rangle$, for all $f, g \in \mathcal{H}L^2(\mu_c)$.

The basic idea to generalize Theorem 3.1 to $\mathcal{H}L^2(\mu_c)$ goes as follows: For fixed polynomials $f, g \in \mathcal{P}(Z_n, n \in \mathbb{N})$ we have

$$f, g \in \overline{\mathcal{P}(Z_1, \dots, Z_d)} =: F_d,$$

for a suitable d. Thus the integral in (5.1) may be written $\int \overline{(\pi_d b)} f g \, \mathrm{d}\mu_c$, where π_d projects on F_d. Clearly this integral can be expressed as an integral over $\mathbb{C}^d$. If we put $b = \mathrm{e}^{-tN}\varphi$ we recover the situation of Theorem 3.1, provided the OU–semigroups are appropriately related. So let us first investigate how e^{-tN} relates to $\pi_d\varphi$, for $\varphi \in \mathcal{H}L^2(\mu_c)$. Let φ be given by the chaos expansion (4.5). It is simple to check that

$$\pi_d\varphi = \sum_{\alpha\in\mathbb{N}_0^d} a_\alpha Z^\alpha. \tag{5.2}$$

This suggests the name "finite variable restriction" for $\pi_d\varphi$. Since e^{-tN} is a diagonal operator on the basis $\{Z^\alpha, \alpha \in \mathbb{N}_c^\infty\}$ formula (5.2) implies

$$\mathrm{e}^{-tN} \circ \pi_d = \pi_d \circ \mathrm{e}^{-tN}. \tag{5.3}$$

The following result summarizes some relations between functions in F_d and functions on $\mathbb{C}^d$, as well as the relation between the corresponding OU–semigroups. Notice that (5.4) specializes the well-known measurable factorization to a "holomorphic factorization".

Lemma 5.1. *For $f \in F_d$ there exists a unique $\tilde{f} \in \mathcal{H}L^2(\mathbb{C}^d, \gamma_c)$ such that*

$$f = \tilde{f}(Z_1, \ldots, Z_d) \quad \mu_c\text{-a.s.} \tag{5.4}$$

The map $J\colon f \mapsto \tilde{f}$ is isometric from $(F_d, \|\cdot\|_{L^2(\mu_c)})$ onto $\mathcal{H}L^2(\mathbb{C}^d, \gamma_c)$. Moreover

$$J \circ \mathrm{e}^{-tN} \circ J^{-1} = \mathrm{e}^{-t\tilde{N}}, \tag{5.5}$$

i.e. the equality $\mathrm{e}^{-tN} f = (\mathrm{e}^{-t\tilde{N}}\tilde{f})(Z_1, \ldots, Z_d)$ holds μ_c-a.s.

With this lemma the first part of Theorem 3.1 generalizes as follows:

Theorem 5.2. *Let e^{-tN} be the OU-semigroup on $\mathcal{H}L^2(\mu_c)$ and let $\varphi \in \mathcal{H}L^p(\mu_c)$ with $p \geq 2$. Put $b := \mathrm{e}^{-tN}\varphi$, with $t \geq t_J := \frac{c}{2}\ln p'$, where $\frac{1}{p} + \frac{1}{p'} = 1$. Then, for all $f, g \in \mathcal{P}(Z_n, n \in \mathbb{N})$:*

$$\left|\int_{W_\mathbb{C}} \bar{b} f g \, \mathrm{d}\mu_c\right| \leq \|\varphi\|_{L^p(\mu_c)} \|f\|_{L^2(\mu_c)} \|g\|_{L^2(\mu_c)}. \tag{5.6}$$

Sketch of proof. Choose d such that $f, g \in \mathcal{P}(Z_1, \ldots, Z_d)$. Put $\tilde{\varphi}_d = J(\pi_d\varphi)$, $\tilde{f} = J(f)$, $\tilde{g} = J(g)$. With Lemma 5.1 we conclude $\Gamma_{\mathrm{e}^{-tN}\varphi}(f, g) = \int_{\mathbb{C}^d} \overline{\mathrm{e}^{-t\tilde{N}}\tilde{\varphi}_d} \tilde{f}\tilde{g} \, \mathrm{d}\gamma_c$. Now (3.2) gives

$$\begin{aligned} |\Gamma_{\mathrm{e}^{-tN}\varphi}(f, g)| &\leq \|\tilde{\varphi}_d\|_{L^p(\gamma_c)} \|\tilde{f}\|_{L^2(\gamma_c)} \|\tilde{g}\|_{L^2(\gamma_c)} \\ &\leq \|\varphi\|_{L^p(\mu_c)} \|f\|_{L^2(\mu_c)} \|g\|_{L^2(\mu_c)}. \end{aligned}$$

□

Remarks. 1. (5.6) implies the continuity of Γ_b because $\mathcal{P}(Z_n, n \in \mathbb{N})$ is dense in $\mathcal{H}L^2(\mu_c)$.

2. Part (ii) of Theorem 3.1 (least regularity) seems not to generalize to $\mathcal{H}L^2(\gamma_c)$: The reverse estimate (3.6) depends on d, and the right side in it goes to zero as $d \to \infty$.

Next let us investigate the Hilbert Schmidt (HS) property of H_b. Since $\mu_c \perp \mu_{2c}$ the spaces $L^2(\mu_c)$ and $L^2(\mu_{2c})$ contain fundamentally different function classes. However, Theorem 4.1 has the important consequence that for $c \leq \tilde{c}$ we can naturally identify $\mathcal{H}L^2(\mu_{\tilde{c}})$ with a subspace in $\mathcal{H}L^2(\mu_c)$ as follows: For a suitable ONB $\{e_1, e_2, \dots\}$ in $L^2([0,T], dx)$ the associated set $\{Z^\alpha, \alpha \in \mathbb{N}_c^\infty\}$ constitutes an OGB in $\mathcal{H}L^2(\mu_{\tilde{c}})$ and in $\mathcal{H}L^2(\mu_c)$. The convergence of (4.6) with parameter $\tilde{c}$ implies convergence of (4.6) with parameter $c < \tilde{c}$. This remarkable relation can be stated in a basis independent way as follows:

Lemma 5.3. *Let $\tilde{c} \geq c > 0$. Then the identity map $I\colon \mathcal{P}(Z) \to \mathcal{P}(Z)$ extends by continuity to a continuous, injective map*

$$\tilde{I}\colon \mathcal{H}L^2(\mu_{\tilde{c}}) \to \mathcal{H}L^2(\mu_c).$$

Remarks. 1. The map $\tilde{I}$ is explicitly given by

$$\tilde{I}\colon \sum_{\alpha \in \mathbb{N}_c^\infty} a_\alpha Z^\alpha \mapsto \sum_{\alpha \in \mathbb{N}_c^\infty} a_\alpha Z^\alpha.$$

Notice that the left side is considered as an orthogonal series in $\mathcal{H}L^2(\mu_{\tilde{c}})$, while the right side is an orthogonal series in $\mathcal{H}L^2(\mu_c)$. We will identify $\mathcal{H}L^2(\mu_{\tilde{c}})$ with its image in $\mathcal{H}L^2(\mu_c)$ under the map $\tilde{I}$ whenever $\tilde{c} \geq c$.

2. For the full space $L^2(\mu_c)$ the identity map on polynomials $\mathcal{P}(X_t, Y_t)$, $t \in [0,T]$ is not continuous. The proof of Lemma 5.3 breaks down in that case, because Hermite–polynomials (replacing the Z^α) with respect to variance $\tilde{c}$ are not orthogonal to each other in $L^2(\mu_c)$ if $\tilde{c} \neq c$.

Since $\mathcal{H}L^2(\mu_c)$ is not a genuine space of functions the notion of a reproducing kernel makes no sense in $\mathcal{H}L^2(\mu_c)$. The construction of an integral representation of type (3.8) along the proof of Theorem 3.2 is therefore not possible. (It does not help to go the Cameron-Martin subspace.) But a generalization of Corollary 3.3 is still possible via finite variable approximations. The following key lemma holds:

Lemma 5.4. *Let $b \in \mathcal{H}L^2(\mu_c)$ be such that H_b is continuous on $\mathcal{H}L^2(\mu_c)$. Then*

$$H_{\pi_d b} = \pi_d \circ H_b \circ \pi_d. \tag{5.7}$$

Let $J\colon f \mapsto \tilde{f}$ *be the isometry defined in Lemma* 5.1, *and put* $b_d := \pi_d b$. *Then*

$$H_{J(b_d)} = J \circ H_{b_d} \circ J^{-1}. \tag{5.8}$$

Remark. Equation (5.7) shows that the "finite variable restriction" $\pi_d \circ H_b \circ \pi_d$ of H_b to the subspace F_d is again a Hankel operator, H_{b_d}. (This statement is false for general orthogonal projections in $\mathcal{H}L^2(\mu_c)$.) Moreover (5.8) shows that H_{b_d} is unitary equivalent to the Hankel operator $H_{\tilde{b}_d}$ on the space $\mathcal{H}L^2(\mathbb{C}^d, \gamma_c)$. These two properties provide a tight relation between Hankel operators on $\mathcal{H}L^2(W_{\mathbb{C}}, \mu_c)$ and Hankel operators on $\mathcal{H}L^2(\mathbb{C}^d, \gamma_c)$.

With these ingredients one can now generalize the HS–characterization (3.9) in a completely satisfying way. Its (slightly technical) proof is based on Lemma 5.4, on Corollary 3.3, and on general properties of HS operators. For details see [3].

Theorem 5.5. *Let* $b \in \mathcal{H}L^2(\mu_c)$. *Then*

$$H_b \textit{ is HS in } \mathcal{H}L^2(\mu_c) \iff b \in \mathcal{H}L^2(\mu_{2c}). \tag{5.9}$$

Moreover, in case H_b *is HS we have* $\|H_b\|^2_{\mathrm{HS}(\mu_c)} = \|b\|^2_{L^2(\mu_{2c})}$.

Remark. The map $\tilde{I}$ is explicitly given by In Corollary 3.3 one may replace $b \in \mathcal{H}L^2(\mathbb{C}^d, \gamma_c)$ by $b \in L^2(\mathbb{C}^d, \gamma_c)$, because $\Gamma_b = 0$ if $b \in \mathcal{H}L^2(\mathbb{C}^d, \gamma_c)^{\perp}$ (the orthogonal complement in $L^2(\mathbb{C}^d, \gamma_c)$). This replacement is not possible in Theorem 5.5 because one cannot identify $L^2(\mu_{2c})$ with a subspace in $L^2(\mu_c)$.

Let us finally generalize the integral representation (3.8). Notice that one cannot simply choose $b \in [b]_c \in \mathcal{H}L^2(\mu_c)$ and consider $b(\omega + \cdot\,)$ with a fixed ω because this function depends significantly on the choice of the representative b (details are given in [3]). So in general the integral kernel $b(\omega + \omega')$ —corresponding to the one in (3.8)— is not well-defined in the Wiener space context. However, if $b \in \mathcal{H}L^2(\mu_{2c}) \subset \mathcal{H}L^2(\mu_c)$ an interesting exception holds which allows to derive the corresponding integral representation:

Theorem 5.6. *Let* $b \in \mathcal{H}L^2(\mu_{2c})$. *Then* $(\omega, \omega') \mapsto b(\omega + \omega')$ *is well-defined as an element in* $L^2(\mu_c \otimes \mu_c)$. *Moreover, the HS–operator*

$$H_b\colon \mathcal{H}L^2(\mu_c) \to \mathcal{H}L^2(\mu_c)$$

is given by

$$H_b f(\omega) = \int_{W_{\mathbb{C}}} b(\omega + \omega') \overline{f(\omega')}\, \mathrm{d}\mu_c(\omega') \quad \mu_c\text{-a.s.} \tag{5.10}$$

Sketch of proof. Choose $b \in [b]_{2c}$. By convolution $\mu_c * \mu_c = \mu_{2c}$ one finds

$$\int |b(\omega+\omega')|^2 \, \mathrm{d}(\mu_c \otimes \mu_c)(\omega,\omega') = \int |b(u)|^2 \, \mathrm{d}\mu_{2c}(u) < \infty. \quad (5.11)$$

Now choose $b_1, b_2 \in [b]_{2c}$ and replace in the previous calculation b by $b_1 - b_2$. This gives

$$\begin{aligned} &\int |b_1(\omega+\omega') - b_2(\omega+\omega')|^2 \, \mathrm{d}(\mu_c \otimes \mu_c)(\omega,\omega') \\ &= \int |b_1(u) - b_2(u)|^2 \, \mathrm{d}\mu_{2c}(u) = 0. \end{aligned} \quad (5.12)$$

(5.11) and (5.12) yield the first assertion. For $f \in F_d$ and $b_d := \pi_d b$ Theorem 3.2 gives $\tilde{b}_d(z + \cdot\,) \in \mathcal{H}L^2(\gamma_c)$, and

$$H_{\tilde{b}_d}\tilde{f}(z) = \int_{\mathbb{C}^d} \tilde{b}_d(z+u)\overline{\tilde{f}(u)} \, \mathrm{d}\gamma_c(u).$$

With $Z = (Z_1, \ldots, Z_d)$ this converts to

$$H_{b_d}f(w) = H_{\tilde{b}_d}\tilde{f}(Z(w)) = \int_{W_{\mathbb{C}}} b_d(w+w')\overline{f(w')} \, \mathrm{d}\mu_c(w'), \quad (5.13)$$

for all $f \in F_d$. By continuity this extends to all $f \in \mathcal{H}L^2(\mu_c)$. A limiting argument (for $d \to \infty$) applied to both sides of (5.13) allows the transition to (5.10). □

References

[1] Bargmann V.: On a Hilbert space of analytic functions and an associated integral transform, Part I, *Comm. Pure Appl. Math.* **24** (1961) 187–214.

[2] Carlen E.: Some integral identities and inequalities for entire functions and their applications to the coherent state transform, *J. Funct. Anal.* **97** (1991) 231–249.

[3] Deck T.: Hankel operators over the complex Wiener space, *Potential Anal.*, to appear.

[4] Deck T., Gross L.: Hankel operators over complex manifolds, *Pacific J. Math.* **205** (2002) 43–97.

[5] Fang S.: On derivatives of holomorphic functions on a complex Wiener space, *J. Math. Kyoto Univ.* **34**(3) (1994) 637–640.

[6] Fang S., Ren J.: Quelques propriétés des fonctions holomorphes sur un espace de Wiener complexe, *C. R. Acad. Sci. Paris Sér. I Math.* **315**(4) (1992) 447–450.

[7] Fang S., Ren J.: Sur le squelette et les dérivées de Malliavin des fonctions holomorphes sur un espace de Wiener complexe, *J. Math. Kyoto Univ.* **33**(3) (1993) 749–764.

[8] Gross L.: Hypercontractivity over complex manifolds, *Acta Math.* **182** (1999) 159–206.

[9] Gross L., Malliavin P.: Hall's Transform and the Segal-Bargmann Map, Festschrift: "Itô's Stochastic Calculus and Probability Theory", N. Ikeda, S. Watanabe, M. Fukushima and H. Kunita, eds., Springer-Verlag 1996, 73–116.

[10] Hankel H.: Über eine besondere Klasse der symmetrischen Determinanten, (Leipziger) Dissertation, Göttingen, 1861.

[11] Hartmann P.: On completely continuous Hankel matrices, *Proc. Amer. Math. Soc.* **9** (1958) 862–866.

[12] Hida T., Kuo H.-H., Potthoff J., Streit L.: *White Noise – An Infinite Dimensional Calculus*, Dordrecht: Kluwer, Dordrecht, 1993.

[13] Hille E., Phillips R.: *Functional Analysis and Semi-Groups*, Amer. Math. Soc., Providence, Rhode Island, 1957.

[14] Holland F., Rochberg R.: Bergman kernels and Hankel forms on generalized Fock spaces, in "Function spaces", K. Jarosz, ed., *Contemporary Mathematics* **232** (1999) 189–200.

[15] Janson S.: *Gaussian Hilbert spaces.* Cambridge Univ. Press, Cambridge, 1998.

[16] Janson S.: On hypercontractivity for multipliers on orthogonal polynomials, *Ark. Mat.* **21** (1983) 97–110.

[17] Janson S.: On complex hypercontractivity, *J. Funct. Anal.* **151** (1997) 270–280.

[18] Janson S.: Hankel operators on Bergman spaces with change of weight, *Math. Scand.* **71** (1992) 267–276.

[19] Janson S., Peetre J.: Weak factorization in periodic Fock space. *Math. Nachr.* **146** (1990) 159–165.

[20] Janson S., Peetre J., Rochberg R.: Hankel Forms and the Fock Space, *Rev. Mat. Iberoamericana* **3** (1987) 61–138.

[21] Janson S., Peetre J., Wallsten R.: A new look on Hankel forms over Fock space, *Studia Math.* **95**(1) (1989) 33–41.

[22] Kronecker L.: Zur Theorie der Elimination einer Variablen aus zwei algebraischen Gleichungen, *Monatsber. Königl. Preuss. Akad. Wiss. (Berlin)* (1881) 535–600.

[23] Malliavin P., Taniguchi S.: Extension holomorphe des fonctionnelles analytiques définies sur un espace de Wiener réel, formule de Cauchy, phase stationnaire. *C. R. Acad. Sci. Paris Sér. I Math.* **322**(3) (1996) 261–265.

[24] Malliavin P., Taniguchi S.: Analytic Functions, Cauchy Formula, and Stationary Phase on a Real Abstract Wiener Space, *J. Funct. Anal.* **143** (1997) 470–528.

[25] Nehari Z.: On bounded bilinear forms, *Ann. of Math.* **65** (1957) 153–162.

[26] Partington J. R.: *An Introduction to Hankel Operators*, Cambridge Univ. Press, Cambridge, 1988.

[27] Peetre J.: *Generalizations of Hankel operators, Nonlinear Analysis, Function spaces and applications*, Vol. 3. Teubner, 1986.

[28] Peetre J.: Hankel forms on Fock space modulo C_N, *Results Math.* **14**(3/4) (1988) 333–339.

[29] Peetre J.: On the S_4-norm of a Hankel form, *Rev. Mat. Iberoamericana* **8**(1) (1992) 121–130.

[30] Peller V. V.: Hankel Operators and Multivariate Stationary Processes, *Proc. Symp. Pure Math.* **51** (1990) 357–371.

[31] Peller V. V., Khrushchev S. V.: Hankel operators, best approximation, and stationary Gaussian processes, *Russian Math. Surveys* **37**(1) (1982) 61–144.

[32] Power S. C.: Hankel Operators on Hilbert Space, *Bull. London Math. Soc.* **12** (1980) 422–442.

[33] Power S. C.: *Hankel Operators on Hilbert Space*, Pitman Research Notes 64, 1982.

[34] Power S. C.: Finite Rank Multivariable Hankel Forms, *Linear Algebra Appl.* **48** (1982) 237–244.

[35] Pustyl'nikov L. D.: Toeplitz and Hankel matrices and their applications, *Russ. Math. Surv.* **39**(4) (1984) 63–98.

[36] Rochberg R.: Higher-Order Hankel Forms and Commutators, In: "Holomorphic Spaces", Axler S. et al., eds., Cambridge 1998.

[37] Rochberg R., Rubel L. A.: A functional equation, *Indiana Univ. Math. J.* **41**(2) (1992) 363–376.

[38] Reed M., Simon B.: *Methods of Modern Mathematical Physics I, Functional analysis*, Academic Press 1980.

[39] Shigekawa I.: Itô–Wiener expansions of holomoprhic functions on the complex Wiener space, In "Stochastic Analysis", E. Meyer et al., eds., 459–473, Academic Press, San Diego, 1991.

[40] Stroethoff K.: Hankel and Toeplitz operators on the Fock space, *Michigan Math. J.* **39**(1) (1992) 3–16.

[41] Sugita H.: Holomorphic Wiener Function, In "New Trends in Stochastic Analysis", Proc. of a Taniguchi Int. Workshop, K. Elworthy, S. Kusuoka, I. Shigekawa, eds., 399–415, World Scientific, 1997.

[42] Sugita H.: Properties of holomorphic Wiener functions – skeleton, contraction and local Taylor expansion, *Prob. Theo. and Rel. Fields* **100** (1994) 117–130.

[43] Sugita H.: Regular version of holomorphic Wiener function. *J. Math. Kyoto Univ.* **34**(4) (1994) 849–857.

[44] Sugita H.: Hu-Meyer's multiple Stratonovich integral and essential continuity of multiple Wiener integrals, *Bull. Sc. Math.* Série 2, **113** (1989) 463–474.

[45] Sugita H., Takanobu S.: Accessibility of infinite dimensional Brownian motion to holomorphically exceptional set, *Proc. Japan Acad. Ser. A Math. Sci.* **71**(9) (1995) 195–198.

[46] Taniguchi S.: Holomorphic functions on balls in an almost complex abstract Wiener space. *J. Math. Soc. Japan* **47**(4) (1995) 655–670.

[47] Wallsten R.: The S_p-criterion for Hankel forms on the Fock space, $0 < p < 1$, *Math. Scand.* **64**(1) (1989) 123–132.

[48] Zhou Z.: The contractivity of the free Hamiltonian semigroup in L^p spaces of entire functions, *J. Funct. Anal.* **96** (1991) 407–425.

[49] Zhu K.: *Operator Theory in Function Spaces*, M. Dekker Inc., New York, Basel, 1990.

MALLIAVIN CALCULUS AND REAL BOTT PERIODICITY

Rémi Léandre
Institut Elie Cartan, Université de Nancy I, 54000, Vandœuvre-les-Nancy, France
leandre@iecn.u-nancy.fr

Abstract We give a "stochastic real Bott periodicity" theorem.

Keywords: Stochastic analysis, Malliavin calculus, K-theory, Bott periodicity

1. Introduction

Let M be a compact manifold. A bundle (complex or real) over M can be trivialized by adding an auxiliary bundle. In K-theory, people consider the algebra A of continuous functionals over M, they consider the injective limit $M(A)$ of matrices $M_n(A) \subset M_{n+1}(A)$ with coefficients in A, and show that a bundle is given by an idempotent. People introduce the Grothendieck group of M in real K-theory $K_{\mathbb{R}}(M)$ or in complex K-theory $K_{\mathbb{C}}(M)$. Its structure is deeply related to the properties of the algebra A of complex or real continuous functionals over M. In particular, we can consider the relative Grothendieck groups $K_{\mathbb{C}}(M,B)$ or $K_{\mathbb{R}}(M,B)$ for a subspace B of M.

The real Bott periodicity theorem says that

$$K_{\mathbb{R}}(M \times I^{n+8}, M \times \partial I^{n+8}) = K_{\mathbb{R}}(M \times I^n, M \times \partial I^n) \tag{1.1}$$

where $I^n \subset \mathbb{R}^n$ is the unit cube and ∂I^n its boundary (see [1, 7, 15, 32]). We refer to the book of Karoubi for material about K-theory ([16]).

The interest for K-theory comes from the Index theorem over a manifold ([3, 4, 6]).

In theoretical physics, people replace the compact manifold M by the space of free loop space over it or the based loop space of it ([12, 31]), study infinite dimensional operators over these infinite dimensional spaces and compute their indices. Computations are purely formals. In particular, physicists consider formal measures over the loop space. A

S. Albeverio et al. (eds.),
Proceedings of the International Conference on Stochastic Analysis and Applications, 37–51.

natural applicant is the Brownian bridge measure. [14, 18, 24, 25, 29]) have studied stochastic regularizations of the operators of physicists.

But in algebra, the auxiliary bundle which is tensorized with the spin bundle where the Dirac operator acts plays a big role (See [8, 10, 11]). Let us remark in these algebraic theories, the auxiliary bundle is given by an idempotent.

On the other hand, there is a beginning of theory of bundles over the Brownian bridge (See [20, 21]). In the transition functionals, some stochastic integrals appear, which are almost surely defined. This shows that the good category in order to study the algebraic K-theory associated to the loop space is to consider idempotent which are almost surely defined (see [24, 25]). We need to consider algebras of functionals which are almost surely defined. In particular, there are a lot of Calculi which were defined before Malliavin Calculus for the flat Brownian motion (See works of Hida, Krée, Albeverio, Elworthy, Fomin,...), but the specificities of the Malliavin Calculus are the followings:

- The set of test functionals is an *algebra.*
- The test functionals are *almost surely* defined.
- It can be applied to *diffusions.*

In [25], we have considered an algebra of complex functionals of Malliavin type over the loop space, and we have defined a complex Grothendieck group in Malliavin sense, and we have established a complex Bott periodicity theorem:

$$K_{\mathbb{C}}(L_x(M)) = K_{\mathbb{C}}(L_x(M) \times B^2, L_x(M) \times S^1) \tag{1.2}$$

where $L_x(M)$ is the based loop space of the compact Riemannian manifold M endowed with the Brownian bridge measure, B^2 is the unit disk in $\mathbb{R}^2$ and S^1 its boundary.

Our motivation is to generalize this theorem in the real case, with an algebra of real functionals of Malliavin type. Our main theorem is:

Theorem 1.1 (Stochastic real Bott periodicity).

$$K_{\mathbb{R}}(L_x(M)\times I^{n+8}, L_x(M)\times \partial I^{n+8}) = K_{\mathbb{R}}(L_x(M)\times I^n, L_x(M)\times \partial I^n) \tag{1.3}$$

The proof follows closely the proof of this theorem for a compact manifold given in the book of M. Karoubi, chapter III (See [16]). This work produces a stochastic example to the considerations of ([16], III). The reader interested in the relation between analysis over loop space and topology can see the two surveys of Léandre about this topic [22] and [31].

2. Stochastic K-Theory and Clifford algebras

Let us consider a compact Riemannian manifold M. Let us introduce the based loop space $L_x(M)$ of continuous applications γ from the circle S^1 into M such that $\gamma(1) = x$. Let Δ be the Laplace-Beltrami operator and $p_t(x,y)$ be the heat kernel. We endow $L_x(M)$ of the Brownian bridge measure. If $0 < s_1 < \cdots < s_r < 1$ and if F is an application from M^r into R, we have

$$\begin{aligned}
&\mathbb{E}\left[F(\gamma(s_1),\dots,\gamma(s_r))\right] \qquad (2.1)\\
&= \frac{1}{p_1(x,x)}\int_{M^r} p_{s_1}(x,x_1)p_{s_2-s_1}(x_1,x_2)\dots p_{1-s_r}(x_r,x)\times\\
&\qquad\qquad \times F(x_1,\dots,x_r)\,\mathrm{d}\pi(x_1)\dots\,\mathrm{d}\pi(x_r)
\end{aligned}$$

if π denotes the Riemannian measure over the Riemannian manifold. Moreover, $s \mapsto \gamma(s)$ is a semi-martingale. Let τ_s be the stochastic parallel transport along a loop. A tangent vector field along a loop is given ([5, 13]) by:

$$X_s = \tau_s H_s \tag{2.2}$$

where $s \mapsto H_s$ is a finite energy path in $T_x(M)$ such that $X_0 = X_1 = 0$. We endow the tangent space with the Hilbert structure:

$$\|X\|_\gamma^2 = \int_0^1 |H_s'|^2\,\mathrm{d}s. \tag{2.3}$$

We have the integration by parts formula for a cylindrical functional F and for H deterministic ([5, 9, 17])

$$\mathbb{E}\left[\langle\,\mathrm{d}F, X\rangle\right] = \mathbb{E}\left[F\,\mathrm{div}\,X\right]. \tag{2.4}$$

where

$$\mathrm{div}\,X = \int_0^1 \langle\tau_s\frac{\mathrm{d}}{\mathrm{d}s}H_s, \delta\gamma_s\rangle + \frac{1}{2}\int_0^1 \langle S_{X_s}, \delta\gamma_s\rangle \tag{2.5}$$

where δ denotes the Itô integral and S the Ricci tensor on the manifold ([5, 9, 17]).

Let us introduce a connection (see [17, 19]) for the introduction of stochastic connections):

$$\nabla_Y X_. = \tau_. D_Y H_. \tag{2.6}$$

where X and Y are two tangent vector fields over $L_x(M)$ and D denotes the operation of H-derivative.

This connection and the integration by parts formula (2.4) allow us to define Sobolev spaces of functionals in Malliavin sense over the based loop space (see [17, 19]):

$$\langle D^k F, X^1, \ldots, X^k \rangle = \int_{[0,1]^k} D^k F(s_1, \ldots, s_k) \frac{\mathrm{d}}{\mathrm{d}s} H^1_{s_1} \ldots \frac{\mathrm{d}}{\mathrm{d}s} H^k_{s_k} \,\mathrm{d}s_1 \ldots \mathrm{d}s_k \tag{2.7}$$

with Sobolev norms

$$\|F\|_{k,p} = \mathbb{E}\left[\left(\int_{[0,1]^k} |D^k F(s_1, \ldots, s_k)|^2 \,\mathrm{d}s_1 \ldots \mathrm{d}s_k\right)^{p/2}\right]^{1/p} \tag{2.8}$$

D^k is defined recursively by:

$$\begin{aligned}\langle D^k F, X^1, \ldots, X^k \rangle &= \langle D \langle D^{k-1} F, X^1, \ldots, X^{k-1} \rangle, X^k \rangle \\ &\quad - \sum \langle D^{k-1} F, X^1, \ldots, \nabla_{X^k} X^i, \ldots, X^{k-1} \rangle\end{aligned} \tag{2.9}$$

Let O be a compact subset of $\mathbb{R}^n$. We define the Sobolev norms of a functional F over $L_x(M) \times O$ as follows: we suppose that the functional can be extended over $L_x(M) \times O'$ where O' is an open subset of $\mathbb{R}^n$ containing O. We define

$$\|F_{L_x(M) \times O'}\|_{r,p} = \sup_{O'} \sum_{k+k'=r} \|D^{k'}_{O'} F\|_{k,p} \tag{2.10}$$

where $D^{k'}_{O'}$ denotes the derivative in O' of order k' and we take the Sobolev norms in the direction of the loop space ([17, 19, 30]) for analogous considerations). We put

$$\|F_{L_x(M) \times O}\|_{k,p} = \inf_{O \subset O'} \|F_{L_x(M) \times O'}\|_{k,p}. \tag{2.11}$$

The algebra A of stochastic functionals over $L_x(M) \times O$ is given by the set of Sobolev norms $\| \cdot \|_{k,p}$ which have to be all finite. We add an extra condition. Let $(\gamma, y) \mapsto F(\gamma, y)$ be the functional over $L_x(M) \times O$. By definition, we suppose that $y \mapsto (\gamma \mapsto F(\gamma, y))$ is continuous and bounded on a neighborhood O' of O for the L^∞ norm over $L_x(M)$. So we consider as extra norm the norm over O' over continuous functionals, bounded, from O' into the Banach space $L^\infty(L_x(M))$ and we consider the infimum of these norms over the open sets O' containing O as in (2.11). We denote this norm $\|F_{X \times O}\|_\infty$. If we consider only this norm, the algebra A is a real Banach algebra. But we consider the whole system of norms $\| \cdot \|_{k,p}$ and $\| \cdot \|_\infty$ and we get a Fréchet algebra denoted $A_{L_x(M) \times O}$. We can proceed in another way: by Tietze theorem, there exist an extension of $y \mapsto \{\gamma \to F(\gamma, y)\}$ to O', which is continuous for the L^∞ norm. In (2.11), we can consider the infimum over $O \subseteq O'$ and over all possible extensions. Let O_1 be a subset of O. We can consider the Fréchet algebra $A_{(L_x(M)\times O)/(L_x(M)\times O_1)}$ of elements of $A_{L_x(M)\times O}$ which are constant $L_x(M) \times O_1$. We get another Fréchet algebra. If we consider only the norm $\|\cdot\|_\infty$, we get a real Banach algebra.

Spaces like

$$L_x(M) \times O \quad \text{and} \quad (L_x(M) \times O)/(L_x(M) \times O_1) \tag{2.12}$$

are denoted by X. To each X, we associate a Fréchet algebra A_X of stochastic functionals, and if we consider only the first condition, i.e., $\|F\|_\infty < \infty$, we get a real Banach algebra. Let us recall recall what is a Clifford algebra (See [16], III.3). Over $\mathbb{R}^{p+q}$, we consider the quadratic form $-x_1^2 - \cdots - x_p^2 + x_{p+1}^2 + \cdots + x_{p+q}^2$. The algebra $C^{p,q}$ is a $\mathbb{R}$-algebra generated by the elements e_i, $i = 1, \ldots, p+q$ subject to the relations:

(i) $(e_i)^2 = -1$ for $1 \le i \le p$,

(ii) $(e_i)^2 = +1$ for $p+1 \le i \le p+q$

(iii) $e_i e_j = -e_j e_i$ for $i \neq j$.

If A is an algebra, we consider the algebra $\mathcal{M}_n(A)$ of $n \times n$ matrices with component in A. We get:

Theorem ([16] p. 133) *The algebras $C^{p+8,q}$, $C^{p,q+8}$, and $\mathcal{M}_{16}(C^{p,q})$ are isomorphic.*

Let E be a bundle over X. By definition, it is given by an idempotent p ($p^2 = p$) belonging to some $\mathcal{M}_n(A_X)$ for some n. We suppose moreover that the dimension of the image of p is almost surely constant if $X = L_x(M)$. We suppose that E is a $C^{p,q}$-vector bundle. It is given by endomorphism of the bundle related to the algebra A_X $\rho(e_i)$ such that:

(i) $\rho(e_i)^2 = -1$ if $i \le p$.

(ii) $\rho(e_i)^2 = 1$ if $p+1 \le i$.

(iii) $\rho(e_i)\rho(e_j) = -\rho(e_j)\rho(e_i)$ if $i \neq j$.

This means that $\rho(e_i)$ can be seen as a matrix with component in the algebra A_X such that almost surely $p\rho(e_i)p = \rho(e_i)$ as random matrices. We would like to lift the structure of $C^{p,q}$-vector bundle of E into a $C^{p,q+1}$-vector bundle. It is given by a gradation η of E, that is an element of $\mathrm{End}(E)$ such that:

(i) $\eta^2 = 1$

(ii) $\eta\rho(e_i) = -\rho(e_i)\eta$.

We call Grad(E) the space of gradations of the stochastic bundle. We will write $\eta_1 \sim \eta_2$ if η_1, η_2 are two gradations which are homotopic by a smooth homotopy. This means that there exists a smooth path in $\mathcal{M}_n(A_X)$ endowed with its Fréchet topology η_t joining η_1 and η_2 belonging to Grad(E). By the Sobolev imbedding theorem, this means that $\eta_.$ belongs to $A_{X\times[1,2]}$ because such a path in $M_n(A_X)$ can be seen as an element of $\mathcal{M}_n(A_{X\times[1,2]})$. This means that in (2.12), we replace O by $O\times[1,2]$. If we consider the space $L_x(M)\times O/L_x(M)\times O_1$, we consider function $F(\gamma, y, t)$ which depends only on t on $L_x(M)\times O_1\times[1,2]$. We consider the set of Sobolev norms over $A_{X\times I}$ ($I=[1,2]$) as it was done in (2.11)

Definition 2.1. $K'^{p,q}(X)$ is the quotient of the free group generated by the triples (E,η_1,η_2) where E is a $C^{p,q}$-stochastic vector bundle and η_1 and η_2 are gradations by the subgroup generated by the relations:

(i) $(E,\eta_1,\eta_2)+(F,\xi_1,\xi_2) = (E\oplus F, \eta_1\oplus\xi_1, \eta_2\oplus\xi_2)$.

(ii) $(E,\eta_1,\eta_2) = 0$ if η_1 is smoothly homotopic to η_2 in Grad(E).

$d(E,\eta_1,\eta_2)$ denotes the class of the triple (E,η_1,η_2).

Let us recall ([16] pp. 143, 144):

$$d(E,\eta_1,\eta_2)+d(E,\eta_2,\eta_1) = 0 \tag{2.13}$$
$$d(E,\eta_1,\eta_2)+d(E,\eta_2,\eta_3) = d(E,\eta_1,\eta_3) \tag{2.14}$$

and that

$$d(E,\eta_1,\eta_2) = d(E,\eta'_1,\eta'_2) \tag{2.15}$$

if η_1 is a gradation homotopic to η'_1 and η_2 is a gradation homotopic to η'_2. Moreover, each element of $K'^{p,q}(X)$ can be seen as $d(E,\eta_1,\eta_2)$ and $d(E,\eta_1,\eta_2) = d(E,\eta'_1,\eta'_2)$ is equivalent with the fact that we can find a triple (T,ζ,ζ) such that the gradations over $E\oplus E'\oplus T$ $\eta_1\oplus\eta'_2\oplus\zeta$ and $\eta_2\oplus\eta'_1\oplus\zeta$ are smoothly homotopics.

We will say that a function from I into A_X is *smooth* if it defines an element of $A_{X\times I}$. If there is no quotient in the definition of X, the definition is clear. If $X = Y/T$, $A_{X\times I}$ denotes the elements of $A_{Y\times I}$ which are almost surely constants on each $T\times\{t\}$. We have a small improvement of Lemma 4.21 of ([16], III):

Lemma 2.2. *Let E be a $C^{p,q}$-bundle over X and let* Grad(E) *be the space of gradations over E (which is included in the inductive limit of*

$\mathcal{M}_n(A_X)$ when $n \to \infty$). If $\eta\colon [0,1] \to \mathrm{Grad}(E)$ is a smooth map, there exists a smooth map $\beta\colon [0,1] \to \mathrm{Aut}(E)$ compatible with the action of $C^{p,q}$ on the the stochastic vector bundle E such that

$$\beta(0) = 1$$
$$\eta(\cdot) \sim \beta(\cdot)\eta(0)\beta(\cdot)^{-1}.$$

Proof. Let us recall we consider as first norm the norm $\|.\|_\infty$ such that if we consider only this norm, A_X is a real Banach algebra. We put

$$\beta_0(t,u) = \frac{1+\eta(t)\eta(u)}{2} \tag{2.16}$$

It is a equal to 1 for $t = u$ and is such that

$$\beta_0(t,u)\eta(u) = \eta(t)\beta_0(t,u)\ . \tag{2.17}$$

Since we consider the norm $\|\cdot\|_\infty$, $\beta_0(t,u)$ is an isomorphism if $|t-u| < I/N$. We deduce a partition $t_k = k/N$ of [0,1] and for $t \in [t_k, t_{k+1}]$, we define

$$\beta_0(t) = \beta_0(t,t_k)\beta_0(t_k,t_{k-1})\ldots\beta(t_1,t_0) \tag{2.18}$$

It satisfies to $\eta(t) = \beta_0(t)\eta(0)\beta_0(t)^{-1}$ but it is not smooth.

We find a smooth homotopy from the function identity over $[t_k, t_{k+1}]$ into an increasing smooth functions which is equal to t_k in a neighborhood of t_k and to t_{k+1} in a neighborhood of t_{k+1}. We call this homotopy $f_{s,k}(t)$ and we choose if $t \in [t_k, t_{k+1}]$

$$\beta_s(t) = \beta_0(f_{s,k}(t), t_k)\beta_0(t_k) \tag{2.19}$$

$\beta_s(t)$ is still an isomorphism and $t \mapsto \beta_1(t)$ is smooth. Moreover,

$$\beta_s(t)\eta(0)\beta_s(t)^{-1} = \eta(f_{s,k}(t)) \tag{2.20}$$

which can be chosen smooth in t and is smoothly homotopic to $\eta(t)$. Moreover, $(s,t) \mapsto \beta_s(t)$ depends smoothly on (s,t) in all Sobolev spaces. □

If we consider a $C^{p,q+1}$-stochastic vector bundle E, we have the map of restriction of scalars ϕ which consider the stochastic bundle E as a $C^{p,q}$ bundle.

By this procedure, we get the notion of *stochastic Grothendieck groups* $K^{p,q}(X)$. It consists of triples (E, F, α) of $C^{p,q+1}$-stochastic bundles E, F and an isomorphism α between E and F considered as $C^{p,q}$-bundles.

(E,F,α) and (E',F',α') are isomorphics if there exists two isomorphism $f\colon E \to E'$ and $g\colon F \to F'$ of $C^{p,q+1}$-stochastic vector bundles such that $\alpha' \circ f = g \circ \alpha$. A triple (E,F,α) is called *elementary* if $E = F$ and α is smoothly homotopic to Id_E in the category of $C^{p,q}$-vector bundles. We define the sum of (E,F,α) and of (E',F',α') as $(E \oplus E', F \oplus F', \alpha \oplus \alpha')$.

$K^{p,q}(X)$ consists of such triples (E,F,α) submitted to the equivalence relation: $(E_1,F_1,\alpha_1) \sim (E_2,F_2,\alpha_2)$ if there exists elementary triples (E'_1,F'_1,α'_1) and (E'_2,F'_2,α'_2) such that $(E_1,F_1,\alpha_1)+(E'_1,F'_1,\alpha'_1)$ and $(E_2,F_2,\alpha_2)+(E'_2,F'_2,\alpha'_2)$ are isomorphic. We get ([16] p. 141) a group denoted $K^{p,q}(X)$. We denote by $d(E,F,\alpha) \in K^{p,q}(X)$ the class of (E,F,α). To $d(E,F,\alpha) \in K^{p,q}(X)$, , we associate $d(E,\eta_1,\eta_2)$ where $\eta_1 = \eta_E$ and $\eta_2 = \alpha^{-1}\eta_F\alpha$ where η_E and η_F are the gradations associated to E and F. This map $K^{p,q}(X) \to K^{'p,q}(X)$ is an isomorphism by Lemma II.2 (see [16] p. 146 for similar statements), because if $\eta_2 \sim \eta'_2$

$$d(E,\eta_1,\eta_2) = d(E,\eta_1,\eta'_2) + d(E,\eta'_2,\eta_2) = d(E,\eta_1,\eta'_2) \tag{2.21}$$

We can give a relative version of these statements. Let $Y \subseteq X$ of the type

$$(L_x(M) \times O_2)/(L_x(M) \times O_1) \tag{2.22}$$

where $O_1 \subseteq O_2 \subseteq O$. We consider a triple (E,η_1,η_2) where E is a $C^{p,q}$ stochastic bundle on X, and η_1 and η_2 are gradations equals on Y.

Definition 2.3. The group $K^{p,q}(X,Y)$ is the quotient of the free abelian groups with basis the triples (E,η_1,η_2) where E is a $C^{p,q}$-stochastic vector bundle on X, and η_1,η_2 are gradations of E with $\eta_{1|Y} = \eta_{2|Y}$, by the subgroup generated by the relations:

(i) $(E,\eta_1,\eta_2) + (F,\zeta_1,\zeta_2) = (E \oplus F, \eta_1 \oplus \zeta_1, \eta_2 \oplus \zeta_2)$.

(ii) $(E,\eta_1,\eta_2) = 0$ if there exists a smooth map $\eta\colon [0,1] \to \mathrm{Grad}(E)$ such that $\eta(0) = \eta_1$, $\eta_2 = \eta(1)$ and such that $\eta(t)$ on Y is equal to η_1.

Let us remark that if Y is a subspace of the type (2.22) of X, X/Y is still of the type (2.22).

Proceeding as in ([16] p. 147), each element of $K^{p,q}(X,Y)$ can be written as $d(T_r,\eta_r,\eta)$ where T_r is a trivial $C^{p,q}$-bundle, η_r a trivial gradation and η is a gradation such that $\eta_{r|Y} = \eta_{|Y}$.

In order to establish that statement, we introduce a sequence of $C^{p,q+1}$-vector spaces M_r satisfying $M_r \oplus M_s = M_{r+s}$ and such that all $C^{p,q+1}$-bundle is isomorphic to a subbundle of the trivial bundle with fiber M_r for some r. We say that $\{M_r\}$ is a cofinal system. It is possible to do that by the classification of Clifford algebras $C^{p,q+1}$ ([16] p. 131-133). Let $\mathrm{Grad}^{p,q}(M_r)$ be the space of gradations of the $C^{p,q}$-vector space M_r. We get

$$K^{p,q}(X) = \operatorname{inj}\lim_r [X, \mathrm{Grad}^{p,q}(M_r)] \tag{2.23}$$

where $[X, \mathrm{Grad}^{p,q}(M_r)]$ denote the set of (smooth) homotopy classes of functions from X into $\mathrm{Grad}^{p,q}(M_r)$ which belong to our functional spaces.

With this formalism, we get that $K^{p,q}(X,Y)$ is the injective limit of the set of (smooth) homotopy classes of maps which belong to our functional spaces from X into $\mathrm{Grad}^{p,q}(M_r)$ such that $\eta = e_{p+q+1}$ on Y.

These remarks show that it is clear that $K^{p,q}(X,Y)$ is canonically isomorphic to $K^{p,q}(X/Y,\{y\})$ where y coincides with the subspace Y.

Let X and Y be spaces of the type (2.12) or (2.22).

Lemma 2.4. *Let E be a stochastic bundle over X. Any section on Y of E can be extended in a section on X of E.*

Proof. We write $Y = Y_1/T$ and $X = X_1/T$ where $Y_1 = L_x(M)\times O$, $X_1 = L_x(M)\times V$ and $T = L_x(M)\times V'$. A section of E restricted to Y can be seen as an element ξ of $A^n_{L_x(M)\times O'}$ which satisfy the required continuity property in $y \in O'$ in $L^\infty(L_x(M))$ and the required differentiability properties for a certain deterministic neighborhood O' of O, such that $p\xi = \xi$ on $L_x(M)\times O$, where p is the projector which gives the bundle E.

Let W be the set of projectors. It is a compact manifold. Namely, the space projectors of rank k is homeomorphic to a closed subspace of the product of the Grassmannian constituted of linear spaces of rank k and of the Grassmannian constituted of the Grassmannian of rank $n-k$, if we work over $\mathbb{R}^n$. Moreover, the Grassmannian are compacts. Let $\{W_i\}$ be a covering of W by small open convex subsets. Let $\{g_i\}$ be a smooth partition of unity associated to $\{W_i\}$. If $p_i \in \operatorname{supp} g_i$, the projection on $\operatorname{Im} p$ is a smooth application denoted $\pi_i(p)$. We put

$$\tilde{\xi} = \sum g_i(p)\pi_i(p)\xi g(y) \tag{2.24}$$

if $g(y)\colon \mathbb{R}^n \to [0,1]$ is a smooth application equal to 0 outside O' and equal to 1 over O and where (γ, y) is the generic element of $L_x(M)\times\mathbb{R}^n$. $\tilde{\xi}$ is equal to ξ on $L_x(M)\times O$ and is equal to a constant on T. □

By using this lemma, we can repeat the arguments of Lemma 2.21, Lemma 2.22 and Proposition 2.24 of ([16], II). In particular, let E be a stochastic bundle over X and let Y be a subspace of the type (2.22), and let $\alpha\colon E \to E$ be an automorphism of E. Let σ' a smooth path in the automorphism of E restricted to Y starting from α restricted to Y, there exist a smooth path σ starting from α in the set of automorphism of E whose restriction to Y is equal to σ'.

This property allows to repeat the proof of Theorem 5.4 of ([16], III) and to exhibit an exact sequence

$$K^{p,q}(X,Y) \to K^{p,q}(X) \to K^{p,q}(Y) \tag{2.25}$$

Moreover, we can repeat these arguments for $K^{p,q}(X/Y,\{y\})$. We get clearly an isomorphism between $K^{p,q}(X/Y,\{y\})$ and $K^{p,q}(X,Y)$ where $\{y\} = Y$ and the explicit exact sequence:

$$0 \to K^{p,q}(X/Y,\{y\}) \to K^{p,q}(X/Y) \to K^{p,q}(\{y\}) \to 0 \tag{2.26}$$

because $\{y\}$ is a retract of X/Y.

Definition 2.5 ([16] p. 61). Let X and Y be two spaces of the considered type. The *relative Grothendieck group* $K(X,Y)$ is the set of classes $d(E,F,\alpha)$ where E, F are stochastic bundles on X and $\alpha \in \mathrm{GL}(A_Y)$ is an isomorphism of the restrictions of these bundles on Y. Moreover,

$$d(E,F,\alpha) = d(E',F',\alpha') \tag{2.27}$$

if there exists two trivial triples (G,G,Id) and (G',G',Id) and two isomorphisms $f\colon E \oplus G \to E' \oplus G'$ and $g\colon F \oplus G \to F' \oplus G'$ such that

$$g_{|Y} \circ (\alpha \oplus I) := (\alpha' \oplus I) \circ f_{|Y}. \tag{2.28}$$

By Lemma 2.4 we get an exact sequence

$$K(X,Y) \to K(X) \to K(Y) \tag{2.29}$$

and if we apply these considerations to $(X/Y,\{y\})$, we get a split exact sequence:

$$0 \to K(X/Y,\{y\}) \to K(X/Y) \to K(\{y\}) \to 0. \tag{2.30}$$

By doing as in ([16], p. 151), we get an isomorphism between $K^{0,0}(X,Y)$ and $K(X,Y)$ by putting

$$g(d(E,\eta_1,\eta_2)) = d(E_1^0, E_2^0, \alpha) \tag{2.31}$$

where $E_i^0 = \ker \frac{1-\eta_i}{2}$ and $\alpha\colon E_1^0|_Y \to E_2^0|_Y$ is the identification map. From the exact sequence (2.30)-(2.26), and since $K(X/Y,\{y\})$ is canonically isomorphic to $K(X,Y)$ we derive that $K^{0,0}(X,Y)$ is canonically isomorphic to $K(X,Y)$.

Moreover, by Morita equivalence ([16] p. 138), the fact that

$$C^{p,q} = \mathcal{M}_{2^q}(C^{p-q,0}) \text{ if } p > q, \tag{2.32}$$

$$C^{p,q} = \mathcal{M}_{2^p}(C^{0,p-q}) \text{ if } p < q, \tag{2.33}$$

$$C^{n,n} = \mathcal{M}_{2^n}(\mathbb{R}), \tag{2.34}$$

and the 8-periodicity of the real Clifford algebra listed previously, we deduce that $K^{p,q}(X,Y)$ only depends, up to isomorphism, on the difference $p-q$ mod 8 ([16], p. 149).

3. Stochastic Bott periodicity

Let (X,Y) two spaces of the considered type, Y being a subspace of X. In order to simplify the exposure, we will take $X = L_x(M) \times I^n$ and $Y = L_x(M) \times \partial I^n$. For each pair (p,q), we define a fundamental homomorphism

$$t\colon K^{p,q+1}(X,Y) \to K^{p,q}(X \times I^1, X \times \{-1,1\} \cup Y \times I^1) \tag{3.1}$$

as follows: we identify I^1 with the semi-circle $\{\exp \mathrm{i}\theta \mid 0 \leq \theta \leq \pi\}$. Let E' be the underlying $C^{p,q}$-bundle defined by the element of $K^{p,q+1}(X,Y)$ considered as a $C^{p,q}$-vector bundle, and let ϵ be the gradation which makes E' in a $C^{p,q+1}$-bundle.

Let $d(E,\eta_1,\eta_2)$ be the element of $K^{p,q+1}(X,Y)$ associated to E'. We define ([16] p. 152)

$$\zeta_i(\theta) = \epsilon\cos(\theta) + \eta_i \sin(\theta) \tag{3.2}$$

and we put

$$t(d(E,\eta_1,\eta_2)) = d(E',\zeta_1,\zeta_2) \tag{3.3}$$

where E' is the pullback bundle of E considered as a $C^{p,q}$-bundle by the map $(x,\theta) \mapsto x$. So E' is a $C^{p,q}$-bundle over $X \times I^1$.

The main goal of this section is to show the following theorem:

Theorem 3.1. *The homomorphism t is an isomorphism:*

$$K^{p,q+1}(X,Y) \simeq K^{p,q}(X \times I^1, X \times \{-1,1\} \cup Y \times I^1). \tag{3.4}$$

From Theorem 3.1, and the considerations of Section 2, we derive Theorem 1.1 as in ([16], p.153). The proof follows closely the corresponding proof of this theorem for a compact space ([16], II.6). By quotienting by Y, we can consider after the considerations of the end of the previous part the case of X/Y and of $\{y\} = y$ and come back to the case where $Y = \varnothing$.

Let E_r^0 be a $C^{p,q+2}$-trivial bundle which constitutes a cofinal system. If $e_1, \ldots, e_{p+q+2}$ are the generators of the real Clifford algebra $C^{p,q+2}$, we write $\xi_r(\theta)$ for the gradation of the trivial bundle E_r^0 deduced from E_r^0 over $X \times I^1$ defined by $\xi_r(\theta) = e_{p+q+1}\cos(\theta) + e_{p+q+2}\sin(\theta)$. Each element of $K^{p,q}(X \times I^1, X \times \{-1,1\})$ can be written ([16], III.6.4) as $d(E_r^0, \xi_r(\theta), \epsilon(\theta))$ where ϵ is a gradation of E_r^0 considered as a $C^{p,q}$-bundle over $X \times I^1$ such that $\epsilon = \xi_r$ on $X \times \{-1,1\}$. This element is equal to 0 if and only if there exists s such that $\epsilon \oplus \xi_s$ is smoothly homotopic to ξ_{r+s}, the homotopy being constant on $X \times \{-1,+1\}$. By using Lemma II.2 and identifying $[-1,1]$ to $[0,\pi]$, we see that each element of $K^{p,q}(X \times I^1, X \times \{0,\pi\})$ can be written as $d(E_r^0, \xi_r(\theta), \epsilon(\theta))$ where $\epsilon(\theta) \sim f(\theta)e_{p+q+1}f(\theta)^{-1}$ for a smooth map from $[0,\pi]$ into $\mathrm{Aut}(E_r^0)$ considered as a $C^{p,q}$-bundle over X such that:

(i) $f(0) = \mathrm{Id}_{E_r^0}$.

(ii) $e_{p+q+1}f(\pi) = -f(\pi)e_{p+q+1}$.

Moreover we introduce the $C^{p,q}$-automorphism $h_r(\theta)$ of the $C^{p,q}$-vector space E_r^0 defined by $h_r(\theta) = \cos(\theta/2) + e_{p+q+1}e_{p+q+2}\sin(\theta/2)$. Its inverse is given by $\cos(\theta/2) - e_{p+q+1}e_{p+q+2}\sin(\theta/2)$. $d(E_r^0, \xi_r(\theta), \epsilon(\theta))$ is zero if and only if there exists an integer s such that $f \oplus h_s$ is smoothly homotopic to h_{r+s} within the automorphism of the trivial $C^{p,q}$-bundle of $E_{r+s}^0 = E_r^0 \oplus E_s^0$ considered as a trivial $C^{p,q}$-bundle over $X \times [0,\pi]$. This is done by smoothing the map $f'(\theta,t)$ in ([16], III, Lemma 6.5) as in Lemma II.2. We introduce the algebra A of $C^{p,q}$-endomorphisms of the trivial bundle over X whose fiber is the Clifford algebra $C^{p,q+2}$ himself. The components are elements of A_X. By adding these Clifford algebras, we get a cofinal system $\{N_r\}$. Over the algebra A, there is an involution $\alpha \mapsto e_{p+q+1}\alpha e_{p+q+1}^{-1} = \overline{\alpha}$.

We denote $\mathrm{GL}_r(A)$ the space of $r \times r$ invertible matrices with entries in A. A matrix g belongs to $\mathrm{GL}_r(A)$ if there exists $g^{-1} \in M_r(A)$ such that $gg^{-1} = g^{-1}g = 1$. We denote

$$\mathrm{GL}_r^-(A) = \{g \in \mathrm{GL}_r(A) \mid \overline{g} = -g\}. \tag{3.5}$$

In particular, $e = e_{p+q+2}e_{p+q+1} \in \mathrm{GL}_1^-(A)$ ([16] p. 164). As based point of $\mathrm{GL}_{2r}^-(A)$, we choose the diagonal matrix $e'^r = \{\delta_{i,j}(-1)^j e\}$. We define

$$\mathrm{GL}^-(A) = \operatorname{inj}\lim_r \mathrm{GL}_{2r}^-(A) \tag{3.6}$$

with respect to the "inclusion" map $g \mapsto g \oplus e'^1$.

Let $\pi_1(\mathrm{GL}_{2r}(A), \mathrm{GL}_{2r}^-(A))$ be the space of smooth homotopy classes of smooth paths $f \colon [0,\pi] \to \mathrm{GL}_{2r}(A)$ such that $f(0) = \mathrm{Id}$ and $f(\pi) \in$

$\mathrm{GL}_{2r}^-(A)$. We denote

$$\pi_1(\mathrm{GL}(A), \mathrm{GL}^-(A)) = \operatorname{inj}\lim_r \pi_1(\mathrm{GL}_{2r}(A), \mathrm{GL}_{2r}^-(A)) \tag{3.7}$$

where the injective limit is taken with respect to the maps $f \mapsto f \oplus h_1$ ([16] p. 165).

The previous considerations can be reformulated and we get:

$$K^{p,q}(X \times [0,\pi], X \times \{0,\pi\}) = \pi_1(\mathrm{GL}(A), \mathrm{GL}^-(A)). \tag{3.8}$$

We can describe the group $K^{p,q+1}(X)$ in terms of the algebra A as in [16], chapter III, 6.8. We consider as cofinal system $\{M_r = N_r \oplus \tilde{N}_r\}$ where $N_r = (C^{p,q+2})^r$ and $\tilde{N}_r$ is the same space where we changed the sign of the action of e_{p+q+2}. We consider M_r as a trivial $C^{p,q+1}$-bundle over X. Let G_r be the set of gradations η of the $C^{p,q+1}$-bundle over X. $\eta^2 = 1$ and $\eta e_i = -e_i \eta$ for $i = 1, \ldots, p+q+1$. $K^{p,q+1}(X)$ can be seen as the set of smooth homotopy classes of such gradations. Let $I_r(A)$ be the subspace of $\mathrm{GL}_{2r}(A)$ constituted of elements g such that $\overline{g} = -g$ and $g^2 = -1$ ([16] p. 166). We identify an element of $I_r(A)$ with the gradation ge_{p+q+1}. We denote

$$I(A) = \operatorname{inj}\lim_r I_r(A) \tag{3.9}$$

We get:

$$K^{p,q+1}(X) = \pi_0(I(A)) \tag{3.10}$$

([16] III, Prop. (6.9), but we identify two elements of $I(A)$ if they are joined by a smooth path in some $I_r(A)$).

To an element g of $I(A)$, we associate the path

$$w(g)\colon f(\theta) = \cos(\theta/2) + g\sin(\theta/2) \tag{3.11}$$

It defines an element of $\pi_1(\mathrm{GL}(A), \mathrm{GL}^-(A))$. This application w is compatible with t ([16], III, Prop. (6.11)).

In order to show Theorem III.1 it is enough to show the following proposition:

Proposition 3.2. $w\colon \pi_0(A) \to \pi_1(\mathrm{GL}(A), \mathrm{GL}^-(A))$ *is an isomorphism.*

The proof follows closely the proof of Theorem 6.12 of ([16], III) (see [32]), by working in a first step in $L^\infty(L_x(M))$, a Banach algebra, and remarking afterwards that the different approximation procedures are compatible with the various Sobolev norms involved.

Remark 3.3. We can give an example of a bundle in our sense over $L_x(M)$. Let Q be a principal bundle with compact structural Lie group G over M. Let V a real linear bundle over G: it is given by an idempotent p on G. We get a map from $L_x(M)$ into G, which is the stochastic parallel transport, denoted τ_1^Q: it belongs to all Sobolev spaces ([19]). The pullback bundle $(\tau_1^Q)^*V$ given by the idempotent $p(\tau_1^Q)$, smooth in Malliavin sense, satisfies to our assumptions.

References

[1] Atiyah M.: K-theory and reality, *Quart. J. Math. Oxford Ser. (2)* **17** (1966) 367–386.

[2] Atiyah M., Hirzebruch F.: Vector bundles and homogeneous spaces, *Proc. Symp. Pure Math.* Vol. III, 7–38, Amer. Math. Soc., Providence, 1961.

[3] Atiyah M., Singer I.: The index of elliptic operators. I. III, *Ann. of Math. (2)* **87** (1968) 484–530, 546–604.

[4] Atiyah M., Segal G.: The index of elliptic operators. II, *Ann. of Math. (2)* **87** (1968) 531–545.

[5] Bismut J.-M.: *Large deviations and the Malliavin Calculus*, Progress in Mathematics **45**, Birkhäuser, Boston, 1984.

[6] Borel A., Serre J.-P.: Le théorème de Riemann-Roch (d'après A. Grothendieck), *Bull. Soc. Math. France* **86** (1958) 97–136.

[7] Bott R.: The stable homotopy of the classical groups, *Ann. of Math. (2)* **70** (1959) 313–337.

[8] Connes A.: Entire cyclic cohomology of Banach algebras and characters of θ-summable Fredholm modules, *K-theory* **1** (1988) 519–548.

[9] Driver B.: A Cameron-Martin type quasi-invariance theorem for Brownian motion on a compact Riemannian manifold, *J. Funct. Anal.* **110** (1992) 272–376.

[10] Getzler E., Szenes A.: On the Chern Character of theta-summable Fredholm module, *J. Funct. Anal.* **84** (1989) 343–357.

[11] Jaffe A., Lesniewski A., Osterwalder K.: Quantum K-theory. I. The Chern Character, *Comm. Math. Phys.* **118** (1988) 1–14.

[12] Jaffe A., Lesniewski A., Weitsman J.: Index of a family of Dirac operators on loop space, *Comm. Math. Phys.* **112** (1987) 75–88.

[13] Jones J.D.S., Léandre R.: L^p-Chern forms on loop spaces, In *"Stochastic analysis" (Durham 1990)*, 103–162, Barlow M., Bingham N., eds. London Math. Soc. Lecture Note Ser., **167**, Cambridge Univ. Press, Cambridge, 1991.

[14] Jones J.D.S., Léandre R.: A stochastic approach to the Dirac operator over the free loop space, *Proc. Steklov Inst. Math.* **217** (1997) 253–282.

[15] Karoubi M.: Algèbres de Clifford et K-théorie, *Ann. Sci. École Norm. Sup. (4)* **1** (1968) 161–270.

[16] Karoubi M.: *K-theory. An introduction*, Grundlehren der Mathematischen Wissenschaften, **226**, Springer-Verlag, Berlin-New York, 1978.

[17] Léandre R.: Integration by parts formulas and rotationally invariant Sobolev calculus on the free loop space, In *"XXVII School of theoretical Physics. Infinite-dimensional geometry in physics (Karpacz, 1992)*. Borowiec A., Gielerak R., eds. *J. Geom. Phys.* **11** (1993) 517–528.

[18] Léandre R.: Brownian motion over a Kähler manifold and elliptic genera of level N. In *"Stochastic analysis and applications in physics" (Funchal 1993)*, Sénéor R., Streit L., eds., 193–217, NATO Adv. Sci. Inst. Ser. C Math. Phys. Sci., **449**, Kluwer Acad. Publ., Dordrecht, 1994.

[19] Léandre R.: Invariant Sobolev calculus on the free loop space, *Act. Appl. Math.* **46** (1997) 267–350.

[20] Léandre R.: Hilbert space of spinor fields over the free loop space. *Rev. Math. Phys.* **9** (1997) 243–277.

[21] Léandre R.: Stochastic gauge transform of the string bundle, *J. Geom. Phys.* **26** (1998) 1–25.

[22] Léandre R.: String structure over the Brownian bridge, *J. Math. Phys.* **40** (1999) 454–479.

[23] Léandre R.: Cover of the Brownian bridge and stochastic symplectic action, *Rev. Math. Phys.* **12** (2000) 91–137.

[24] Léandre R.: Quotient of a loop group and Witten genus, *J. Math. Phys.* **42** (2001) 1364–1383.

[25] Léandre R.: A stochastic approach to the Euler-Poincaré characteristic of a quotient of a loop group, *Rev. Math. Phys.* **13** (2001) 1307–1322.

[26] Léandre R.: Analysis on loop spaces and topology, *Math. Notes* **72** (2002) 212–229.

[27] Léandre R.: Wiener analysis and cyclic cohomology, In "Stochastic Analysis And Mathematical Physics (Santiago, Chile, 2002), Rebolledo R. Zambrini, to appear.

[28] Léandre R.: Stochastic K-theory and stochastic Bott periodicity, Preprint Institut Elie Cartan Nancy 2001/52.

[29] Léandre R., Roan S.-S.: A stochastic approach to the Euler-Poincaré number of the loop space of a developable orbifold, *J. Geom. Phys.* **16** (1995) 71–98.

[30] Taubes C.: S^1 actions and elliptic genera, *Comm. Math. Phys.* **112** (1989) 455–526.

[31] Witten Ed.: The Index of Dirac operator in loop space, In "*Elliptic curves and modular forms in algebraic topology (Princeton, 1986)*", P.S. Landweber ed., 161–181, *Lecture Notes in Math.* **1326**, Springer, Berlin, 1988.

[32] Wood R.: Banach algebras and Bott periodicity, *Topology* **4** (1965/1966) 371–389.

HEAT EQUATION ASSOCIATED WITH LÉVY LAPLACIAN

Nobuaki Obata
Graduate School of Information Sciences
Tohoku University
Sendai 980-8579 Japan
obata@math.is.tohoku.ac.jp

Habib Ouerdiane
Département de Mathématiques
Faculté de Sciences
Université de Tunis El Manar
Campus Universitaire, Tunis 1060 Tunisia
habib.ouerdiane@fst.rnu.tn

Abstract A solution to the heat equation associated with the Lévy Laplacian is studied by means of nuclear spaces of infinite dimensional entire functions. In particular, evolution of positive distributions and relation to the quadratic quantum white noise are discussed in a unified manner.

1. Introduction

In the famous books [17, 18] P. Lévy introduced and studied an infinite dimensional generalization of the classical Laplace operator:

$$\Delta_L = \lim_{N\to\infty} \frac{1}{N} \sum_{n=1}^{N} \frac{\partial^2}{\partial x_n^2}. \tag{1.1}$$

This operator, called the *Lévy Laplacian*, possesses many peculiar properties and has been studied by many authors from various aspects. For example, formulating as a differential operator acting on functions on a Hilbert space, Feller [10], Polishchuk [28] and others (see the references cited therein) studied differential equations such as boundary problems in detail and Obata [19] gave a group-theoretical characterization. In recent years, more attention has been paid to the Lévy Laplacian act-

S. Albeverio et al. (eds.),
Proceedings of the International Conference on Stochastic Analysis and Applications, 53–68.

ing on functions on a nuclear space for its rich structure. A somehow unexpected relation to the Gross Laplacian was found by Kuo–Obata–Saito [15]. A connection between Yang–Mills equations and heat equations associated with the Lévy Laplacian, first pointed out by Aref'eva–Volovich [6], has become an important research topic, see e.g., Accardi [1], Accardi–Bogachev [2], Accardi–Gibilisco–Volovich [3]. Using a particular domain constructed from Lévy's normal functions, Chung–Ji–Saitô [7] solved a heat equation associated with the Lévy Laplacian by means of an analytic one-parameter group $e^{z\Delta_L}$, see also Saitô [29]. Recently, using an idea of Poisson analysis, Saitô–Tsoi [31] found a new space where the Lévy Laplacian is formulated as a selfadjoint operator. In this direction further progress has been made by Saitô [30] and Kuo–Obata–Saitô [16].

In this paper, we focus on the heat equation associated with the Lévy Laplacian acting on functions on a real nuclear space E. Thus we are interested in the Cauchy problem:

$$\frac{\partial}{\partial t} F(t,\xi) = \Delta_L F(t,\xi), \qquad F(0,\xi) = F_0(\xi), \tag{1.2}$$

where the initial condition F_0 is a certain function on a nuclear space E. When $F_0(\xi)$ is the Fourier transform of a measure μ on E' which is invariant under a certain shift operator, a solution to (1.2) was explicitly obtained by Accardi–Roselli–Smolyanov [5]. Another interesting function $F(t,\xi)$ satisfying the heat equation was constructed by Obata [22] from a normal-ordered white noise equation involving the quadratic quantum white noises. The main purpose of this paper is to show that the above two classes of solutions are obtained in a unified manner without assuming that $\{x_1, x_2, \dots\}$ is an orthogonal coordinate system, which is a traditional assumption in the definition of the Lévy Laplacian (1.1). Furthermore, employing the recent framework of infinite dimensional holomorphic functions due to Gannoun–Hachaichi–Ouerdiane–Rezgui [11], we obtain an evolution of a positive distribution driven by the Lévy Laplacian. It is noted that our approach is independent of Gaussian analysis and seems appropriate for analysis of the Lévy Laplacian.

2. Preliminaries

In this section we assemble some basic notation and results on entire functions on nuclear spaces, for more details see [11].

2.1 Entire functions with θ-exponential growth

We begin with a general notation. For a complex Banach space $(B, \|\cdot\|)$ we denote by $H(B)$ the space of entire functions on B, i.e., continuous functions $B \to \mathbb{C}$ whose restrictions to every affine line of B are entire holomorphic on $\mathbb{C}$. We classify such entire functions by growth rates. Let θ be a Young function, i.e., $\theta : \mathbb{R}_+ \to \mathbb{R}_+$ is a continuous, convex, increasing function such that $\theta(0) = 0$ and

$$\lim_{x \to \infty} \frac{\theta(x)}{x} = \infty. \tag{2.1}$$

For $m > 0$ we define the *space of entire functions on B with θ-exponential growth of finite type m* by

$$\mathrm{Exp}\,(B, \theta, m) = \{f \in H(B)\,;\, \|f\|_{\theta,m} \equiv \sup_{u \in B} |f(u)| \mathrm{e}^{-\theta(m\|u\|)} < \infty\}.$$

If θ is a Young function,

$$\theta^*(x) = \sup_{t \geq 0} (tx - \theta(t)), \qquad x \geq 0, \tag{2.2}$$

becomes also a Young function. This is called the *polar function* of θ and plays a role in duality argument.

2.2 Nuclear spaces of entire functions

Let N be a complex nuclear Fréchet space whose topology is defined by a family of increasing Hilbertian norms $\{|\cdot|_p\,, p \in \mathbb{N}\}$. The space N can be represented as $N = \cap_{p \in \mathbb{N}} N_p$, where N_p is the Hilbert space obtained by completing N with respect the norm $|\cdot|_p$. Denote by N_{-p} the topological dual space of N_p. Then by general duality theory the dual space N' can be expressed as $N' = \cup_{p \in \mathbb{N}} N_{-p}$. Because of the nuclearity of the space N, the strong topology of N' coincides with the inductive limit topology.

It is easily verified that $\{\mathrm{Exp}\,(N_{-p}, \theta, m)\}$ forms a projective system of Banach spaces as $p \to \infty$ and $m \downarrow 0$. We then define the *space of entire functions on N' with θ-exponential growth of minimal type* by

$$\mathcal{F}_\theta(N') = \bigcap_{p \in \mathbb{N}, m > 0} \mathrm{Exp}\,(N_{-p}, \theta, m). \tag{2.3}$$

Similarly, $\{\mathrm{Exp}\,(N_p, \theta, m)\}$ forms an inductive system of Banach spaces as $p \to \infty$ and $m \to \infty$, and we define the *space of entire functions on N with θ-exponential growth of (arbitrarily) finite type* by

$$\mathcal{G}_\theta(N) = \bigcup_{p \in \mathbb{N}, m > 0} \mathrm{Exp}\,(N_p, \theta, m). \tag{2.4}$$

If θ and φ are two Young functions which are equivalent at infinity, i.e., $\lim_{x \to \infty} \theta(x)/\varphi(x) = 1$, we have $\mathcal{F}_\theta(N') = \mathcal{F}_\varphi(N')$ and $\mathcal{G}_\theta(N) = \mathcal{G}_\varphi(N)$.

2.3 Taylor series map

Each $f \in \mathcal{F}_\theta(N')$ and $g \in \mathcal{G}_\theta(N)$ admit Taylor series expansions:

$$f(x) = \sum_{n=0}^{\infty} \langle x^{\otimes n}, f_n \rangle, \qquad x \in N',$$

$$g(\xi) = \sum_{n=0}^{\infty} \langle g_n, \xi^{\otimes n} \rangle, \qquad \xi \in N.$$

Characterization of these spaces in terms of Taylor expansion is useful. The correspondences $f \mapsto \vec{f} = (f_n)_{n\geq 0}$ and $g \mapsto \vec{g} = (g_n)_{n\geq 0}$ are called the *Taylor series map* (at zero) and denoted by $\mathcal{T}$.

Given a Young function θ, we put

$$\theta_n = \inf_{r>0} \frac{e^{\theta(r)}}{r^n}.$$

Then we define the Hilbert space $F_{\theta,m}(N_p)$ by

$$F_{\theta,m}(N_p) = \left\{ \vec{f} = (f_n)_{n\geq 0} \,;\, f_n \in N_p^{\odot n},\, \|\vec{f}\|_{\theta,p,m} < \infty \right\},$$

where $N_p^{\odot n}$ is the n-fold symmetric tensor power of N_p and

$$\|\vec{f}\|^2_{\theta,p,m} = \sum_{n=0}^{\infty} \theta_n^{-2} m^{-n} |f_n|_p^2, \qquad \vec{f} = (f_n).$$

Then, equipped with the projective limit topology,

$$F_\theta(N) = \bigcap_{p\in\mathbb{N}, m>0} F_{\theta,m}(N_p)$$

becomes a nuclear Fréchet space. In a similar manner, one defines

$$G_{\theta,m}(N_{-p}) = \{ \vec{\Phi} = (\Phi_n)_{n\geq 0} \,;\, \Phi_n \in N_{-p}^{\odot n},\, \|\vec{\Phi}\|_{\theta,-p,m} < \infty \},$$

where

$$\|\vec{\Phi}\|^2_{\theta,-p,m} = \sum_{n=0}^{\infty} (n!\theta_n)^2 m^n |\Phi_n|^2_{-p}.$$

Then we put

$$G_\theta(N') = \bigcup_{p\in\mathbb{N}, m>0} G_{\theta,m}(N_{-p}),$$

which is equipped with the inductive limit topology. By definition the power series spaces $F_\theta(N)$ and $G_\theta(N')$ are dual each other with the canonical bilinear form defined by

$$\langle\langle \vec{\Phi}, \vec{f} \rangle\rangle = \sum_{n=0}^{\infty} n!\, \langle \Phi_n, f_n \rangle. \tag{2.5}$$

Theorem 2.1 (Gannoun-Hachaichi-Ouerdiane-Rezgui [11]). *The Taylor series map $\mathcal{T}$ induces two topological isomorphisms:*

$$\mathcal{T} : \mathcal{F}_\theta(N') \longrightarrow F_\theta(N) \quad \textit{and} \quad \mathcal{T} : \mathcal{G}_{\theta^*}(N) \longrightarrow G_\theta(N'), \tag{2.6}$$

where θ^* is the polar function of θ.

2.4 Laplace Transform

Let $\mathcal{F}_\theta^*(N')$ denote the strong dual space of $\mathcal{F}_\theta(N')$. We shall obtain its concise description.

Let $\Phi \in \mathcal{F}_\theta^*(N')$. By the adjoint map $\mathcal{T}^* : F_\theta^*(N) \to \mathcal{F}_\theta^*(N')$, which is also an isomorphism by (2.6), we obtain $\mathcal{T}^{*-1}\Phi \in F_\theta^*(N)$. On the other hand, $F_\theta^*(N)$ is identified with $G_\theta(N')$ through (2.5). Let $\vec{\Phi} = (\Phi_n) \in G_\theta(N')$ be the element corresponding to $\mathcal{T}^{*-1}\Phi$. Then, for $f \in \mathcal{F}_\theta(N')$ we have

$$\Phi(f) = \langle\langle \vec{\Phi}, \vec{f} \rangle\rangle = \sum_{n=0}^{\infty} n!\, \langle \Phi_n, f_n \rangle, \tag{2.7}$$

where $\vec{f} = (f_n) = \mathcal{T} f$.

For $\xi \in N$ we define the exponential function $\mathrm{e}^\xi : N' \to \mathbb{C}$ by $\mathrm{e}^\xi(x) = \mathrm{e}^{\langle x, \xi \rangle}$, $x \in N'$. It is proved with the help of (2.1) that $\mathrm{e}^\xi \in \mathcal{F}_\theta(N')$ for all $\xi \in N$. The *Laplace transform* of $\Phi \in \mathcal{F}_\theta^*(N')$ is defined by

$$(\mathcal{L}\Phi)(\xi) \equiv \hat{\Phi}(\xi) = \Phi(\mathrm{e}^\xi), \qquad \xi \in N. \tag{2.8}$$

Since the Taylor expansion of e^ξ is given by

$$\mathrm{e}^\xi(x) = \mathrm{e}^{\langle x, \xi \rangle} = \sum_{n=0}^{\infty} \left\langle x^{\otimes n}, \frac{\xi^{\otimes n}}{n!} \right\rangle,$$

we have

$$\mathcal{T}(\mathrm{e}^\xi) = \left(1, \xi, \dots, \frac{\xi^{\otimes n}}{n!}, \dots \right).$$

Therefore, by (2.7) we see that (2.8) becomes

$$(\mathcal{L}\Phi)(\xi) = \hat{\Phi}(\xi) = \sum_{n=0}^{\infty} n! \left\langle \Phi_n, \frac{\xi^{\otimes n}}{n!} \right\rangle = \sum_{n=0}^{\infty} \langle \Phi_n, \xi^{\otimes n} \rangle.$$

Thus, taking Theorem 2.1 into account, we come to the following

Theorem 2.2 (Gannoun–Hachaichi–Ouerdiane–Rezgui [11]). *The Laplace transform induces a topological isomorphism*

$$\mathcal{L} : \mathcal{F}_\theta^*(N') \to \mathcal{G}_{\theta^*}(N), \tag{2.9}$$

where θ^* is the polar function of θ.

2.5 Integral representation of positive distributions

We assume that $N = E + iE$, where E is a real nuclear Fréchet space. Then an element $f \in \mathcal{F}_\theta(N')$ is called *positive* if $f(x + i0) \geq 0$ for all $x \in E'$. We denote by $\mathcal{F}_\theta(N')_+$ the set of positive functions. An element $\Phi \in \mathcal{F}_\theta^*(N')$ is called *positive* if $\Phi(f) \geq 0$ for all $f \in \mathcal{F}_\theta(N')_+$. The cone of positive elements in $\mathcal{F}_\theta^*(N')$ is denoted by $\mathcal{F}_\theta^*(N')_+$. We always assume that E' is equipped with the Borel σ-field.

Theorem 2.3 (Ouerdiane–Rezgui [27]). *For each $\Phi \in \mathcal{F}_\theta^*(N')_+$ there exists a unique positive Radon measure $\mu = \mu_\Phi$ on the space E' such that*

$$\Phi(f) = \int_{E'} f(x + i0)\, d\mu(x), \qquad f \in \mathcal{F}_\theta(N'). \tag{2.10}$$

In that case there exist $q > 0$ and $m > 0$ such that the measure μ is carried by the space E_{-q} and

$$\int_{E_{-q}} \mathrm{e}^{\theta(m|x|_{-q})}\, d\mu(x) < \infty. \tag{2.11}$$

Conversely, such a positive finite measure μ on the space E' defines a positive distribution $\Phi \in \mathcal{F}_\theta^*(N')_+$ by formula (2.10).

Note that the Fourier transform of μ_Φ and the Laplace transform of Φ is related as

$$\mathcal{F}\mu_\Phi(\xi) = \int_{E'} \mathrm{e}^{\mathrm{i}\langle x,\xi\rangle}\, d\mu_\Phi(x) = \Phi(\mathrm{e}^{\mathrm{i}\xi}) = \mathcal{L}\Phi(i\xi), \qquad \xi \in E. \tag{2.12}$$

3. The Lévy Laplacian

3.1 Definition in general

Let E be a real nuclear Fréchet space as before. A function $F : E \to \mathbb{R}$ is called of class $C^2(E)$ if there exist two continuous maps $\xi \mapsto F'(\xi) \in E'$ and $\xi \mapsto F''(\xi) \in \mathcal{L}(E, E')$, $\xi \in E$, such that

$$F(\xi + \eta) = F(\xi) + \langle F'(\xi), \eta\rangle + \frac{1}{2}\langle F''(\xi)\eta, \eta\rangle + \epsilon(\eta), \qquad \xi, \eta \in E,$$

where the error term satisfies:

$$\lim_{t\to 0} \frac{\epsilon(t\eta)}{t^2} = 0, \qquad \eta \in E.$$

In view of the nuclear kernel theorem $\mathcal{L}(E, E') \cong (E \otimes E)' \cong \mathcal{B}(E, E)$ we use the common symbol $F''(\xi)$ for all:

$$\langle F''(\xi)\eta, \eta\rangle = \langle F''(\xi), \eta \otimes \eta\rangle = F''(\xi)(\eta, \eta) = D_\eta D_\eta F(\xi),$$

where D_η is the Fréchet derivative in the direction η, i.e.,

$$(D_\eta F)(\xi) = \lim_{\lambda\to 0} \frac{F(\xi + \lambda\eta) - F(\xi)}{\lambda}.$$

A $\mathbb{C}$-valued function $F : E \to \mathbb{C}$ is a member of $C^2(E)$ if so are its real and imaginary parts. In that case, $F'(\xi) \in N'$ and $F''(\xi) \in (N \otimes N)'$.

Fix an arbitrary infinite sequence $\{e_n\}_{n=1}^\infty \subset E$. We shall assume additional properties later, though. The *Lévy Laplacian* is defined for $F \in C^2(E)$ by

$$\Delta_L F(\xi) = \lim_{N\to\infty} \frac{1}{N} \sum_{n=1}^{N} \langle F''(\xi)e_n, e_n\rangle, \qquad \xi \in E,$$

whenever the limit exists. Let $\mathcal{D}_L(E)$ be the space of all $F \in C^2(E)$ for which $\Delta_L F(\xi)$ exists for all $\xi \in E$. It is noted that the Lévy Laplacian depends on the choice of the sequence $\{e_n\}$ as well as its arrangement.

3.2 Cesàro mean

We prepare a notation. Let $\{e_n\}_{n=1}^\infty \subset E$ be an arbitrary sequence as in the previous subsection. Recall that $N = E + iE$. We denote by $(E\otimes E)'_L$ (resp. $(N\otimes N)'_L$) the set of all $f \in (E\otimes E)'$ (resp. $f \in (N\otimes N)'$) which admit the limit

$$\langle f\rangle_L = \lim_{N\to\infty} \frac{1}{N} \sum_{n=1}^{N} \langle f, e_n \otimes e_n\rangle.$$

Although not explicitly written, $(E \otimes E)'_L$ and $(N \otimes N)'_L$ depend on the choice of $\{e_n\}$. Obviously, $f \in (N \otimes N)'_L$ if and only if its real and imaginary parts belong to $(E \otimes E)'_L$.

By definition we have the following

Lemma 3.1. *A function $F \in C^2(E)$ belongs to $\mathcal{D}_L(E)$ if and only if $F''(\xi) \in (N \otimes N)'_L$ for all $\xi \in E$. In that case,*

$$\Delta_L F(\xi) = \langle F''(\xi)\rangle_L.$$

Let E'_L (resp. N'_L) denote the set of all $a \in E'$ (resp. $a \in N'$) such that $a \otimes a \in (E \otimes E)'_L$ (resp. $a \otimes a \in (N \otimes N)'_L$), i.e., the limit

$$\langle a \otimes a \rangle_L = \lim_{N \to \infty} \frac{1}{N} \sum_{n=1}^{N} \langle a, e_n \rangle^2$$

exists. For a real $a \in E'_L$ we also write $\|a\|_L^2 = \langle a \otimes a \rangle_L$. It is clear that $E'_L \subset N'_L$ but $N'_L = E'_L + iE'_L$ does not necessarily hold.

Lemma 3.2. *For $a, b \in E'_L$ it holds that*

$$\limsup_{N \to \infty} \frac{1}{N} \sum_{n=1}^{N} |\langle a \otimes b, e_n \otimes e_n \rangle| \le \|a\|_L \|b\|_L .$$

Proof. Note the Schwarz inequality

$$\sum_{n=1}^{N} |\langle a \otimes b, e_n \otimes e_n \rangle| \le \left(\sum_{n=1}^{N} \langle a, e_n \rangle^2 \right)^{1/2} \left(\sum_{n=1}^{N} \langle b, e_n \rangle^2 \right)^{1/2},$$

from which the assertion follows immediately. □

Lemma 3.3. *Let $a, b \in E'_L$.*

(1) *If $a + b \in E'_L$, then $\|a + b\|_L \le \|a\|_L + \|b\|_L$.*
(2) *If $\|b\|_L = 0$, then $a + b \in E'_L$ and $\|a + b\|_L = \|a\|_L$.*

Proof. We note the obvious identity:

$$\sum_{n=1}^{N} \langle a + b, e_n \rangle^2 = \sum_{n=1}^{N} \langle a, e_n \rangle^2 + \sum_{n=1}^{N} \langle b, e_n \rangle^2 + 2 \sum_{n=1}^{N} \langle a \otimes b, e_n \otimes e_n \rangle .$$

Then the assertions are immediate from Lemma 3.2. □

3.3 Eigenfunctions

Lemma 3.4. *Let $p \in \mathcal{D}_L(E)$ with $p'(\xi) \in N'_L$ for all $\xi \in E$. Then $e^p \in \mathcal{D}_L(E)$ and*

$$\Delta_L e^{p(\xi)} = (\langle p''(\xi) \rangle_L + \langle p'(\xi) \otimes p'(\xi) \rangle_L)\, e^{p(\xi)} .$$

Proof. The assertion is immediate from

$$D_{e_n}^2 e^{p(\xi)} = \left\{ \langle p''(\xi) e_n, e_n \rangle + \langle p'(\xi), e_n \rangle^2 \right\} e^{p(\xi)},$$

which is verified by a direct computation. □

Now we show two typical classes of eigenfunctions of Δ_L.

Proposition 3.5. (1) *For $a \in N'_L$ it holds that*

$$\Delta_L e^{\langle a,\xi\rangle} = \langle a \otimes a\rangle_L \, e^{\langle a,\xi\rangle}. \tag{3.1}$$

(2) *Let $f \in (N \otimes N)'_L$ be symmetric. If $\langle (f \otimes_1 \xi) \otimes (f \otimes_1 \xi)\rangle_L = 0$ for all $\xi \in E$, it holds that*

$$\Delta_L e^{\langle f,\xi\otimes\xi\rangle} = 2\langle f\rangle_L e^{\langle f,\xi\otimes\xi\rangle}.$$

Proof. (1) is immediate from Lemma 3.4. We prove (2). Put $p(\xi) = \langle f, \xi \otimes \xi\rangle$. Note that

$$\langle p'(\xi), e_n\rangle = 2\langle f, \xi \otimes e_n\rangle = 2\langle f \otimes_1 \xi, e_n\rangle,$$

where $f \otimes_1 \xi$ is contraction of degree one and is defined as above. Then

$$\langle p'(\xi) \otimes p'(\xi)\rangle_L = 4\langle (f \otimes_1 \xi) \otimes (f \otimes_1 \xi)\rangle_L = 0$$

by assumption. On the other hand, $\langle p''(\xi)\rangle_L = 2\langle f\rangle_L$. Hence the assertion follows from Lemma 3.4. □

3.4 Derivation property

It is widely known as one of the peculiar properties that the Lévy Laplacian is a derivation, i.e., behaves like a first order differential operator. This property, however, depends on the domain as shown in the next proposition. A similar fact was already pointed out by Accardi–Obata [4].

Proposition 3.6. *Let $F_1, F_2 \in \mathcal{D}_L(E)$. If $\langle F'(\xi) \otimes G'(\xi)\rangle_L = 0$ for all $\xi \in E$, then*

$$\Delta_L(F_1F_2) = (\Delta_L F_1)F_2 + F_1(\Delta_L F_2).$$

The proof is straightforward. A function $H \in C^2(E)$ is called *Lévy-harmonic* if $\Delta_L H(\xi) = 0$ for all $\xi \in E$. Then we have immediately the following

Corollary 3.7. *Let $F \in \mathcal{D}_L(E)$ and H a Lévy-harmonic function. If $\langle F'(\xi) \otimes H'(\xi)\rangle_L = 0$ for all $\xi \in E$, then*

$$\Delta_L(FH) = (\Delta_L F)H.$$

4. Heat Equation

4.1 Cauchy problem

In general, the Cauchy problem associated with the Lévy Laplacian is stated as follows:

$$\frac{\partial F}{\partial t} = \gamma \Delta_L F, \qquad F(0,\xi) = F_0(\xi), \tag{4.1}$$

where $\gamma \in \mathbb{C}$ is a constant, the initial condition F_0 is a suitable function on E and t runs over an interval including 0. Note that (4.1) involves both heat type and Schrödinger type equations associated with the Lévy Laplacian.

The formal solution F of (4.1) is given by

$$F(t,\xi) = (\mathrm{e}^{t\gamma\Delta_L} F_0)(\xi) = \sum_{n=0}^{\infty} \frac{(t\gamma)^n}{n!}(\Delta_L^n F_0)(\xi).$$

However, the convergence is always in question. For particular initial conditions the convergence is proved by Chung–Ji–Saitô [7]. We do not go into this direction.

As a general remark we only mention the following

Proposition 4.1. *Let $p, q \in \mathcal{D}_L(E)$ and assume that*

$$\Delta_L p(\xi) \equiv \alpha, \qquad \Delta_L q(\xi) \equiv 0,$$

and

$$\langle p'(\xi) \otimes p'(\xi) \rangle_L = \langle p'(\xi) \otimes q'(\xi) \rangle_L \equiv 0, \qquad \langle q'(\xi) \otimes q'(\xi) \rangle_L \equiv \beta,$$

where $\alpha, \beta \in \mathbb{C}$ are constant numbers. Let $\gamma \in \mathbb{C}$ be another constant. Then,

$$F_t(\xi) = F(t,\xi) = \mathrm{e}^{t\gamma(\alpha+\beta)} \mathrm{e}^{p(\xi)+q(\xi)}, \qquad t \in \mathbb{R}, \quad \xi \in E, \tag{4.2}$$

satisfies the Cauchy problem associated with the Lévy Laplacian:

$$\frac{\partial}{\partial t} F(t,\xi) = \gamma \Delta_L F(t,\xi), \qquad F(0,\xi) = \mathrm{e}^{p(\xi)+q(\xi)}. \tag{4.3}$$

Proof. By Lemma 3.4 we have

$$\begin{aligned} \Delta_L \mathrm{e}^{p(\xi)+q(\xi)} = {} & [\langle p''(\xi) \rangle_L + \langle q''(\xi) \rangle_L \\ & + \langle (p'(\xi) + q'(\xi)) \otimes (p'(\xi) + q'(\xi)) \rangle_L]\, \mathrm{e}^{p(\xi)+q(\xi)}. \end{aligned}$$

By assumption we have

$$\Delta_L e^{p(\xi)+q(\xi)} = (\alpha+\beta)\, e^{p(\xi)+q(\xi)},$$

from which the assertion is immediate. □

In order to check the condition in the above theorem such results as in Lemmas 3.2 and 3.3 are useful. Typically we take

$$p(\xi) = \langle f, \xi\otimes\xi\rangle, \qquad q(\xi) = \langle a, \xi\rangle,$$

and consider (4.2), see Proposition 3.5. From the next subsection on, we shall show that superposition of (4.2) gives a solution to the Cauchy problem (4.1) with an interesting initial condition.

4.2 Shift-invariance

Recall that the Lévy Laplacian Δ_L depends on an arbitrarily fixed sequence $\{e_n\}_{n=1}^{\infty} \subset E$. We now consider the shift operator S associated with this sequence. Assume that there exists a continuous operator $S : E \to E$ such that $Se_n = e_{n+1}$ for all n. It would be more natural to do the converse. Given a continuous operator S and a fixed $e_1 \in E$, we may construct the sequence $\{e_n\}$ by $e_n = S^{n-1}e_1$.

Proposition 4.2. *The Lévy Laplacian is invariant under the shift* S*, i.e.,*

$$\Delta_L(F\circ S) = (\Delta_L F)\circ S, \qquad F\in\mathcal{D}_L(E).$$

Proof. By a direct computation we have

$$\langle (F\circ S)''(\xi)e_n, e_n\rangle = \langle F''(S\xi)Se_n, Se_n\rangle = \langle F''(S\xi)e_{n+1}, e_{n+1}\rangle.$$

Since the Cesàro mean is invariant under the shift, the assertion follows. □

4.3 Evolution of positive distributions

For $x\in E'$ we put $q_x(\xi) = e^{i\langle x,\xi\rangle}$. Then, if $x \in E'_L$ we have

$$\Delta_L q_x(\xi) = -\langle x\otimes x\rangle_L q_x(\xi) = -\|x\|_L^2 q_x(\xi).$$

We shall consider a superposition of such q_x.

Note that the adjoint S^* is a continuous operator from E' into itself.

Theorem 4.3. *Let* $\Phi_0 \in \mathcal{F}_\theta^*(N')_+$ *and* μ *the corresponding Radon measure on* E'*, see Theorem 2.3. If* μ *is invariant under* S^**, then* $x\in E'_L$ *for* μ*-a.e.* x *and*

$$F_t(\xi) = F(t,\xi) = \int_{E'} e^{-t\|x\|_L^2} e^{i\langle x,\xi\rangle} d\mu(x), \qquad \xi\in E, \quad t\ge 0, \tag{4.4}$$

is a solution to the Cauchy problem:

$$\frac{\partial F}{\partial t} = \Delta_L F, \qquad F(0,\xi) = \mathcal{L}\Phi_0(i\xi) = \int_{E'} e^{i\langle x,\xi\rangle} d\mu(x). \tag{4.5}$$

Proof. For simplicity we put

$$G(x) = \langle x, e_1\rangle^2, \qquad x \in E'.$$

Then $G \in L^1(E',\mu)$. In fact, taking Theorem 2.3 into account, we note that

$$\begin{aligned}\int_{E'} |G(x)|\, d\mu(x) &= \int_{E_{-q}} \langle x, e_1\rangle^2 e^{-\theta(m|x|_{-q})} e^{\theta(m|x|_{-q})}\, d\mu(x)\\ &\le \int_{E_{-q}} |x|_{-q}^2 |e_1|_q^2 e^{-\theta(m|x|_{-q})} e^{\theta(m|x|_{-q})}\, d\mu(x).\end{aligned} \tag{4.6}$$

Since $\sup_{t\ge 0} t^2 e^{-\theta(mt)} < \infty$ by the assumptions on θ, (4.6) is finite as desired. Now we recall the assumption that μ is invariant under the measurable transformation S^*. Then applying the ergodic theorem (see e.g., [9, Chapter VIII]), we see that

$$\tilde{G}(x) = \lim_{N\to\infty} \frac{1}{N} \sum_{n=1}^{N} G(S^{*(n-1)}x)$$

converges for μ-a.e. $x \in E'$. Moreover, the convergence holds also in the L^1-sense and $\tilde{G} \in L^1(E',\mu)$. On the other hand, since

$$\sum_{n=1}^{N} G(S^{*(n-1)}x) = \sum_{n=1}^{N} \langle S^{*(n-1)}x, e_1\rangle^2 = \sum_{n=1}^{N} \langle x, S^{n-1}e_1\rangle^2 = \sum_{n=1}^{N} \langle x, e_n\rangle^2,$$

we have $\tilde{G}(x) = \langle x\otimes x\rangle_L = \|x\|_L^2$. Consequently, a measurable function $x \mapsto \|x\|_L^2$ is defined μ-a.e. $x \in E'$ and belongs to $L^1(E',\mu)$. Then, one can check easily that (4.4) is a solution to (4.5) by the Lebesgue convergence theorem. □

In the usual definition of Δ_L, the sequence $\{e_n\} \subset E$ is assumed to have some particular properties, typically, to be a complete orthonormal basis for a certain Hilbert space. We note that in Theorem 4.3 such additional assumptions are not required. However, the idea of proof is essentially due to Accardi–Roselli–Smolyanov [5] and Accardi–Obata [4].

4.4 A relation with quadratic quantum white noises

In this subsection we take a concrete nuclear triple:

$$E = \mathcal{S}(\mathbb{R}) \subset H = L^2(\mathbb{R}) \subset E' = \mathcal{S}'(\mathbb{R}).$$

As before, we set $N = E + iE$. For $s \geq 0$ consider

$$p_s(\xi) = \langle 1_{[0,s]}\xi, \xi \rangle = \int_0^s \xi(u)^2 \mathrm{d}u, \qquad \xi \in E.$$

Then,

$$\|p_s'(\xi)\|_L^2 = 4\|1_{[0,s]}\xi\|_L^2 = 4 \lim_{N\to\infty} \frac{1}{N} \sum_{n=1}^{N} \left(\int_0^s e_n(u)\xi(u)\mathrm{d}u \right)^2$$

and

$$\Delta_L p_s(\xi) = 2\langle 1_{[0,s]}\tau \rangle_L = 2 \lim_{N\to\infty} \frac{1}{N} \sum_{n=1}^{N} \int_0^s e_n(u)^2 \mathrm{d}u,$$

where $\tau \in \mathcal{S}'(\mathbb{R} \times \mathbb{R})$ is the trace. With the help of Proposition 4.1 we come to the following

Lemma 4.4. *Let $s \geq 0$. If $\|1_{[0,s]}\xi\|_L = 0$ for all $\xi \in E$ and $\langle 1_{[0,s]}\tau \rangle_L = s$, then*

$$f_t(\xi) = \mathrm{e}^{2st}\mathrm{e}^{q_s(\xi)}, \qquad t \in \mathbb{R}, \quad \xi \in E,$$

satisfies the heat equation:

$$\frac{\partial}{\partial t} f_t(\xi) = \Delta_L f_t(\xi).$$

Let $\{a_t, a_t^*\}_{t\in\mathbb{R}}$ be the quantum white noise, namely, a_t is a continuous linear operator on $\mathcal{F}_\theta(N')$ defined by

$$a_t \mathrm{e}^\xi = \xi(t)\mathrm{e}^\xi, \qquad \xi \in N, \quad t \in \mathbb{R},$$

and a_t^* is the dual operator. Consider the normal-ordered white noise differential equation

$$\frac{d\Xi}{dt} = (a_t^2 + a_t^{*2}) \diamond \Xi, \qquad \Xi(0) = I. \tag{4.7}$$

This is a "singular" quantum stochastic differential equation beyond the traditional Itô theory. By general theory [21] there exists a unique solution to (4.7) in $\mathcal{L}(\mathcal{F}_\theta(N'), \mathcal{F}_\theta^*(N'))$. Let $\{\Psi_t\} \subset \mathcal{F}_\theta^*(N')$ be the "classical" stochastic process corresponding to the "quantum" stochastic process $\{\Xi_t\}$ defined by $\Psi_t = \Xi_t \mathrm{e}^0$, where $\mathrm{e}^0(\xi) \equiv 1$. Then by a direct computation we have

$$\mathrm{e}^{q_s(\xi)} = \mathcal{L}\Psi_s(\xi), \qquad s \geq 0, \quad \xi \in E.$$

Summing up,

Theorem 4.5. *Let* $\{\Psi_s\} \subset \mathcal{F}^*_\theta(N')$ *be the classical stochastic process corresponding to the quantum stochastic process determined by* (4.7). *Let* $s \geq 0$. *Assume that* $\|1_{[0,s]}\xi\|_L = 0$ *for all* $\xi \in E$ *and* $\langle 1_{[0,s]}\rangle_L = s$. *Then*

$$F_t(\xi) = \mathrm{e}^{2st}\mathcal{L}\Psi_s(\xi), \qquad t \geq 0, \quad \xi \in E,$$

satisfies the heat equation:

$$\frac{\partial}{\partial t}F = \Delta_L F, \qquad F(0,\xi) = \mathcal{L}\Psi_s(\xi). \tag{4.8}$$

Let T be a compact interval equipped with a finite measure ν. If the assumption in Theorem 4.5 is true for all $s \in T$, then

$$F_t(\xi) = \int_T \mathrm{e}^{2st}\mathcal{L}\Psi_s(\xi)\,\nu(ds), \qquad t \geq 0, \quad \xi \in E,$$

satisfies the heat equation (4.8) with an initial condition:

$$F(0,\xi) = \int_T \mathcal{L}\Psi_s(\xi)\,\nu(ds).$$

This draws out an essence of [22, Theorem 6].

Remark 4.6. Recall that for $\Phi \in \mathcal{F}^*_\theta(N')$, the Laplace transform $\mathcal{L}\Phi$ belongs to $\mathcal{G}_{\theta^*}(N)$. In particular, $\mathcal{L}\Phi \in C^2(E)$. Let $\mathcal{D}_L$ denote the space of all $\Phi \in \mathcal{F}^*_\theta(N')$ such that $\mathcal{L}\Phi \in \mathcal{D}_L(E)$ and $\Delta_L\mathcal{L}\Phi \in \mathcal{G}_{\theta^*}(N)$. Then the *Lévy Laplacian* $\tilde{\Delta}_L$ is defined by

$$\tilde{\Delta}_L\Phi = \mathcal{L}^{-1}\Delta_L\mathcal{L}\Phi, \qquad \Phi \in D_L.$$

This $\tilde{\Delta}_L$ is essentially the same as the Lévy Laplacian formulated within white noise theory, see e.g., Kuo [14] and references cited therein.

Acknowledgments

For the first author Supported in part by JSPS Grant-in-Aid for Scientific Research No. 12440036.

References

[1] Accardi L.: Yang-Mills equations and Lévy-Laplacians, in "Dirichlet Forms and Stochastics Processes (Z. M. Ma, M. Röckner and J. A. Yan, Eds.)," pp. 1–24, Walter de Gruyter, 1995.

[2] Accardi L., Bogachev V. I.: The Ornstein-Uhlenbeck process associated with the Lévy Laplacian and its Dirichlet form, *Prob. Math. Stat.* **17** (1997) 95–114.

[3] Accardi L., Gibilisco P., Volovich I. V.: The Lévy Laplacian and the Yang-Mills equations, *Rend. Accad. Sci. Fis. Mat. Lincei* **4** (1993) 201–206.

[4] Accardi L., Obata, N.: Derivation property of the Lévy Laplacian, *RIMS Kokyuroku* **874** (1994) 8–19.

[5] Accardi L., Roselli P., Smolyanov O. G.: The Brownian motion generated by the Lévy-Laplacian, *Mat. Zametki* **54** (1993) 144–148.

[6] Aref'eva I. Ya., Volovich I.: Higher order functional conservation laws in gauge theories, in "Generalized Functions and their Applications in Mathematical Physics," Proc. Internat. Conf., Moscow, 1981 (Russian).

[7] Chung D. M., Ji U. C., Saitô K.: Cauchy problems associated with the Lévy Laplacian in white noise analysis, *Infin. Dimens. Anal. Quantum Probab. Relat. Top.* **2** (1999) 131–153.

[8] Cochran W. G., Kuo H.-H., Sengupta A.: A new class of white noise generalized functions, *Infin. Dimens. Anal. Quantum Probab. Relat. Top.* **1** (1998) 43–67.

[9] Dunford N., Schwartz J. T.: *Linear Operators, Part I: General Theory*, Wiley Classical Library Edition, 1988.

[10] Feller M. N.: Infinite-dimensional elliptic equations and operators of Lévy type, *Russian Math. Surveys* **41** (1986) 119–170.

[11] R. Gannoun, R. Hachaichi, Ouerdiane H., Rezgui A.: Un théorème de dualité entre espaces de fonctions holomorphes à croissance exponentielle, *J. Funct. Anal.* **171** (2000) 1–14.

[12] Gel'fand I. M., Vilenkin N. Ya.: *Generalized Functions, Vol. 4*, Academic Press, 1964.

[13] Krée P., Ouerdiane H.: Holomorphy and Gaussian analysis, Prépublication de l'Institut de Mathématique de Jussieu. C.N.R.S. Univ. Paris 6, 1995.

[14] Kuo H.-H.: *White Noise Distribution Theory*, CRC Press, 1996.

[15] Kuo H.-H., Obata N., Saitô K.: Lévy Laplacian of generalized functions on a nuclear space, *J. Funct. Anal.* **94** (1990) 74–92.

[16] Kuo H.-H., Obata N., Saitô K.: Diagonalization of the Lévy Laplacian and related stable processes, *Infin. Dimens. Anal. Quantum Probab. Relat. Top.* **5** (2002) 317–331.

[17] Lévy P.: *Leçons d'Analyse Fonctionnelle*, Gauthier–Villars, Paris, 1922.

[18] Lévy P.: *Problèmes Concrets d'Analyse Fonctionnelle*, Gauthier–Villars, Paris, 1951.

[19] Obata N.: A characterization of the Lévy Laplacian in terms of infinite dimensional rotation groups, *Nagoya Math. J.* **118** (1990) 111–132.

[20] Obata N.: *White Noise Calculus and Fock Space*, Lect. Notes in Math. Vol. 1577, Springer–Verlag, 1994.

[21] Obata N.: Wick product of white noise operators and quantum stochastic differential equations, *J. Math. Soc. Japan* **51** (1999) 613–641.

[22] Obata N.: Quadratic quantum white noises and Lévy Laplacian, *Nonlinear Analysis, Theory, Methods and Applications* **47** (2001) 2437–2448.

[23] Ouerdiane H.: Fonctionnelles analytiques avec conditions de croissance et application à l'analyse gaussienne, *Japan. J. Math.* **20** (1994) 187–198.

[24] Ouerdiane H.: Noyaux et symboles d'opérateurs sur des fonctionnelles analytiques gaussiennes, *Japan. J. Math.* **21** (1995) 223–234.

[25] Ouerdiane H.: Algèbre nucléaires et équations aux dérivées partielles stochastiques, *Nagoya Math. J.* **151** (1998) 107–127.

[26] Ouerdiane H.: Distributions gaussiennes et applications aux équations aux dérivées partielles stochastiques, in "*Proc. International conference on Mathematical Physics and stochastics Analysis (in honor of L. Streit's 60th birthday),*" S. Albeverio et al. Eds., World Scientific, 2000.

[27] Ouerdiane H., Rezgui A.: Représentations intégrales de fonctionnelles analytiques, *Can. Math. Soc. Conf. Proc.* **28** (2000) 283–290.

[28] Polishchuk E. M.: *Continual Means and Boundary Value Problems in Function Spaces*, Birkhäuser, 1988.

[29] Saitô K.: A (C_0)-group generated by the Lévy Laplacian II, *Infin. Dimens. Anal. Quantum Probab. Relat. Top.* **1** (1998) 425–437.

[30] Saitô K.: A stochastic process generated by the Lévy Laplacian, *Acta Appl. Math.* **63** (2000) 363–373.

[31] Saitô K., Tsoi, A. H.: The Lévy Laplacian as a self-adjoint operator, in "*Quantum Information*", Hida T., Saitô K., Eds., pp. 159–171, World Scientific, 1999.

ZETA-REGULARIZED TRACES VERSUS THE WODZICKI RESIDUE AS TOOLS IN QUANTUM FIELD THEORY AND INFINITE DIMENSIONAL GEOMETRY

Sylvie Paycha
Département de Mathématiques, Complexe des Cézeaux, 63177 Aubière Cedex, France
sylvie.paycha@math.univ-bpclermont.fr

Abstract We compare two candidates for a "trace" in infinite dimensions, namely the Wodzicki residue and ζ-regularized "traces". We discuss their relevance when trying to extend some classical notions in finite dimensions to an infinite dimensional context appropriate for quantum field theory. This is a review article based on [6, 5, 18, 19].

Keywords: Zeta-regularized trace, Wodzicki residue, quantum field theory, infinite dimensional geometry

Introduction

From a path integral point of view, quantization brings into play formal integrals on an infinite dimensional configuration space. Thus, going from some classical field theory to the corresponding quantum field theory using path integral quantization procedures (we leave aside the case of a statistical model), involves stepping from a finite dimensional setting to an infinite dimensional setting. This step is the source of many a difficulty, a major one being that one typically needs to work with the algebra of classical pseudo-differential operators (or even with the more general log-polyhomogeneous pseudo-differential operators), instead of the algebra of ordinary matrices. In particular, giving a meaning to the partition function defined by such a path integral requires extending ordinary determinants –and hence traces– of matrices, to determinants –and hence "traces"– of pseudo-differential operators.

Not only does the trace on ordinary matrices underly the notion of determinant but it also provides a useful tool to describe the geometry of finite dimensional manifolds and vector bundles; Chern-Weil invariants

S. Albeverio et al. (eds.),
Proceedings of the International Conference on Stochastic Analysis and Applications, 69–84.

are a striking example of useful geometric concepts defined via the trace. A natural question is therefore what to replace the ordinary trace and determinant by in order to get a grasp on the geometry of some infinite dimensional manifolds such as configuration spaces arising in quantum field theory.

The Wodzicki residue [22] yields a trace on the algebra of classical pseudo-differential operators on a closed manifold and Chern-Weil calculus on finite rank vector bundles extends without problem to a class of infinite rank vector bundles using the Wodzicki residue. However, since the Wodzicki residue vanishes on finite rank operators, it "does not see" finite dimensional objects and is therefore not a good candidate to extend the finite dimensional setting. Moreover, it yields vanishing characteristic classes in the case of loop manifolds [14].

On the other hand, ζ-regularized traces investigated in different contexts in [6, 18, 17, 5] are linear functionals on the algebra of classical pseudo-differential operators that coincide with the usual trace on finite rank operators so that at first glance they could appear as good candidates to extend the finite dimensional setting. Also, the ζ-determinant built from ζ-regularized traces, which extends the ordinary determinant on matrices, is widely used in the physics literature. In general ζ-regularized traces are (despite their name) not tracial and do not commute with exterior differentiation, and ζ-determinants are not multiplicative, thus giving rise to various discrepancies that can in turn be expressed in terms of the Wodzicki residue. We show how these discrepancies, which we refer to as *tracial anomalies*, come into the way in quantum field theory and in infinite dimensional geometry, and discuss possible ways to circumvent them.

The article is organized as follows. In sections 1 and 2, we recall the notion of ζ-regularized trace, Wodzicki residue and their associated determinants, the ζ-determinant and an exotic determinant. Section 3 is devoted to the analysis of tracial and determinant anomalies, namely discrepancies arising from the use of ζ-regularized traces and determinants. In section 4 we discuss how these discrepancies can arise in quantum field theory and we show in section 5, how they come up in infinite dimensional geometry. We point out to possible ways of avoiding these discrepancies to define Chern-Weil type invariants in the infinite dimensional context. In section 6, we give two examples of these Chern-Weil type forms.

Notations. Let $C\ell(M,E)$ denote the algebra of classical pseudo-differential operators (P.D.O.s) acting on smooth sections of some hermitian vector bundle $\pi : E \to M$ based on a closed connected Riemannian manifold M of dimension n. Let $C\ell_{\mathrm{ord}\leq\alpha}(M,E)$ denote the subset of classical P.D.O.s or order $\leq \alpha$, let $C\ell^*(M,E)$, resp. $\mathrm{Ell}(M,E)$, resp. $\mathrm{Ell}^*(M,E)$ denote the subsets of invertible, resp. elliptic, resp. invertible elliptic classical P.D.O.s, and finally let $\mathrm{Ell}^{*\,\mathrm{adm}}_{\mathrm{ord}>0}(M,E)$ denote the subset of admissible, invertible elliptic classical P.D.O.s with positive order. Admissibility is a technical assumption on the spectrum of the operator, which allows to build complex powers and the logarithm of the operator. The order of P.D.O.s is denoted by small letters, e.g., the order of A is denoted by a.

1. Weighted traces versus Wodzicki residue

Given $A \in C\ell(M,E)$ of order a, $Q \in \mathrm{Ell}^{*\,\mathrm{adm}}_{\mathrm{ord}>0}(M,E)$ of order q, for $\mathrm{Re}\, z > \frac{a+n}{q}$, the operator AQ^{-z} is trace-class and the map $z \to \mathrm{tr}(AQ^{-z})$ extends to a meromorphic function on the complex plane with simple pole at $z=0$:

$$\mathrm{tr}(AQ^{-z}) = \mathrm{tr}^Q(A) + \frac{\mathrm{res}(A)}{q \cdot z} + \mathrm{O}(z). \tag{1.1}$$

The residue in (1.1) is proportional to the *Wodzicki residue* $\mathrm{res}(A)$ of A which can be defined independently of the choice of Q by the following local formula [22]:

$$\mathrm{res}(A) := \frac{1}{(2\pi)^n} \int_M \left(\int_{ST_x^* M} \mathrm{tr}_x(\sigma_{-n}(A)) d\xi \right) d\mu(x) \tag{1.2}$$

where $ST_x^*M := \{\xi \in T_x^*M, \|\xi\| = 1\}$ is the tangent cosphere of M at point x, $d\mu(x)$ the Riemannian volume element on M and $\sigma_{-n}(A)$ the homogeneous component of degree $-n$ of the symbol of A. Although the symbol of A is only defined locally, the above integral does not depend on the local representation chosen to define it.

We call $\mathrm{tr}^Q(A)$ the *Q-weighted trace* of A. When A is trace-class, the residue term vanishes in the expression $\mathrm{tr}(AQ^{-z})$, so that this expression converges to $\mathrm{tr}^Q(A) = \mathrm{tr}(A)$ when $z \to 0$. Hence weighted traces extend to P.D.O.s the ordinary trace on trace-class operators, whereas the Wodzicki residue vanishes on such operators.

Wodzicki [22] showed that $A \to \mathrm{res}(A)$ defines a trace on the algebra $C\ell(M,E)$ i.e.

$$\forall A, B \in C\ell(M,E), \quad \mathrm{res}([A,B]) = 0$$

and that it is the only trace (up to some multiplicative factor) on that algebra when $n > 1$.

In contrast, linear functionals on $C\ell(M,E)$ obtained from weighted traces $A \mapsto \mathrm{tr}^Q(A)$ do not define traces on the algebra of classical pseudo-differential operators i.e. $\mathrm{tr}^Q([A,B]) \neq 0$ in general and they depend on the choice of the weight. Only on very special subalgebras (see [10, 7]) do they define cyclic traces.

Both the Wodzicki residue and weighted traces have a covariance property:

$$\mathrm{res}(C^{-1}AC) = \mathrm{res}(A), \quad \forall A, B \in C\ell(M,E), \tag{1.3}$$

and

$$\mathrm{tr}^{C^{-1}QC}(C^{-1}AC) = \mathrm{tr}^{Q}(A), \ \forall A \in C\ell(M,E), \ \forall C \in C\ell^*(M,E), \tag{1.4}$$

the first one resulting from the tracial property of the Wodzicki residue. We refer to [6] for the second one.

If we want to recover the finite dimensional setting as a special case of the infinite dimensional one, weighted traces turn out to be very useful. If instead we want to focus on the purely infinite dimensional properties, we might as well work with the Wodzicki residue which singles them out, picking the divergent part and leaving aside the finite part.

2. The ζ-determinant versus an exotic determinant

Just as the ordinary determinant on invertible matrices is built from the trace by

$$\det A = \exp \int_0^1 \mathrm{tr}(\gamma^{-1} d\gamma)$$

where $\{\gamma(t), t \in [0,1]\}$ is a path of invertible matrices from the identity matrix to A, one can build an exotic determinant [22], see also [9]:

$$\det{}^{\mathrm{res}} A := \exp \int_0^1 \mathrm{res}(\gamma^{-1} d\gamma)$$

where $\{\gamma(t) \in C\ell_0^*(M,E), t \in [0,1]\}$ is a path of invertible zero order classical P.D.O.s from the identity operator to A. These definitions make sense because of the cyclicity of the traces involved. In both cases the exponentiated integral defining the determinant is independent of the choice of path since it is trivial on loops. In the case of the exotic determinant, the triviality follows from the vanishing of the residue on finite rank operators. For the same reason, the exotic determinant is identically equal to 1 on finite rank operators and also on Fredholm determinants $1 + A$ where A is trace-class since the Wodzicki residue vanishes on trace-class operators.

The above definitions do not apply to the case of weighted traces since they are not tracial on the algebra of classical P.D.O.s; only if when restricting to subalgebras of P.D.O.s on which such regularized traces are cyclic can one define the corresponding determinant as above [7].

Instead, we adopt another point of view, based on the fact that in finite dimensions det$=\exp\circ\mathrm{tr}\circ\log$. Extending this definition to the infinite dimensional setting leads to the ζ-*determinant* used by physicists. We first need to introduce the logarithm of a classical P.D.O. For $A\in \mathrm{Ell}^{*\,\mathrm{adm}}_{\mathrm{ord}>0}(M,E)$, we set $\log A := \frac{d}{dz}_{|z=0}A^z$, which depends on the spectral cut one chooses to define the complex power A^z. On zero-order operators, the logarithm is defined by $\log A = \frac{i}{2\pi}\int_\Gamma \log\lambda(A-\lambda I)^{-1}d\lambda$, where Γ is a contour around the spectrum of A avoiding a given spectral cut.

Although the logarithm of a classical P.D.O. is not classical, the bracket $[\log Q, A]$ and the difference $\frac{\log Q_1}{q_1}-\frac{\log Q_2}{q_2}$ of two such logarithms are classical P.D.O.s. Recall that q_i denotes the order of Q_i.

For an admissible elliptic operator $A\in \mathrm{Ell}^{\mathrm{adm}}_{\mathrm{ord}>0}(M,E)$ of positive order with non zero eigenvalues, the function $\zeta_A(z):=\mathrm{tr}(A^{-z})$ is holomorphic at $z=0$ and hence $z\to \mathrm{tr}(\log A A^{-z})$ converges when $z\to 0$ to the limit $\mathrm{tr}^A(\log A) := -\zeta_A'(0)$. The ζ-determinant of A is defined by:

$$\det{}_\zeta(A) := \exp\left(-\zeta_A'(0)\right) = \exp \mathrm{tr}^A(\log A). \tag{2.1}$$

In fact physicists (see e.g. [1, 12]) often implicitly work with *weighted determinants* $\det^Q(A) := \exp \mathrm{tr}^Q(\log A)$, or *relative determinants* investigated in [7]:

$$\frac{\det^Q(A)}{\det_\zeta(Q)} = \exp \mathrm{tr}^Q\left(\log A - \log Q\right),$$

which correspond to quotients of a weighted determinant with the ζ-determinant of a fixed reference operator (the weight Q here). Weighted and ζ-determinants are related by a Wodzicki residue

$$\det{}_\zeta(A) = \det{}^Q(A)\exp\left(-\frac{a}{2}\,\mathrm{res}\left(\frac{\log Q}{q}-\frac{\log A}{a}\right)^2\right)$$

with the convention of notation for the order of an operator adopted at the beginning of the article.

Because weighted traces coincide with ordinary trace on trace-class operators, both the weighted and the ζ-determinant coincide with the Fredholm determinant on Fredholm operators.

Inner automorphisms of $\mathrm{Ell}^*_{\mathrm{ord}>0}(M,E)$ leave the ζ-determinant invariant. Indeed, let A be an operator in $\mathrm{Ell}^{*\,\mathrm{adm}}_{\mathrm{ord}>0}(M,E)$ and let $C\in$

$C\ell(M,E)$ be invertible, then CAC^{-1} lies in $\mathrm{Ell}^*_{\mathrm{ord}>0}(M,E)$ and is also admissible since an inner automorphism on P.D.Os induces an inner automorphism on leading symbols $\sigma_L(CAC^{-1}) = \sigma_L(C)\sigma_L(A)\sigma_L(C)^{-1}$ and hence leaves both the spectra of the operator and of its leading symbol unchanged. Moreover, using the fact that, given $Q \in \mathrm{Ell}^*_{\mathrm{ord}>0}(M,E)$ admissible, we have $\log CAC^{-1} = \log A$ and $\mathrm{tr}^{CQC^{-1}}(C(\log A)C^{-1}) = \mathrm{tr}^Q(\log A)$, a fact which can easily be deduced from covariance properties of weighted traces extended to logarithms similar to (1.4) (see [6]), it follows that:

$$\det{}_\zeta(CAC^{-1}) = \det{}_\zeta(A). \tag{2.2}$$

3. Tracial and determinant anomalies in terms of the Wodzicki residue

As briefly mentioned in sections 1 and 2, ζ-regularized traces and ζ-determinants present some discrepancies, namely the coboundary of a weighted trace given by

$$\partial\,\mathrm{tr}^Q(A,B) := \mathrm{tr}^Q([A,B]) \quad \forall A,B \in C\ell(M,E),$$

its dependence on the weight Q and a lack of multiplicativity for the ζ-determinant

$$\det{}_\zeta(AB) \neq \det{}_\zeta(A)\det{}_\zeta(B).$$

These discrepancies can be described in terms of Wodzicki residues via a formula established by Kontsevich and Vishik [10]:

$$\mathrm{Res}_{z=0}\left(\mathrm{tr}(A(z))\right) = -\frac{1}{a'(0)}\,\mathrm{res}(A(0)) \tag{3.1}$$

where $A(z)$ is a holomorphic family of operators on a domain $D \subset \mathbb{C}$ such that $a'(0) \neq 0$, $a(z)$ being the order of $A(z)$.

• **Tracial anomalies.** From (2.1) one can derive the following expressions for what we shall call *weighted trace anomalies* [15, 6, 18, 17]:

$$\partial\,\mathrm{tr}^Q(A,B) = -\frac{1}{q}\,\mathrm{res}\left([\log Q, A]B\right), \tag{3.2}$$

$$\mathrm{tr}^{Q_1}(A) - \mathrm{tr}^{Q_2}(A) = -\,\mathrm{res}\left(A\left(\frac{\log Q_1}{q_1} - \frac{\log Q_2}{q_2}\right)\right) \tag{3.3}$$

where $A,B \in C\ell(M,E)$, $Q \in \mathrm{Ell}^{*\,\mathrm{adm}}_{\mathrm{ord}>0}(M,E)$ of order q, $Q_i \in \mathrm{Ell}^{*\,\mathrm{adm}}_{\mathrm{ord}>0}(M,E)$ with q_i the order of Q_i, $i = 1,2$.

Furthermore, given a smooth family of weights $Q \in \mathrm{Ell}^{*\,\mathrm{adm}}_{\mathrm{ord}>0}(M,E)$ with constant order q parametrized by some manifold B we have [6, 18, 17]:

$$[\mathrm{d}\,\mathrm{tr}^Q](A) = -\frac{1}{q}\,\mathrm{res}(A\,\mathrm{d}\log Q) \quad \forall A \in C\ell(M,E). \tag{3.4}$$

Weighted traces extend to the space $\Omega(B, C\ell(M,E))$ of $C\ell(M,E)$-valued forms on a manifold B setting:

$$\mathrm{tr}^Q(\alpha\otimes A) := \mathrm{tr}^Q(A)\alpha \quad \forall\alpha\in\Omega(B), \quad \forall A\in C\ell(M,E).$$

Formula (3.4) then generalizes to a local formula on P.D.O.-valued forms ([6, 17]): for all $\alpha \in \Omega(B, C\ell(M,E))$,

$$\left(\mathrm{d}\circ\mathrm{tr}^Q - \mathrm{tr}^Q\circ d\right)(\alpha) = \frac{(-1)^{|\alpha|+1}}{q}\,\mathrm{res}(\alpha\,\mathrm{d}\log Q). \tag{3.5}$$

• **Multiplicative anomaly[10].** Another type of anomaly which is closely related to weighted trace anomalies is the multiplicative anomaly of ζ-determinants. The Fredholm determinant is multiplicative but the ζ-determinant is not, this leading to an anomaly

$$F_\zeta(A,B) := \frac{\det_\zeta(AB)}{\det_\zeta(A)\det_\zeta(B)}$$

which reads [10, 7]:

$$\begin{aligned}\log F_\zeta(A,B) = &\frac{1}{2a}\,\mathrm{res}\left(\left(\log A - \frac{a}{a+b}\log(AB)\right)^2\right)\\ &+\frac{1}{2b}\,\mathrm{res}\left(\left(\log B - \frac{b}{a+b}\log(AB)\right)^2\right)\\ &+\mathrm{tr}^{AB}\left(\log(AB) - \log A - \log B\right)\end{aligned} \tag{3.6}$$

for any two operators $A, B \in \mathrm{Ell}^{*\,\mathrm{adm}}_{\mathrm{ord}>0}(M,E)$ of order a and b, respectively. Specializing to $B = A^*$, the adjoint of A for the L^2 structure induced by a Riemannian metric on M and a hermitian one on E, in general we have $F_\zeta(A, A^*) \neq 0$ and hence:

$$\det_\zeta(A^*A) \neq |\det_\zeta(A)|^2.$$

Weighted determinants are generally not multiplicative either and their multiplicative anomaly can be expressed using a Campbell-Hausdorff formula for P.D.O.s. Indeed, it was shown in [16] that given two operators $A, B \in \mathrm{Ell}^{*\,\mathrm{adm}}_{\mathrm{ord}>0}$ with scalar symbols, for any integer K,

$$\log(AB) - \log A - \log B - \sum_{k=2}^{K} C^{(k)}(\log A, \log B) \in Cl_{1-K}(M, E)$$

where $C^{(k)}(P, Q)$ is a linear combination of Lie monomials of degree k in P and Q of the type $(AdP)^{\alpha_1}(AdQ)^{\beta_1}(AdP)^{\alpha_2}\cdots(AdQ)^{\beta_j}Q$. Here we have set $AdM(N) := [M, N]$. Applying a Q-weighted trace yields the following expression for the weighted determinant multiplicative anomaly $F^Q(A, B) := \frac{\det^Q(AB)}{\det^Q(A)\det^Q(B)}$:

$$\begin{aligned}\log F^Q(A, B) &= \mathrm{tr}^Q(\log(AB)) - \mathrm{tr}^Q(\log A) - \mathrm{tr}^Q(\log B) \\ &= \sum_{k=2}^{\infty} \mathrm{tr}^Q\left(C^{(k)}(\log A, \log B)\right) \qquad (3.7)\end{aligned}$$

where we have used the continuity of the weighted trace on the Fréchet space $C\ell_{\mathrm{ord}\leq 0}(M, E)$. Let us now check that this a priori infinite sum is in fact finite. It was shown in [7] that formula (3.2) can be extended to logarithms:

$$\partial\, \mathrm{tr}^Q(\log A, C) = -\frac{1}{q}\, \mathrm{res}\left([\log Q, A_Q]C\right) \quad \forall C \in C\ell(M, E) \qquad (3.8)$$

where we have set $A_Q := \log A - \frac{a}{q}\log Q$ which lies in $C\ell_{\mathrm{ord}\leq 0}(M, E)$. Here a is the order of A and q the order of Q. For k large enough $C = C^{(k)}(\log A, \log B) \in C\ell_{\mathrm{ord}<-n}(M, E)$ so that, inserting it in (3.8) we find that $\partial\, \mathrm{tr}^Q(\log A, C) = -\frac{1}{q}\, \mathrm{res}\left([\log Q, A_Q]C\right)$ vanishes since the Wodzicki residue vanishes on operators of order $< -n$. Hence all but a finite number of terms vanish in the above asymptotic expansion (3.7).

4. How tracial anomalies can lead to discrepancies in quantum field theory

Tracial anomalies can cause some trouble when trying to quantize the classical action of a physical system with some symmetry property which one would like to preserve at the quantized level.

From a path integral point of view, quantization requires building (heuristic) integrals of the type:

$$\int_{\mathcal{C}} e^{-A(\phi)}\mathcal{D}[\phi]$$

where the formal integration $\mathcal{D}[\phi]$ is carried out on a configuration space $\mathcal{C}$ which is typically an infinite dimensional manifold modelled on a space of sections $\Gamma(M, E)$ of some vector bundle E based on a closed manifold M, $\mathcal{A}$ being the action describing the classical action. Anomalies in quantum field theory occur while quantizing because of the lack of invariance of the formal measure $\mathcal{D}[\phi]$ under some symmetry group of the action $\mathcal{A}$. These are broadly speaking described as logarithmic variations of jacobian determinants arising from a transformation of the formal measure under the action of some symmetry group of the classical action. Before even defining its logarithmic derivative, it is often the very definition of the determinant function which causes problems. Chiral anomalies coming from determinants of Dirac operators on even dimensional manifolds are typically described using the language of determinant bundles, where determinant functions are replaced by sections of determinant line bundles. The idea is then to choose some canonical flat section and to build a determinant function as a quotient of a section of the determinant bundle with this flat section.

We briefly describe the geometric setting in the case of a smooth family of elliptic operators $\{A_b^+ : C^\infty(M, E^+) \to C^\infty(M, E^-), b \in B\}$ parametrized by some manifold B where $E := E^+ \oplus E^-$ is some *fixed* super vector bundle based on a *fixed* closed Riemannian manifold M. Alternatively, one considers $C\ell(M, E)$-valued zero-forms $A := \begin{bmatrix} 0 & A^+ \\ A^- & 0 \end{bmatrix}$ on B of formally self-adjoint elliptic operators (using an inner product built from one on the bundle E and the volume measure on the manifold M), so that $A^- = (A^+)^*$. We refer the reader to [3] for a more general setting involving families of manifolds. One associates to this family a determinant line bundle à la Quillen [20]

$$\mathcal{L}_A := (\Lambda^{\max} \operatorname{Ker} A)^* \otimes (\Lambda^{\max} \operatorname{Coker} A)$$

where $\Lambda^{\max} \operatorname{Ker} A$ denotes the maximal exterior power of the finite dimensional kernel of A^+ and $\Lambda^{\max} (\operatorname{Coker}(A))$ of its cokernel which is also finite dimensional. It can be equipped with the Quillen metric defined in terms of a ζ-determinant:

$$\|\operatorname{Det} A^+\|_Q := \sqrt{\det_\zeta(A^- A^+)} \tag{4.1}$$

for a section $\operatorname{Det} A^+$ of $\mathcal{L}_A$, at a point where it is invertible. It can further be equipped with the Bismut-Freed [3] connection (which is compatible with the Quillen metric) defined in terms of weighted traces:

$$\left(\operatorname{Det} A^+\right)^{-1} \nabla^{\operatorname{Det}} \operatorname{Det} A^+ := \operatorname{tr}^{\Delta^+} \left(\left(A^+\right)^{-1} dA^+ \right).$$

Here we have set $\Delta^+ := A^- A^+$ which serves as a weight denoted by Q in the previous sections.

Discrepancies arise when computing the curvature Ω^{Det}; if it were not for the weight Δ^+ in the formula for the connection, one would expect the curvature of the Bismut-Freed connection to vanish. However:

$$\begin{aligned}(\mathrm{Det}\, A^+)^{-1}\,\Omega^{\mathrm{Det}}\,\mathrm{Det}\, A^+ &= \mathrm{d}\,\mathrm{tr}^{\Delta^+}\left((A^+)^{-1}\,\mathrm{d}A^+\right)\\ &= (\mathrm{d}\,\mathrm{tr}^{\Delta^+})[(A^+)^{-1}\,\mathrm{d}A^+] + \mathrm{tr}^{\Delta^+}[(A^+)^{-1}\,\mathrm{d}A^+(A^+)^{-1}dA^+]. \qquad (4.2)\end{aligned}$$

The first term on the r.h.s. of (4.2) corresponds to a tracial anomaly described in (3.4) and the second one to the one described in (3.2). This curvature, which according to those formulas can be expressed in terms of the Wodzicki residue, is therefore local in the sense of (1.2). It can be interpreted as an obstruction to the Wess-Zumino consistency relations for chiral anomalies [5].

5. Chern-Weil type classes on infinite rank vector bundles

Let us first briefly recall how one can build Chern-Weil cohomology classes in finite dimensions using the usual trace on matrices. Let $\pi : \mathcal{E} \to B$ be a finite rank vector bundle equipped with a connection ∇; its curvature is a $\mathrm{Hom}(\mathcal{E})$ valued two form $\Omega \in \Omega^2(B, \mathrm{Hom}(\mathcal{E}))$. Thus for any $k \in \mathbb{N}/\{0\}$, the trace $\mathrm{tr}(\Omega^k)$ defines a $2k$-form on B. Writing $\nabla = \mathrm{d}+\theta$ in some local trivialization, the induced connection on $\mathrm{Hom}(\mathcal{E})$ reads $\nabla^{\mathrm{Hom}} = d + [\theta, \cdot]$. Using the cyclicity of the trace on matrices, we have for any $k \in \mathbb{N}/\{0\}$:

$$\begin{aligned} d\,\mathrm{tr}(\Omega^k) &= \mathrm{tr}(d\Omega^k)\\ &= \mathrm{tr}(d\Omega^k) + \mathrm{tr}([\theta, \Omega^k])\\ &= \mathrm{tr}([\nabla, \Omega^k])\\ &= \sum_{i=1}^{k} \mathrm{tr}(\Omega^i[\nabla, \Omega]\Omega^{k-i}) = 0 \qquad (5.1)\end{aligned}$$

where we have used Bianchi identity $[\nabla, \Omega] = 0$ in the last identity.

Thus $\mathrm{tr}(\Omega^k)$ defines a de Rham cohomology class and similar arguments using the cyclicity of the trace together with the Bianchi identity, show that the corresponding cohomology class does not depend on the choice of connection. Since the vector bundle has finite rank, for any analytic function f, the form $\mathrm{tr}(f(\Omega))$ on B reduces to the trace of some polynomial expression $\sum_{i=0}^{K} a_k \Omega^k$ in powers of Ω to which we can extend the above construction by linearity. Chern-Weil classes are cohomology classes of forms of the type $\mathrm{tr}(f(\Omega))$ where f is an analytic function (see e.g. [4]).

The question naturally arises how to extend this construction to the infinite dimensional setting. There the finite rank vector bundle with structure group $\mathrm{GL}_n(\mathbb{C})$ is replaced by an infinite rank $\mathcal{E}$ vector bundle with structure group $C\ell^*(M,E)$ where as before, $\pi\colon E \to M$ is a finite rank vector bunlde based on a closed Riemannian manifold M. Locally $\mathcal{E}$ can be seen as a product of some open subset U of B with a space $\Gamma(M,E)$ of sections of E. The connection ∇ on the finite rank vector bundle, which locally reads $\nabla = d + \theta$ where θ is a $gl_n(\mathbb{C})$-valued one form on some open subset U of B is replaced by a connection for which θ is a $C\ell(M,E)$ valued one form on some open subset U of B. Such a connection, which we refer to as *a P.D.O connection*, induces a connection on an infinite rank bundle $C\ell(\mathcal{E})$ based on B which is locally of the form $U \times C\ell(M,E)$ where U is some open subset of B. Because transition maps on $\mathcal{E}$ are given by invertible classical P.D.O.s, the notion of P.D.O. connection is independent of the choice of trivialization. The curvature being tensorial, the curvature of a P.D.O. connection will be a $C\ell(\mathcal{E})$-valued two form on B.

If M is a point, then $C\ell^*(M,E)$ coincides with $\mathrm{GL}_n(\mathbb{C})$, the vector bundle reduces to a finite rank vector bundle, and the P.D.O. connection to a usual connection.

A natural idea to build Chern-Weil type forms on P.D.O. bundles described above is to replace the ordinary trace on matrices by a trace on P.D.Os, namely the Wodzicki residue. The Wodzicki residue extends to $C\ell(\mathcal{E})$ -valued forms on B since it is invariant under changes of trivializations given by conjugation by invertible P.D.O.s. Thus, given a P.D.O. connection ∇, one can build $C\ell(\mathcal{E})$-valued $2k$-forms $\mathrm{res}(\Omega^k)$. The cyclicity of the residue together with the Bianchi identity yield closed forms in a similar way to (5.1):

$$\begin{aligned} d\,\mathrm{res}(\Omega^k) &= \mathrm{res}(d\Omega^k) \\ &= \mathrm{res}(d\Omega^k) + \mathrm{res}([\theta,\Omega^k]) \\ &= \mathrm{res}([\nabla,\Omega^k]) \\ &= \sum_{i=1}^{k} \mathrm{res}(\Omega^i[\nabla,\Omega]\Omega^{k-i}) = 0 \end{aligned}$$

where we have written as before $\nabla = d + \theta$ in some local trivialization. Similar arguments then show that the corresponding de Rham class is independent of the choice of connection: we call such a class a *residue Chern-Weil* type class. On a loop group, which is trivializable, one can choose a trivial connection so that there the residue Chern-Weil classes vanish. But more surprisingly, residue Chern-Weil classes also vanish on manifolds of loops $\mathrm{Map}(S^1,N)$ as it was shown in [14] using a natural connection built from a connection on N. The residue vanishes because the corresponding curvature is a multiplication operator. This raises the question whether one has not left out some interesting information taking the residue instead of weighted traces to define Chern-Weil type cohomology classes.

In view of their covariance property (1.4), weighted traces also extend to sections of the bundle $C\ell(\mathcal{E})$, provided one equips the bundle with a *weight*, i.e. a section of invertible admissible elliptic operator of positive order. These properties, which make sense in a local trivialization, are preserved under a change of trivialization since they are conserved under conjugation by an invertible zero order P.D.O.

Here obstructions occur when trying to build closed forms using weighted traces instead of Wodzicki residues because of the tracial anomalies described in section 3. Indeed, two basic features of the Wodzicki residue which were used in the finite dimensional setting (see (5.1)) fail to hold here:

$$\mathrm{d} \circ \mathrm{tr}^Q \neq \mathrm{tr}^Q \circ \mathrm{d}$$

and

$$\partial \, \mathrm{tr}^Q := \mathrm{tr}^Q \left([\,\cdot\,,\,\cdot\,] \right) \neq 0.$$

On one hand, the non cyclicity of regularized traces leads to interesting BRST cocycles of Dirac operators investigated in [13], the locality of which is related to the locality of the Wodzicki residue by formula (3.2). On the other hand, this lack of cyclicity is an obstacle when trying to extend finite dimensional geometric constructions involving traces. One possible way to circumvent this obstacle is to choose classes of weights and connections for which the obstructions $d\,\mathrm{tr}^Q$ and $\partial\,\mathrm{tr}^Q$ vanish. This was carried out in [14] where Chern-Weil classes on certain loop groups were built, restricting the class of connections and the class of weights. This approach seems, in many cases, to lead to vanishing Chern classes. Further investigations along this line of thought were made in [19].

Another way to circumvent the difficulties related to tracial anomalies is to modify the underlying geometry so as to compensate for the anomalies. This type of idea was carried out in [3] to define the Chern character associated to a family of Dirac operators on a compact spin manifold, by introducing one parameter families of super connections which can be seen as deformations of the initial superconnection [2] (this procedure is described in more detail below).

Yet another idea suggested later in [19] is to pick another type of trace, namely the integral over the cotangent unit sphere of the (ordinary) trace of the leading symbol of the pseudo-differential operator. Since the leading symbol reduces to a matrix when the underlying manifold reduces to a point, these traces boil down to the usual trace on matrices in the finite dimensional setting so that the Chern-Weil classes one builds this way can also be seen as generalizations of the finite dimensional ones.

To sum up, there does not yet seem to be a canonical way of generalizing Chern-Weil classes to an infinite dimensional setting and different approaches seem to lead to different types of characteristic classes.

6. Examples

Let us illustrate this discussion by two examples, one in the context of loop groups, the other one in the context of the family index theorem. What follows is based on previous joint work, namely [6] for loop groups and [18] for families of Dirac operators.

i) *Loop groups*. Following Freed [8], one can equip the complex manifold $H_e^{\frac{1}{2}}(S^1, G)$ of based $H^{\frac{1}{2}}$-loops with values in a semi-simple Lie group G (with Lie algebra denoted by $Lie\, G$) of compact type with a left invariant Levi-Civita ∇ connection which turns out to be a classical P.D.O. connection (see [14] for a careful investigation of P.D.O. connections in this context). Using a left invariant weight defined on the Lie algebra $H_0^{\frac{1}{2}}(S^1, Lie\, G) \simeq H_0^{\frac{1}{2}}(S^1, \mathbb{C}) \otimes Lie\, G$ by $Q = Q_0 \otimes \mathrm{Id}_{Lie\, G}$ (we call such weights *scalar*) we showed [6] that Freed's first Chern from could be interpreted as a weighted first Chern form $\mathrm{tr}^Q(\Omega)$ and that it coincides with a coboundary $\partial\, \mathrm{tr}^Q(A, B)$ where Q is some scalar weight, A and B some classical P.D.O.s.

ii) *Families of Dirac operators*. Given a finite rank complex super vector bundle $E \to B$ equipped with a connection ∇ with curvature $\Omega = \nabla^2$, the part of degree 2 of the Chern character coincides with minus the first Chern form:

$$\mathrm{ch}(\nabla)_{[2]} = \left[\mathrm{str}(e^{-\nabla^2})\right]_{[2]} = \left[\sum_{k=0}^{\infty} \frac{(-1)^k}{k!}\, \mathrm{str}(\nabla^{2k})\right]_{[2]} = -\,\mathrm{str}(\Omega). \quad (6.1)$$

On the other hand, letting ∇^{Det} denote the connection induced on the associated determinant bundle $\mathrm{Det}\, E = \Lambda^{\max} E$, we have for a section $\mathrm{Det}\, L$ of $\mathrm{Det}\, E$:

$$\mathrm{Det}\, L^{-1} \nabla^{\mathrm{Det}}\, \mathrm{Det}\, L = \frac{1}{2}\, \mathrm{str}(L^{-1}[\nabla, L])$$

(str stands for supertrace) at any point where this section is invertible. Differentiating this expression yields its curvature Ω^{Det}:

$$\begin{aligned}\operatorname{Det} L^{-1}\Omega^{\mathrm{Det}} \operatorname{Det} L &= \frac{1}{2} d\ \operatorname{str}(L^{-1}[\nabla, L]) \\ &= -\frac{1}{2}\operatorname{str}(L^{-1}\nabla L L^{-1}\nabla L) + \frac{1}{2}\operatorname{str}(L^{-1}[\Omega, L]) \\ &= -\operatorname{str}(\Omega) \qquad (6.2)\end{aligned}$$

where we use the traciality of the supertrace in the last identity. Combining (6.1) and (6.2) we finally get:

$$\Omega^{\mathrm{Det}} = -\operatorname{str}(\nabla^2) = \operatorname{ch}(\nabla)_{[2]}. \qquad (6.3)$$

Given a fibration $\pi\colon \mathbb{M} \to B$ of spin manifolds with fibre M_b above $b \in B$ (in section 4, this fibration was trivial), a vector bundle $\mathbb{E} \to \mathbb{M}$ on $\mathbb{M}$, there is an associated infinite rank vector bundle $\mathcal{E} = \pi_* \mathbb{E}$ the fibre above $b \in B$ of which is a space of sections of the finite rank bundle $E_b = \mathbb{E}_{|M_b} \to M_b$ (see e.g. [4] for a precise construction). A family of Dirac operators $\{D_b, b \in B\}$ acting on sections of $E_b \to M_b$ yields a weight D on $\mathcal{E}$ and hence a weighted vector bundle bundle $(\mathcal{E}, D)$. Letting $C\ell(\mathcal{E})$ be the bundle of algebras of classical pseudo-differential operators based on B with fibre above $b \in B$ given by $C\ell(M_b, E_b)$, the weight D can be seen as a section of $C\ell(\mathcal{E})$ (see [18]). There is an associated Quillen determinant line bundle $\mathcal{L}_D$.

A P.D.O. connection ∇ on $\mathcal{E}$ (i.e., a connection on $\mathcal{E}$ which is locally of the form $\nabla = d + \theta$ where θ is a $C\ell(\mathcal{E})$) induces a connection (due to Bismut and Freed [3]) on ∇^{Det} on $\mathcal{L}_D$, which in turn gives rise to a curvature denoted by Ω^{Det} (compare with section 4 where $\mathbb{M} = M \times B \to B$ was a trivial fibration of manifolds).

On the other hand, following Bismut [2], this weight D can be used to build a one parameter family of superconnections (this notion was first introduced in [21]) $\{\nabla^D_\varepsilon, \varepsilon > 0\}$ on the super-bundle $\mathcal{E}$. Bismut and Freed [3] showed that the expression $\operatorname{str}\left(e^{-(\nabla^D_\varepsilon)^2}\right)_{[2]}$ converges when $\varepsilon \to 0$ to what one can interpret as the second degree part $\operatorname{ch}^D(\nabla)_{[2]}$ of a Chern character on the super bundle $\mathcal{E}$ and we have [3]:

$$\Omega^{\mathrm{Det}} = \lim_{\varepsilon \to 0} \operatorname{str}\left(e^{-(\nabla^D_\varepsilon)^2}\right)_{[2]} = \operatorname{ch}^D(\nabla)_{[2]}.$$

However, because of the tracial anomalies (3.2) and (3.4), the part of degree 2 of the Chern character, in contrast to the finite dimensional case, only coincides with minus the first Chern form up to a residue term [18]:

$$\Omega^{\mathrm{Det}} = -\,\mathrm{str}^Q(\Omega) + \text{Wodzicki residues} \\ \text{involving the operators } Q, [\nabla, Q] \text{ and } \Omega \tag{6.4}$$

where str^Q stands for Q-weighted supertrace which is defined in a simar way to a Q-weighted trace up to the introduction of a grading inside the trace (see [5]). Here we have set $Q := |D|$.

If one of these three forms has a local expression in terms of the integral over the manifold of some form involving the curvature, then all have since the Wodzicki residue by which they differ has a local description. In the Bismut-Freed setting, the choice of connection ∇ yields a curvature Ω given by a multiplication operator so that –as we pointed out in Section 1– the expression $\mathrm{str}^Q(\Omega)$ presents some locality feature. Bismut and Freed give, via some supersymmetric cancellation computations, a local description of this expression in terms of finite dimensional Chern forms corresponding to the underlying geometric data.

References

[1] Arnlind J., Mickelsson J.: Trace extensions, determinant bundles and gauge group cocycles, `arXiv: hep-th/0205126` (2002).

[2] Bismut J.-M.: Localization formulae, superconnections and the index theorem for families, *Comm. Math. Phys.* **103**, (1986) 127–166.

[3] Bismut J.-M., Freed D.: The analysis of elliptic families I, *Comm. Math. Phys.* **106** (1986) 159–176.

[4] Berline N., Getzler E., Vergne M.: *Heat Kernels and Dirac Operators*, Springer-Verlag, Berlin, 1998.

[5] Cardona A., Ducourtioux C., Paycha S.: From tracial anomalies to anomalies in Quantum Field Theory, *Comm. Math. Phys.***242** (2003) 31–65.

[6] Cardona A., Ducourtioux C., Magnot J.-P., Paycha S.: Weighted traces on algebras of pseudodifferential operators and geometry on loop groups, *Infin. Dimens. Anal. Quantum Probab. Relat. Top.* **5**, 4 (2002) 503–540.

[7] Ducourtioux C.: Weighted traces on pseudo-differential operators and associated determinants, Ph.D thesis, Mathematics Department, Université Blaise Pascal, 2001 (unpublished).

[8] Freed D.: The geometry of loop groups, *J. Diff. Geom.* **28** (1988) 223–276.

[9] Kassel Ch.: Le résidu non commutatif [d'après Wodzicki, *Séminaire Bourbaki* **708** (1989).

[10] Kontsevich M., Vishik S.: Determinants of elliptic pseudodifferential operators, Max Planck Preprint (1994); Geometry of determinants of elliptic operators, in Funct. Anal. on the Eve of the 21st Century Vol. **I** (eds. S.Gindikin, J.Lepowski, R.L.Wilson) Progress in Mathematics (1994).

[11] Lesch M.: On the non commutative residue for pseudo-differential operators with log-polyhomogeneous symbols, *Annals of Global Analysis and Geometry* **17** (1998) 151–187.

[12] Langmann E., Mickelsson J.: Elementary derivation of the chiral anomaly, *Lett. Math. Phys.* **6** (1996) 45–54.

[13] Langmann E., Mickelsson J., Rydh S.: Anomalies and Schwinger terms in NCG field theory models, *J. Math. Phys.* **42** (2001) 4779–4793.

[14] Magnot J.-P.: The geometry of loop spaces, Ph.D thesis, Mathematics Department, Université Blaise Pascal, 2002 (unpublished).

[15] Melrose R., Nistor V.: Homology of pseudodifferential operators I. Manifolds with boundary, Preprint `funct-an/9606005`, Oct. 98.

[16] Okikiolu K.: The multiplicative anomaly for determinants of elliptic operators; The Campbell-Hausdorff theorem for elliptic operators and a related trace formula, *Duke Math. J.* **79** (1995) 723–750; 687–722.

[17] Paycha S.: Renormalized traces as a looking glass into infinite-dimensional geometry, *Infin. Dimens. Anal. Quantum Probab. Relat. Top.* **4**, 2 (2001) 221–266.

[18] Paycha S., Rosenberg S.: Curvature on determinant bundles and first Chern forms, *J. Geom. Phys.* **45** (2003) 393–429.

[19] Paycha S., Rosenberg S.: Traces and characteristic classes on loop spaces, In *Infinite dimensional groups and manifolds*, Proceedings of the 70th meeting of theretical physicists and mathematicans held in Strasbourg, May 23–25, 2002. Edited by T.Wurzbacher. IRMA Lectures in mathematics and Theoretical Physics. Walter de Gruyter and Co., Berlin, to appear in 2004.

[20] Quillen D.: Determinants of Cauchy-Riemann operators over a Riemann surface, *Funktsional. Anal. i Prilozhen.* **19** (1985) 37–41.

[21] Quillen D.: Superconnections and the Chern character, *Topology* **24** (1985) 89–95.

[22] Wodzicki M.: *Non-commutative residue*, Lecture Notes in Mathematics, **1289**, Berlin, Springer-Verlag, 1987.

QUANTUM STOCHASTIC CALCULUS APPLIED TO PATH SPACES OVER LIE GROUPS

Nicolas Privault

Département de Mathématiques, Université de La Rochelle, avenue Michel Crépeau, 17042 La Rochelle, France

nprivaul@univ-lr.fr

Abstract Quantum stochastic calculus is applied to the proof of Skorokhod and Weitzenböck type identities for functionals of a Lie group-valued Brownian motion. In contrast to the case of $\mathbb{R}^d$-valued paths, the computations use all three basic quantum stochastic differentials.

Keywords: Quantum stochastic calculus, Lie group-valued Brownian motion. *Mathematics Subject Classification.* 60H07, 81S25, 58J65, 58C35.

1. Introduction

Quantum stochastic calculus [4, 7], and anticipating stochastic calculus [6] have been linked in [5], where the Skorokhod isometry is formulated and proved using the annihilation and creation processes. On the other hand, a Skorokhod type isometry has been constructed in [3] for functionals on the path space over Lie groups. This isometry yields in particular a Weitzenböck type identity in infinite-dimensional geometry. We refer to [1, 2] for the case of path spaces over Riemannian manifolds.

We will prove such a Skorokhod type isometry formula on the path space over a Lie group, using the conservation operator which is usually linked to stochastic calculus for jump processes. In this way we will recover the Weitzenböck formula established in [3]. This provides a link between the non-commutative settings of Lie groups and of quantum stochastic calculus.

This paper is organised as follows. In Sect. 2 we recall how the Skorokhod isometry can be derived from quantum stochastic calculus in the case of $\mathbb{R}^d$-valued Brownian motion. In Sect. 3 the gradient and divergence operators of stochastic analysis on path groups are introduced,

S. Albeverio et al. (eds.),
Proceedings of the International Conference on Stochastic Analysis and Applications, 85–94.

and the Skorokhod type isometry of [3] is stated. The proof of this isometry is given in Sect. 4 via quantum stochastic calculus on the path space over a Lie group. Sect. 4 ends with a remark on the links between vanishing of torsion and quantum stochastic calculus.

2. Skorokhod isometry on the path space over $\mathbb{R}^d$

In this section we recall how the Skorokhod isometry is linked to quantum stochastic calculus. Let $(B(t))_{t\in\mathbb{R}_+}$ denote an $\mathbb{R}^d$-valued Brownian motion on the Wiener space W with Wiener measure μ. Let

$$\mathcal{S} = \{G = g(B(t_1), \ldots, B(t_n)) \mid g \in \mathcal{C}_b^\infty((\mathbb{R}^d)^n),\ t_1, \ldots, t_n > 0\},$$

and

$$\mathcal{U} = \Big\{\sum_{i=1}^n u_i G_i \ \Big|\ G_i \in \mathcal{S},\ u_i \in L^2(\mathbb{R}_+;\mathbb{R}^d),\ i = 1, \ldots, n,\ n \geq 1\Big\}.$$

Let $D\colon L^2(W) \to L^2(W \times \mathbb{R}_+;\mathbb{R}^d)$ be the closed operator given by

$$D_t G = \sum_{i=1}^n 1_{[0,t_i]}(t)\nabla_i g(B(t_1), , \ldots, B(t_n)), \quad t \geq 0,$$

for $G = g(B(t_1), \ldots, B(t_n)) \in \mathcal{S}$, and let δ denote its adjoint. Thus

$$\mathbb{E}[G\delta(u)] = \mathbb{E}[\langle DG, u\rangle], \quad G \in \mathrm{Dom}(D),\ u \in \mathrm{Dom}(\delta),$$

where $\mathrm{Dom}(D)$ and $\mathrm{Dom}(\delta)$ are the respective domains of D and δ. We let $\langle \cdot, \cdot \rangle$ denote the scalar product in both $L^2(\mathbb{R}_+;\mathbb{R}^d)$ and $L^2(W \times \mathbb{R}_+;\mathbb{R}^d)$, and let $(\cdot, \cdot)$ denote the scalar product on $\mathbb{R}^d$. Since D is a derivation, we have the divergence relation

$$\delta(uG) = G\delta(u) - \langle u, DG\rangle, \quad G \in \mathcal{S},\ u \in \mathcal{U}.$$

Given $u \in L^2(\mathbb{R}_+;\mathbb{R}^d)$, the quantum stochastic differentials $\mathrm{d}a^-(t)$ and $\mathrm{d}a^+(t)$ are defined from

$$a_u^- G = \int_0^\infty u(t)\mathrm{d}a^-(t)G = \langle DG, u\rangle, \quad G \in \mathrm{Dom}(D), \tag{2.1}$$

and

$$a_v^+ G = \int_0^\infty v(t)\mathrm{d}a^+(t)G = \delta(vG), \quad G \in \mathrm{Dom}(D). \tag{2.2}$$

They satisfy the Itô table

$\cdot$	$\mathrm{d}t$	$\mathrm{d}a_v^-(t)$	$\mathrm{d}a_v^+(t)$
$\mathrm{d}t$	0	0	0
$\mathrm{d}a_u^+(t)$	0	0	0
$\mathrm{d}a_u^-(t)$	0	0	$(u(t), v(t))\mathrm{d}t$

with $\mathrm{d}a_u^-(t) = u(t)\mathrm{d}a^-(t)$ and $\mathrm{d}a_v^+(t) = v(t)\mathrm{d}a^+(t)$. Using the Itô table we have

$$a_u^- a_v^+ = \int_0^\infty \int_0^t \mathrm{d}a_v^+(s)\mathrm{d}a_u^-(t) + \int_0^\infty \int_0^t \mathrm{d}a_u^-(s)\mathrm{d}a_v^+(t) + \int_0^\infty u(t)v(t)\mathrm{d}t$$

and

$$a_v^+ a_u^- = \int_0^\infty \int_0^t \mathrm{d}a_u^-(s)\mathrm{d}a_v^+(t) + \int_0^\infty \int_0^t \mathrm{d}a_v^+(s)\mathrm{d}a_u^-(t),$$

which implies the canonical commutation relation

$$a_u^- a_v^+ = \langle u, v\rangle + a_v^+ a_u^- . \tag{2.3}$$

This relation and its proof can be abbreviated as

$$\mathrm{d}[a_u^-, a_v^+](t) = [\mathrm{d}a_u^-(t), \mathrm{d}a_v^+(t)] = (u(t), v(t))\mathrm{d}t,$$

where $[\cdot, \cdot]$ denotes the commutator of operators. Relation (2.3) is easily translated back to the Skorokhod isometry:

$$\begin{aligned} \mathbb{E}\left[\delta(uF)\delta(vG)\right] &= \langle a_u^+ F, a_v^+ G\rangle = \langle F, a_u^- a_v^+ G\rangle \\ &= \langle u\otimes F, v\otimes G\rangle + \langle F, a_v^+ a_u^- G\rangle \\ &= \langle u\otimes F, v\otimes G\rangle + \langle a_v^- F, a_u^- G\rangle \\ &= \mathbb{E}[\langle u, v\rangle FG] \\ &\quad + \mathbb{E}\left[\int_0^\infty \int_0^\infty (u(t)\otimes D_sF, D_tG\otimes v(s))\mathrm{d}s\mathrm{d}t\right], \end{aligned}$$

$F, G \in \mathcal{S}$, $u, v \in L^2(\mathbb{R}_+;\mathbb{R}^d)$, which implies

$$\mathbb{E}[\delta(h)^2] = \mathbb{E}[\|h\|^2_{L^2(\mathbb{R}_+;\mathbb{R}^d)}] + \mathbb{E}\left[\int_0^\infty \int_0^\infty (D_sh(t), (D_th(s))^*)\mathrm{d}s\mathrm{d}t\right], \ h\in\mathcal{U},$$

where $(D_th(s))^*$ denotes the adjoint of $D_th(s)$ in $\mathbb{R}^d\otimes\mathbb{R}^d$. In this note we carry over this method to the proof of a Skorokhod type isometry on the path space over a Lie group, using the calculus of all the annihilation, creation and gauge (or conservation) process and the Itô table

$\cdot$	$\mathrm{d}t$	$\mathrm{d}a_v^-(t)$	$\mathrm{d}a_v^+(t)$	$q(t)\mathrm{d}\Lambda(t)$
$\mathrm{d}t$	0	0	0	0
$\mathrm{d}a_u^+(t)$	0	0	0	0
$\mathrm{d}a_u^-(t)$	0	0	$(u(t), v(t))\mathrm{d}t$	$q^*(t)u(t)\mathrm{d}a^-(t)$
$p(t)\mathrm{d}\Lambda(t)$	0	0	$p(t)v(t)\mathrm{d}a^+(t)$	$p(t)q(t)\mathrm{d}\Lambda(t)$

where $(q(t))_{t\in\mathbb{R}_+}$ is a (bounded) measurable operator process on $\mathbb{R}^d$ and $q(t)\mathrm{d}\Lambda(t)$ is defined from

$$\int_0^\infty q(t)\mathrm{d}\Lambda(t)F = \delta(q(\cdot)D.F),$$

for $F \in \mathrm{Dom}(D)$ such that $(q(t)D_tF)_{t\in\mathbb{R}_+} \in \mathrm{Dom}(\delta)$.

3. Skorokhod isometry on the path space over a Lie group

Let G be a compact connected d-dimensional Lie group with associated Lie algebra $\mathcal{G}$ identified to $\mathbb{R}^d$ and equipped with an Ad-invariant scalar product on $\mathbb{R}^d \simeq \mathcal{G}$, also denoted by $(\cdot, \cdot)$. The commutator in $\mathcal{G}$ is denoted by $[\cdot, \cdot]$. Let $\mathrm{ad}(u)v = [u, v]$, $u, v \in \mathcal{G}$, with $\mathrm{Ad}\,\mathrm{e}^u = \mathrm{e}^{\mathrm{ad}\,u}$, $u \in \mathcal{G}$.

The Brownian motion $(\gamma(t))_{t\in\mathbb{R}_+}$ on G is constructed from $(B(t))_{t\in\mathbb{R}_+}$ via the Stratonovich differential equation

$$\begin{cases} d\gamma(t) = \gamma(t) \odot \mathrm{d}B(t) \\ \gamma(0) = e_{\mathsf{G}}, \end{cases}$$

where e_{G} is the identity element in G. Let $\mathbf{P}(\mathsf{G})$ denote the space of continuous G-valued paths starting at e_G, with the image measure of the Wiener measure by $I\colon (B(t))_{t\in\mathbb{R}_+} \mapsto (\gamma(t))_{t\in\mathbb{R}_+}$. Let

$$\tilde{\mathcal{S}} = \{F = f(\gamma(t_1), \ldots, \gamma(t_n)) \mid f \in \mathcal{C}_b^\infty(\mathsf{G}^n)\},$$

and

$$\tilde{\mathcal{U}} = \Big\{\sum_{i=1}^n u_i F_i \;\Big|\; F_i \in \tilde{\mathcal{S}},\ u_i \in L^2(\mathbb{R}_+; \mathcal{G}),\ i = 1, \ldots, n,\ n \geq 1\Big\}.$$

Definition 3.1. For $F = f(\gamma(t_1), \ldots, \gamma(t_n)) \in \tilde{\mathcal{S}}$, $f \in \mathcal{C}_b^\infty(\mathsf{G}^n)$, we let $\tilde{D}F \in L^2(W \times \mathbb{R}_+; \mathcal{G})$ be defined as

$$\langle \tilde{D}F, v\rangle = \frac{\mathrm{d}}{\mathrm{d}\varepsilon} f\Big(\gamma(t_1)\mathrm{e}^{\varepsilon\int_0^{t_1} v(s)\mathrm{d}s}, \ldots, \gamma(t_n)\mathrm{e}^{\varepsilon\int_0^{t_n} v(s)\mathrm{d}s}\Big)\Big|_{\varepsilon=0}, \quad v \in L^2(\mathbb{R}_+, \mathcal{G}).$$

In other terms, $\tilde{D}$ acts as a natural gradient on the cylindrical functionals on $\mathbf{P}(\mathsf{G})$ with

$$\tilde{D}_t F = \sum_{i=1}^n \partial_i f(\gamma(t_1), \ldots, \gamma(t_n)) 1_{[0,t_i]}(t), \quad t \geq 0.$$

Let $\tilde{\delta}$ denote the adjoint of $\tilde{D}$, that satisfies

$$\mathbb{E}[F\tilde{\delta}(v)] = \mathbb{E}[\langle \tilde{D}F, v\rangle], \quad F \in \tilde{\mathcal{S}},\ v \in L^2(\mathbb{R}_+; \mathcal{G}), \tag{3.1}$$

(that $\tilde{\delta}$ exists and satisfies (3.1) can be seen as a consequence of Lemma 4.1 below). Given $v \in L^2(\mathbb{R}_+; \mathcal{G})$ we define

$$q_v(t) = \int_0^t \mathrm{ad}(v(s))\mathrm{d}s, \quad t > 0.$$

Definition 3.2 ([3]). The covariant derivative of $u \in L^2(\mathbb{R}_+;\mathcal{G})$ in the direction $v \in L^2(\mathbb{R}_+;\mathcal{G})$ is the element $\nabla_v u$ of $L^2(\mathbb{R}_+;\mathcal{G})$ defined as follows:

$$\nabla_v u(t) = q_v(t)u(t) = \int_0^t \mathrm{ad}(v(s))u(t)\mathrm{d}s, \quad t > 0.$$

In the following we will distinguish between ∇_v which acts on $L^2(\mathbb{R}_+;\mathcal{G})$ and $q_v(t)$ which acts on $\mathcal{G}$ and is needed in the quantum stochastic integrals to follow.

The operators $q_v(t)$ and ∇_v are antisymmetric on $\mathcal{G}$ and $L^2(\mathbb{R}_+;\mathcal{G})$ respectively, because $(\cdot\,,\,\cdot)$ is Ad-invariant.

The Skorokhod isometry on the path space over G holds for the covariant derivative ∇. The definition of ∇_v extends to $\tilde{\mathcal{U}}$, as

$$\nabla_v(uF)(t) = u(t)\langle \tilde{D}F, v\rangle + Fq_v(t)u(t), \quad t > 0,\ F \in \tilde{\mathcal{S}},\ u \in L^2(\mathbb{R}_+;\mathcal{G}).$$

Let $\nabla_s(uF)(t) \in \mathcal{G}\otimes\mathcal{G}$ be defined as $(i,j = 1,\ldots,d)$

$$\langle e_i \otimes e_j, \nabla_s(uF)(t)\rangle = \langle u(t), e_j\rangle\langle e_i, \tilde{D}_sF\rangle + 1_{[0,t]}(s)F\langle e_j, \mathrm{ad}(e_i)u(t)\rangle.$$

In this context the following isometry has been proved in [3].

Theorem 3.3 ([3]). *We have for $h \in \tilde{\mathcal{U}}$:*

$$\mathbb{E}[\tilde{\delta}(h)^2] = \mathbb{E}[\|h\|^2_{L^2(\mathbb{R}_+;\mathcal{G})}] + \mathbb{E}\left[\int_0^\infty\int_0^\infty (\nabla_s h(t), (\nabla_t h(s))^*)\mathrm{d}t\mathrm{d}s\right]. \quad (3.2)$$

The proof in [3] is clear and self-contained, however its calculations involve a number of coincidences which are apparently not related to each other. In this paper we provide a short proof which offers some explanation for these. Let the analogs of (2.1)-(2.2) be defined as

$$\tilde{a}_u^- F = \int_0^\infty \mathrm{d}\tilde{a}_u^-(t)F = \langle \tilde{D}F, u\rangle, \quad F \in \tilde{\mathcal{S}},$$

and

$$\tilde{a}_u^+ F = \int_0^\infty \mathrm{d}\tilde{a}_u^+(t)F = \tilde{\delta}(uF), \quad F \in \tilde{\mathcal{S}},$$

$u \in L^2(\mathbb{R}_+;\mathcal{G})$, i.e. $\mathrm{d}\tilde{a}_u^-(t)F = u(t)\tilde{D}_tF\mathrm{d}t$.

Our proof relies on

(a) the relation

$$\mathrm{d}\tilde{a}_u^-(t) = \mathrm{d}a_u^-(t) + q_u(t)\mathrm{d}\Lambda(t), \quad t > 0,\ u \in L^2(\mathbb{R}_+;\mathcal{G}), \quad (3.3)$$

see Lemma 4.1 below,

(b) the commutation relation between $\tilde{a}_u^-$ and $\tilde{a}_v^+$ which is analogous to (2.3) and is proved via quantum stochastic calculus in the following lemma.

Lemma 3.4. *We have on $\tilde{\mathcal{S}}$, with $u, v \in L^2(\mathbb{R}_+; \mathcal{G})$:*

$$\tilde{a}_u^- \tilde{a}_v^+ - \tilde{a}_v^+ \tilde{a}_u^- = \langle u, v\rangle + \tilde{a}^-_{\nabla_v u} + \tilde{a}^+_{\nabla_u v}. \tag{3.4}$$

Proof. Using the quantum Itô table, relation (3.3), Lemma 4.2 below and the fact that $q_v^*(t) = -q_v(t)$, we have

$$\begin{aligned}
&\mathrm{d}\tilde{a}_u^-(t) \cdot \mathrm{d}\tilde{a}_v^+(t) - \mathrm{d}\tilde{a}_v^+(t) \cdot \mathrm{d}\tilde{a}_u^-(t) \\
&\quad = (\mathrm{d}a_u^-(t) + q_u(t)\mathrm{d}\Lambda(t)) \cdot (\mathrm{d}a_v^+(t) - q_v(t)\mathrm{d}\Lambda(t)) \\
&\qquad - (\mathrm{d}a_v^+(t) - q_v(t)\mathrm{d}\Lambda(t)) \cdot (\mathrm{d}a_u^-(t) + q_u(t)\mathrm{d}\Lambda(t)) \\
&\quad = (u(t), v(t))\mathrm{d}t + q_v(t)u(t)\mathrm{d}a^-(t) - q_u(t)q_v(t)\mathrm{d}\Lambda(t) \\
&\qquad + q_u(t)v(t)\mathrm{d}a^+(t) + q_v(t)q_u(t)\mathrm{d}\Lambda(t) \\
&\quad = (u(t), v(t))\mathrm{d}t + \nabla_v u(t)\mathrm{d}a^-(t) \\
&\qquad + q_{\nabla_v u}(t)\mathrm{d}\Lambda(t) + \nabla_u v(t)\mathrm{d}a^+(t) - q_{\nabla_u v}(t)\mathrm{d}\Lambda(t) \\
&\quad = (u(t), v(t))\mathrm{d}t + \mathrm{d}\tilde{a}^-_{\nabla_v u}(t) + \mathrm{d}\tilde{a}^+_{\nabla_u v}(t).
\end{aligned}$$

□

This commutation relation can be interpreted to give a proof of the Skorokhod isometry (3.2):

Proof of Theorem 3.3. Applying Lemma 3.4 we have

$$\begin{aligned}
&\mathbb{E}\left[\tilde{\delta}(uF)\tilde{\delta}(vG)\right] = \langle \tilde{a}_u^+ F, \tilde{a}_v^+ G\rangle = \langle F, \tilde{a}_u^- \tilde{a}_v^+ G\rangle \\
&= \langle u \otimes F, v \otimes G\rangle + \langle \tilde{a}_v^- F, \tilde{a}_u^- G\rangle + \langle F\nabla_v u, \tilde{D}G\rangle + \langle \tilde{D}F, G\nabla_u v\rangle \\
&= \mathbb{E}[\langle u, v\rangle FG] \\
&+ \mathbb{E}\left[\int_0^\infty (F\nabla_s u(t) + \tilde{D}_s F \otimes u(t), G(\nabla_t v(s))^* + v(s) \otimes \tilde{D}_t G)\mathrm{d}s\mathrm{d}t\right] \\
&= \mathbb{E}[\langle u, v\rangle FG] + \mathbb{E}\left[\int_0^\infty \langle \nabla_s(uF)(t), (\nabla_t(vG)(s))^*\rangle \mathrm{d}s\mathrm{d}t\right],
\end{aligned}$$

for $F, G \in \tilde{\mathcal{S}}$, $u, v \in L^2(\mathbb{R}_+; \mathcal{G})$ □

We mention that a consequence of Th. 3.3 is the following Weitzenböck type identity, cf. [3], which extends the Shigekawa identity [8] to path spaces over Lie groups:

Theorem 3.5 ([3]). *We have for* $u \in \tilde{\mathcal{U}}$*:*

$$\mathbb{E}[\tilde{\delta}(u)^2] + \mathbb{E}\left[\|du\|^2_{L^2(\mathbb{R}_+;\mathcal{G})\wedge L^2(\mathbb{R}_+;\mathcal{G})}\right] \\ = \mathbb{E}[\|u\|^2_{L^2(\mathbb{R}_+)}] + \mathbb{E}\left[\|\nabla u\|^2_{L^2(\mathbb{R}_+;\mathcal{G})\otimes L^2(\mathbb{R}_+;\mathcal{G})}\right]. \qquad (3.5)$$

The next section is devoted to two lemmas that are used to prove (3.3).

4. Quantum stochastic differentials on path space

The following expression for $\tilde{D}$ using quantum stochastic integrals can be viewed as an intertwining formula between $\tilde{D}$, D and I.

Lemma 4.1. *We have for* $v \in L^2(\mathbb{R}_+;\mathcal{G})$*:*

$$\mathrm{d}\tilde{a}_v^-(t) = \mathrm{d}a_v^-(t) + q_v(t)\mathrm{d}\Lambda(t), \qquad t > 0.$$

Proof. The process $t \mapsto \gamma(t)\mathrm{e}^{\int_0^t v(s)\mathrm{d}s}$ satisfies the following stochastic differential equation in the Stratonovich sense:

$$\mathrm{d}\left(\gamma(t)\mathrm{e}^{\int_0^t v(s)\mathrm{d}s}\right) = \gamma(t)\mathrm{e}^{\int_0^t v(s)\mathrm{d}s}\left(\odot\,\mathrm{Ad}\,\mathrm{e}^{-\int_0^t v(s)\mathrm{d}s}\mathrm{d}B(t) + v(t)\mathrm{d}t\right) \\ = \gamma(t)\mathrm{e}^{\int_0^t v(s)\mathrm{d}s}\left(\odot\mathrm{e}^{-q_v(t)}\mathrm{d}B(t) + v(t)\mathrm{d}t\right), \quad t > 0.$$

Let $I_1(u)$ denote the Wiener integral of $u \in L^2(\mathbb{R}_+;\mathbb{R}^d)$ with respect to $(B(t))_{t\in\mathbb{R}_+}$, and let

$$G = g\left(I_1(u_1),\ldots,I_1(u_n)\right) \in \mathcal{S}, \quad \text{and} \quad F = f(\gamma(t_1),\ldots,\gamma(t_n)) \in \tilde{\mathcal{S}}.$$

Since $\exp(q_v(t))\colon \mathcal{G} \longrightarrow \mathcal{G}$ is isometric, we have from the Girsanov theorem:

$$\mathbb{E}\left[f\left(\gamma(t_1)\mathrm{e}^{\int_0^{t_1} v(s)\mathrm{d}s},\ldots,\gamma(t_n)\mathrm{e}^{\int_0^{t_n} v(s)\mathrm{d}s}\right)G\right] = \mathbb{E}\left[F\mathrm{e}^{I_1(v)-\frac{1}{2}\|v\|^2}\Theta_v G\right],$$

where

$$\Theta_v G = g\left[\int_0^\infty u_1(s)\mathrm{e}^{q_v(s)}\mathrm{d}B(s) - \langle u_1, v\rangle,\ldots,\int_0^\infty u_n(s)\mathrm{e}^{q_v(s)}\mathrm{d}B(s) - \langle u_n, v\rangle\right].$$

From the derivation property of D and the divergence relation $\delta(vG) = G\delta(v) - \langle v, DG\rangle$ we have

$$\begin{aligned}
\frac{\mathrm{d}}{\mathrm{d}\varepsilon}\Theta_{\varepsilon v}G|_{|\varepsilon=0} &= \sum_{i=1}^{n} \partial_i g\left(I_1(u_1),\dots,I_1(u_n)\right)\frac{\mathrm{d}}{\mathrm{d}\varepsilon}\Theta_{\varepsilon v}I_1(u_i)|_{|\varepsilon=0} \\
&= \sum_{i=1}^{n} \partial_i g\left(I_1(u_1),\dots,I_1(u_n)\right)\left(-\langle v, u_i\rangle + \int_0^\infty q_v^*(s)u_i(s)\mathrm{d}B(s)\right) \\
&= -\sum_{i=1}^{n} \partial_i g\left(I_1(u_1),\dots,I_1(u_n)\right)\left(-\langle v, DI_1(u_i)\rangle + \delta(\nabla_v DI_1(u_i))\right) \\
&= -\langle v, Dg\left(I_1(u_1),\dots,I_1(u_n)\right)\rangle \\
&\quad -\sum_{i=1}^{n} \partial_i g\left(I_1(u_1),\dots,I_1(u_n)\right)\delta(\nabla_v DI_1(u_i)) \\
&= -\langle v, DG\rangle - \sum_{i=1}^{n}\delta(\partial_i g\left(I_1(u_1),\dots,I_1(u_n)\right)\nabla_v DI_1(u_i)) \\
&\quad - \sum_{i,j=1}^{n} \partial_j\partial_i g\left(I_1(u_1),\dots,I_1(u_n)\right)\left(\langle u_j, \nabla_v u_i\rangle + \langle u_i, \nabla_v u_j\rangle\right) \\
&= -\langle v, DG\rangle - \delta(\nabla_v DG),
\end{aligned}$$

since ∇_v is antisymmetric. Hence

$$\begin{aligned}
&\mathbb{E}[\langle \tilde{D}F, v\rangle G] \\
&= \frac{\mathrm{d}}{\mathrm{d}\varepsilon}\mathbb{E}\left[f\left(\gamma(t_1)\mathrm{e}^{\varepsilon\int_0^{t_1} h(s)\mathrm{d}s},\dots,\gamma(t_n)\mathrm{e}^{\varepsilon\int_0^{t_n} h(s)\mathrm{d}s}\right)G\right]\Big|_{\varepsilon=0} \\
&= \frac{\mathrm{d}}{\mathrm{d}\varepsilon}\mathbb{E}\left[F\mathrm{e}^{\varepsilon I_1(v)-\frac{1}{2}\varepsilon^2\|v\|^2}\Theta_{\varepsilon v}G\right]\Big|_{\varepsilon=0} \\
&= \mathbb{E}\left[F(GI_1(v) - \langle v, DG\rangle - \delta(\nabla_v DG))\right].
\end{aligned}$$

For $G = 1$ this implies the well known identity $\delta(v) = I_1(v)$, hence

$$\begin{aligned}
\mathbb{E}[\langle \tilde{D}F, v\rangle G] &= \mathbb{E}\left[F(G\delta(v) - \langle v, DG\rangle - \delta(\nabla_v DG))\right] \\
&= \mathbb{E}\left[F(\delta(vG) - \delta(\nabla_v DG))\right] \\
&= \mathbb{E}\left[F\left(a_v^+ - \int_0^\infty q_v(t)\mathrm{d}\Lambda(t)\right)G\right] \\
&= \mathbb{E}\left[G\left(a_v^- + \int_0^\infty q_v(t)\mathrm{d}\Lambda(t)\right)F\right]. \qquad \square
\end{aligned}$$

It follows from the proof of Lemma 4.1 that $\tilde{D}$ admits an adjoint $\tilde{\delta}$ that satisfies

$$\tilde{\delta}(uF) = a_v^+ F - \int_0^\infty q_v(t)\mathrm{d}\Lambda(t)F, \quad F \in \tilde{\mathcal{S}},$$

and

$$\mathbb{E}[F\tilde{\delta}(u)] = \mathbb{E}[\langle \tilde{D}F, u\rangle], \quad F \in \tilde{\mathcal{S}},\ u \in \tilde{\mathcal{U}}.$$

Letting

$$\tilde{a}_u^+ F = \int_0^\infty \mathrm{d}\tilde{a}_u^+(t) F = \tilde{\delta}(uF),$$

we have

$$\mathrm{d}\tilde{a}_u^+(t) = \mathrm{d}a_u^+(t) - q_u(t)\mathrm{d}\Lambda(t).$$

The following Lemma shows that

$$[\nabla_v, \nabla_u] = \nabla_{\nabla_v u} - \nabla_{\nabla_u v}.$$

This means that the Lie bracket $\{u, v\}$ associated to the gradient ∇ on $L^2(\mathbb{R}_+; \mathcal{G})$ via $[\nabla_u, \nabla_v] = \nabla_{\{u,v\}}$ satisfies $\{u, v\} = \nabla_u v - \nabla_v u$, i.e. the connection defined by ∇ on $L^2(\mathbb{R}_+; \mathcal{G})$ has vanishing torsion.

Lemma 4.2. *We have*

$$[q_v(t), q_u(t)] = q_{\nabla_v u}(t) - q_{\nabla_u v}(t), \quad t > 0,\ u, v \in L^2(\mathbb{R}_+; \mathcal{G}).$$

Proof. The Jacobi identity on $\mathcal{G}$ shows that

$$\begin{aligned}
[q_v(t), q_u(t)] &= \left[\int_0^t \mathrm{ad}(v(s))\mathrm{d}s, \int_0^t \mathrm{ad}(u(s))\mathrm{d}s\right] \\
&= \mathrm{ad}\left(\left[\int_0^t v(s)\mathrm{d}s, \int_0^t u(s)\mathrm{d}s\right]\right) \\
&= \int_0^t \int_0^s \mathrm{ad}([v(\tau), u(s)])\mathrm{d}\tau\mathrm{d}s - \int_0^t \int_0^s \mathrm{ad}([u(\tau), v(s)])\mathrm{d}\tau\mathrm{d}s \\
&= \int_0^t \mathrm{ad}(q_v(s)u(s))\mathrm{d}s - \int_0^t \mathrm{ad}(q_u(s)v(s))\mathrm{d}s \\
&= q_{\nabla_v u}(t) - q_{\nabla_u v}(t).
\end{aligned}$$

□

The Lie derivative on $\mathbb{P}(\mathsf{G})$ in the direction $u \in L^2(\mathbb{R}_+; \mathcal{G})$, introduced in [3], can be written $\tilde{a}_u^-$ in our context. Finally we show that the vanishing of torsion discovered in [3] can be obtained via quantum stochastic calculus. Precisely, the Lie bracket $\{u, v\}$ associated to $\tilde{a}_v^-$ via $[\tilde{a}_u^-, \tilde{a}_v^-] = \tilde{a}_{\{u,v\}}^-$ satisfies $\{u, v\} = \nabla_u v - \nabla_v u$, i.e. the connection defined by ∇ on $\mathbb{P}(\mathsf{G})$ also has a vanishing torsion.

Proposition 4.3. *We have on $\tilde{\mathcal{S}}$, with $u, v \in L^2(\mathbb{R}_+; \mathcal{G})$:*

$$\tilde{a}_u^- \tilde{a}_v^- - \tilde{a}_v^- \tilde{a}_u^- = \tilde{a}_{\nabla_v u}^- - \tilde{a}_{\nabla_u v}^-.$$

Proof. Using Lemma 4.2, the quantum Itô table implies

$$\begin{aligned}
&\mathrm{d}\tilde{a}_u^-(t) \cdot \mathrm{d}\tilde{a}_v^-(t) - \mathrm{d}\tilde{a}_v^-(t) \cdot \mathrm{d}\tilde{a}_u^-(t) \\
&= \left(da_u^-(t) + q_u(t)\mathrm{d}\Lambda(t)\right) \cdot \left(\mathrm{d}a_v^-(t) + q_v(t)\mathrm{d}\Lambda(t)\right) \\
&\quad - \left(\mathrm{d}a_v^-(t) + q_v(t)\mathrm{d}\Lambda(t)\right) \cdot \left(\mathrm{d}a_u^-(t) + q_u(t)\mathrm{d}\Lambda(t)\right) \\
&= q_v(t)u(t)\mathrm{d}a^-(t) + q_u(t)q_v(t)\mathrm{d}\Lambda(t) \\
&\quad - q_v(t)u(t)\mathrm{d}a^-(t) - q_v(t)q_u(t)\mathrm{d}\Lambda(t) \\
&= \nabla_v u(t)\mathrm{d}a^-(t) - q_{\nabla_v u}(t)\mathrm{d}\Lambda(t) - (\nabla_u v(t)\mathrm{d}a^-(t) - q_{\nabla_u v}(t)\mathrm{d}\Lambda(t)) \\
&= \mathrm{d}\tilde{a}_{\nabla_v u}^-(t) - \mathrm{d}\tilde{a}_{\nabla_u v}^-(t),
\end{aligned}$$

from Lemma 4.1. □

References

[1] Cruzeiro A.B., Malliavin P.: Renormalized differential geometry on path space: Structural equation, curvature, *J. Funct. Anal.*, **139** (1996) 119–181.

[2] Driver B.: A Cameron-Martin type quasi-invariance theorem for Brownian motion on a compact Riemannian manifold, *J. Funct. Anal.*, **110**, 2 (1992) 272–376.

[3] Fang S., Franchi J.: Platitude de la structure riemannienne sur le groupe des chemins et identité d'énergie pour les intégrales stochastiques [Flatness of Riemannian structure over the path group and energy identity for stochastic integrals], *C. R. Acad. Sci. Paris Sér. I Math.* **321**, 10 (1995) 1371–1376.

[4] Hudson R.L., Parthasarathy K.R.: Quantum Itô's formula and stochastic evolutions, *Comm. Math. Phys.* **93**, 3 (1984) 301–323.

[5] Lindsay J.M.: Quantum and non-causal stochastic calculus, *Probab. Theory Related Fields* **97** (1993) 65–80.

[6] Nualart D., Pardoux E.: Stochastic calculus with anticipative integrands, *Probab. Theory Related Fields* **78** (1988) 535–582.

[7] Parthasarathy K.R.: *An Introduction to Quantum Stochastic Calculus*, Birkhäuser, 1992.

[8] Shigekawa I.: de Rham-Hodge-Kodaira's decomposition on an abstract Wiener space, *J. Math. Kyoto Univ.* **26**, 2 (1986) 191–202.

MARTINGALE APPROXIMATION FOR SELF-INTERSECTION LOCAL TIME OF BROWNIAN MOTION

Margarida de Faria
CCM, Universidade da Madeira, P 9000 Funchal
ccm@uma.pt

Anis Rezgui
ZiF, Univ. Bielefeld, D 33615 Bielefeld
rezgui@physik.uni-bielefeld.de

Ludwig Streit
CCM, Universidade da Madeira, P 9000 Funchal,
BiBoS, Univ. Bielefeld, D 33615 Bielefeld
streit@physik.uni-bielefeld.de

Abstract Given that each term of the multiple Wiener integral expansion for the renormalized self-intersection local time of higher dimensional Brownian motion converges in law to another, independent Brownian motion we resum the leading, martingale parts of these terms in closed form and also represent this sum as a stochastic integral.

Keywords: Brownian motion, white noise
Primary 60J65; Secondary 60H40

1. Introduction

An informal definition of the self-intersection local time of d-dimensional Brownian motion is given in terms of an integral over Donsker's δ–function.

$$L_T = \int_{\Delta_T} dt_2 dt_1 \delta(B_{t_2} - B_{t_1}),$$

where B_t is a d-dimensional Brownian motion and $\Delta_T = \{(t_1, t_2) : 0 < t_1 < t_2 < T\}$.

S. Albeverio et al. (eds.),
Proceedings of the International Conference on Stochastic Analysis and Applications, 95–106.

To make sense of this integral one can invoke a regularization such as

$$L_T^\epsilon = \int_{\Delta_T} dt_2 dt_1 \delta_\epsilon(B_{t_2} - B_{t_1}) \tag{1.1}$$

where $\delta_\epsilon(x) = \frac{1}{(2\pi\epsilon)^{d/2}} e^{-\frac{x^2}{2\epsilon}}$ for $x \in \mathbb{R}^d$, and define L_T as the limit when ϵ goes to zero.

If the dimension is $d \geq 2$, $\lim_{\epsilon \to 0^{++}} \mathbb{E}(L_T^\epsilon) = +\infty$; Varadhan, in [20] renormalizes L_T^ϵ by substracting its expectation and proves for $d = 2$ that the limit exists in mean square.

Hence we consider the centered self-intersection local time defined by

$$L_{T,c}^\epsilon = \int_{\Delta_T} dt_2 dt_1 \delta_{\epsilon,c}(B_{t_2} - B_{t_1})$$

where $\delta_{\epsilon,c}(B_{t_2} - B_{t_1}) = \delta_\epsilon(B_{t_2} - B_{t_1}) - \mathbb{E}\Big(\delta_\epsilon(B_{t_2} - B_{t_1})\Big)$, this is the so called Varadhan renormalization.

For $d \geq 3$, a further multiplicative renormalization $r_d(\epsilon)$ is required for the existence of a limiting process. M. Yor in [19] shows, using the near passage regularization

$$\delta(B_{t_2} - B_{t_1}) \to \delta(B_{t_2} - B_{t_1} + \epsilon)$$

for $d = 3$, that

$$r_3(\epsilon)(L^\epsilon - \mathbb{E}(L^\epsilon)) \xrightarrow{\mathcal{L}} c\beta$$

with β a Brownian motion independent of B.

This can be understood in the light of the fact that each term in the (renormalized) chaos expansion or multiple Wiener integral expansion for the self-intersection local time converges in law to a Brownian motion, for any $d > 2$ as shown in [4], using the fact that the dominant part of each of these multiple Wiener integrals is in fact a martingale. With a view towards convergence results for the renormalized local time in dimensions $d > 3$, it is desirable to resum these dominant terms, i.e. to split the local time into a dominant martingale part and a subdominat remainder. We do this in Theorems 1 and 2, in Theorem 3 we express the two terms as stochastic integrals via the Clark-Ocone formula.

Recent investigations of self-intersection local time have used white noise analysis [1, 4, 5, 21]. So, before announcing the main results of this paper, let us briefly recall some tools from white noise analysis and some of these results on self-intersection local time.

1.1 Tools from White Noise Analysis

We quote some white noise analysis concepts as introduced in [4], referring to [9] for a systematic presentation.

Consider a white noise space $(S'(\mathbb{R})^d, \mathcal{B}, \mu)$, where $\mathcal{B}$ is the weak Borel σ-algebra of $S'(\mathbb{R})^d$, and μ is the centered Gaussian measure whose covariance is given by the inner product of $L^2(\mathbb{R})^d$, in the sense that the vector valued white noise has the characteristic function

$$C(\mathbf{f}) = \mathbb{E}(e^{i\langle\omega,\mathbf{f}\rangle}) = \int_{S'(\mathbb{R})^d} d\mu[\omega] e^{i\langle\omega,\mathbf{f}\rangle} = e^{-\frac{1}{2}\langle\mathbf{f},\mathbf{f}\rangle},$$

where $\langle\omega,\mathbf{f}\rangle = \sum_{i=1}^d \langle\omega_i, f_i\rangle$ and $f_i \in S(\mathbb{R},\mathbb{R})$.

Then a realization of a vector of independent Brownian motions B_i, $i = 1, \ldots, d$, is given by

$$B_i(t) = \langle\omega_i, \mathbf{1}_{[0,t]}\rangle = \int_0^t \omega_i(s) ds.$$

Hence we consider independent d-tuples of Gaussian white noise $\omega = (\omega_1, \ldots, \omega_d)$ and d-tuples of test functions $\mathbf{f} = (f_1, \ldots, f_d) \in S(\mathbb{R},\mathbb{R}^d)$, and use the following multi-index notation:

$$\vec{n}! = \prod_1^d n_i!$$

$$\langle\mathbf{f},\mathbf{f}\rangle = \sum_{i=1}^d \int dt f_i^2(t)$$

$$\langle F_{\vec{n}}, \mathbf{f}^{\otimes\vec{n}}\rangle = \int d^n t F_{\vec{n}}(t_1, \ldots, t_n) \bigotimes_{i=1}^d f_i^{\otimes n_i}(t_1, \ldots, t_n)$$

and similarly for $\langle :\omega^{\otimes\vec{n}}:, F_{\vec{n}}\rangle$ where for d-tuples of white noise the Wick product $:\cdots:$ (see [9]) generalizes to

$$:\omega^{\otimes\vec{n}}: := \bigotimes_{i=1}^d :\omega_i^{\otimes n_i}: .$$

The Hilbert space

$$(L^2) = L^2(d\mu)$$

is canonically isomorphic to the d-fold tensor product of Fock spaces of symmetric square integrable functions:

$$(L^2) \simeq \Big(\bigoplus_{k=0}^\infty \operatorname{Sym} L^2(\mathbb{R}^k, k! d^k t)\Big)^{\otimes d} = \mathcal{F}.$$

For a general element φ of (L^2) this implies the chaos expansion

$$\varphi(\omega) = \sum_{\vec{n}=0}^{\infty} \langle : \omega^{\otimes \vec{n}} :, F_{\vec{n}} \rangle,$$

the norm of φ is given by

$$\|\varphi\|^2_{(L^2)} = \sum_{\vec{n}}^{\infty} \vec{n}! |F_{\vec{n}}|^2_{2,n}$$

with kernel functions F in $\mathcal{F}$ and where $|\ |_{2,n}$ is the norm in $L^2(\mathbb{R}^n, dt)$.
Given $\xi \in S(\mathbb{R})^d$, let us consider the Wick exponential

$$\begin{aligned} : \exp\langle \omega, \xi \rangle : &\equiv \exp\left(\langle \omega, \xi \rangle - \frac{1}{2}(\xi, \xi) \right) \\ &= \sum_{\vec{n}} \frac{1}{\vec{n}!} \langle : \omega^{\otimes \vec{n}} :, \xi^{\otimes \vec{n}} \rangle, \qquad \omega \in S'(\mathbb{R})^d. \end{aligned}$$

The S-transform plays an important role in the study of stochastic processes in the white noise framework, see for example [3][9]; we define the S-transform of φ in (L^2) as

$$S\varphi(\xi) \equiv \ll \varphi, : \exp\langle ., \xi \rangle : \gg = \sum_{\vec{n}} (\varphi_{\vec{n}}, \xi^{\otimes \vec{n}})_{2,n}.$$

In particular, for Hermitean operators A in $L^2(\mathbb{R})$, we can define the "second quantization" of A as an operator $\Gamma(A)$ in (L^2) given by

$$S\Gamma(A)\Phi(\cdot) = S\Phi(A\cdot) \tag{1.2}$$

for $\Phi \in (L^2)$.

1.2 Self-Intersection Local Time in Terms of White Noise

The regularization of the self-intersection local time given by (1) has the following chaos expansion, for $d \geq 3$.

Proposition 1.1 ([1]). *For any t, $\epsilon > 0$, $L^{\epsilon}_{t,c}$ has kernel functions $F \in \mathcal{F}$ given by*

$$\begin{aligned} F_{\epsilon,\vec{n}}(s_1, \ldots, s_n) = (-1)^{\frac{n}{2}} &\left[\chi(\chi+1)(2\pi)^{d/2} 2^{\frac{n}{2}} \frac{\vec{n}}{2}! \right]^{-1} \theta(u)\theta(t-v) \\ &\times \left[(v-u+\epsilon)^{-\chi} + (t+\epsilon)^{-\chi} - (v+\epsilon)^{-\chi} - (t-u+\epsilon)^{-\chi} \right] \end{aligned}$$

if all n_i are even, and zero otherwise, with

$$v(s_1,\cdots,s_n) \equiv \max(s_1,\ldots,s_n),$$

$$u(s_1,\ldots,s_n) \equiv \min(s_1,\ldots,s_n), \qquad \chi \equiv \frac{n+2}{2} - 2.$$

θ is the Heaviside function.

One can then divide each chaos into a martingale part

$$M^{\epsilon}_{\vec{n}} = \int_{[0,t]^n} d^n s (v-u+\epsilon)^{-\chi} : w^{\otimes \vec{n}}(s) :$$

and a remainder

$$N^{\epsilon}_{\vec{n}} = \int_{[0,t]^n} d^n s \left((t+\epsilon)^{-\chi} - (v+\epsilon)^{-\chi} - (t-u+\epsilon)^{-\chi}\right) : w^{\otimes \vec{n}}(s) :$$

which is less singular in the limit $\varepsilon \to 0$. We note that one could have included the part $N^{\epsilon}_{\vec{n}}$ arising from the second term in the integrand, i.e. from $(v+\epsilon)^{-\chi}$, without losing the martingale property:

$$M'^{\epsilon}_{\vec{n}} = \int_{[0,t]^n} d^n s \left((v-u+\epsilon)^{-\chi} - (v+\epsilon)^{-\chi}\right) : w^{\otimes \vec{n}}(s) :$$

is also a martingale since the varable t does not appear in the integrand. After renormalization the difference vanishes in (L^2) as $\varepsilon \to 0$.

For the $\vec{n}$th order chaos,

$$K^{\epsilon}_{\vec{n}} = \alpha_{\vec{n}} \left(M^{\epsilon}_{\vec{n}} + N^{\epsilon}_{\vec{n}}\right) ,$$

$\vec{n} \in \mathbb{N}^d - \{0\}$, it was shown in [4], that for any $d \geq 3$,

$$r_d(\epsilon) K^{\epsilon}_{\vec{n}} \xrightarrow{\mathcal{L}} c_{\vec{n}} \beta_{\vec{n}} \tag{1.3}$$

where $\beta_{\vec{n}}$ are one dimensional Brownian motions independent of the initial one and among each other, with

$$c^2_{\vec{n}} = k^2_n \alpha^2_{\vec{n}} ,$$

$$k^2_n = \begin{cases} n(n-1) & d=3 \\ \frac{n!(d-4)!}{(n+d-5)!} & d>3 , \end{cases}$$

$$\alpha_{\vec{n}} = (-1)^{n/2} \left[\chi(\chi+1)(2\pi)^{d/2} 2^{n/2} \frac{\vec{n}}{2}!\right]^{-1} ,$$

$$r_d(\epsilon) = \begin{cases} |\log \epsilon|^{-1/2} & d=3 \\ \epsilon^{\frac{d-3}{2}} & d>3 . \end{cases}$$

It has been proved in [5] that the variances of these Brownian motions sum up to the variance of the renormalized self-intersection local time,

$$T \sum_{\vec{n}, n \neq 0} c_{\vec{n}}^2 = \lim_{\epsilon \to 0} \mathbb{E}\Big[(r_d(\epsilon) L_{T,c}^{\epsilon})^2 \Big]$$

and that the right hand side can be calculated in closed form for all $d \geq 3$, using a Clark-Ocone formula for L^{ϵ}. For higher order self-intersections chaos decompositions and their renormalized limits are discussed in [14] and [11].

2. Statement of the results

If we inspect the kernels given in the above proposition we can see that the dominant terms of the chaos expansion for the local time are martingales. Hence it is desirable to sum up these dominant terms. This is indeed possible.

Theorem 2.1. *Let $\epsilon, T > 0$ and denote by θ_T the multiplication operator by $1_{[0,T]}$.*
Then

$$L_{T,c}^{\epsilon} = M_{T,c}^{\epsilon} + R_{T,c}^{\epsilon}$$

where

$$M_{T,c}^{\epsilon} = \int_0^{\infty} dt_2 \int_0^{t_2} dt_1 \Gamma(\theta_T) \delta_{\epsilon,c}(B_{t_2} - B_{t_1}) \tag{2.1}$$

is a martingale, and

$$R_{T,c}^{\epsilon} = - \int_{\epsilon}^{+\infty} dt_2 \int_0^T dt_1 \delta_{t_2,c}(B_T - B_{t_1}).$$

The next theorem gives us the corresponding chaos expansions, for $d \geq 3$.

Theorem 2.2. *Let $d > 2$, for any $\epsilon > 0$ and $\vec{n} \in \mathbb{N}^d$, $\vec{n} \neq \vec{0}$, $M_{T,c}^{\epsilon}$ and $R_{T,c}^{\epsilon}$ have respectively kernel functions $\mathcal{M}$ and $\mathcal{R}$ in $\mathcal{F}$ given by*

$$\begin{aligned}
\mathcal{M}_{\vec{n}}^{\epsilon} &= (-1)^{n/2} \Big[\kappa(\kappa+1)(2\pi)^{d/2} 2^{n/2} \frac{\vec{n}}{2}! \Big]^{-1} \\
&\qquad \times \theta(u)\theta(T-v) \big[(v-u+\epsilon)^{-\kappa} - (v+\epsilon)^{-\kappa} \big] \\
\mathcal{N}_{\vec{n}}^{\epsilon} &= (-1)^{n/2} \Big[\kappa(\kappa+1)(2\pi)^{d/2} 2^{n/2} \frac{\vec{n}}{2}! \Big]^{-1} \\
&\qquad \times \theta(u)\theta(T-v) \big[(T+\epsilon)^{-\kappa} - (T-u+\epsilon)^{-\kappa} \big]
\end{aligned}$$

if all n_i are even, and zero otherwise, with

$$v = v(s_1, \cdots, s_n) = \max_i s_i,$$

$$u = u(s_1, \ldots, s_n) = \min_i s_i, \quad \kappa = n + d/2 - 2,$$

and θ the Heaviside function.

We recall that if $(\Phi_t)_{t\geq 0}$ is a stochastic process or even a generalized one (such as $(\Phi_t)_{t\geq 0} \subset \mathcal{G}^{-1}$, the Potthoff-Timpel space [16]), it is shown in [2] that for each $t \geq 0$, Φ_t can be written as

$$\Phi_t = \mathbb{E}\Phi_t + I(m_{\Phi_t}) \tag{2.2}$$

This is the so called generalized Clark-Ocone formula, where $m_{\Phi_t} \in \mathbb{R}^d \otimes L^2(\mathbb{R}) \otimes \mathcal{G}^{-1}$ such that

$$m^i_{\Phi_t} = \Gamma(\theta_.)\partial^i_.\Phi \tag{2.3}$$

and

$$I(m_{\Phi_t}) = \sum_{i=1}^{d} I_i(m^i_{\Phi_t})$$

where $I_i(m^i_{\Phi_t})$ is in $\mathcal{G}^{-1}$, defined by

$$\ll I_i(m^i_{\Phi_t}), \varphi \gg \; = \; \ll m^i_{\Phi_t}, \partial^i_.\varphi \gg \qquad \text{for every } \varphi \in \mathcal{G}^{+1}.$$

$\partial^i_.\varphi \in L^2(\mathbb{R}) \otimes \mathcal{G}^{+1}$ is the Hida-gradient of φ, see also [2] for an explicit formula. In our next theorem, we give the explicit formula of the integrand by applying the Clark-Ocone formula 2.3.

Theorem 2.3. *For T, $\epsilon > 0$ we have*

$$M^\epsilon_{T,c} = \frac{1}{(2\pi)^{d/2}} \int_0^{+\infty} dt_2 \int_0^{T\wedge t_2} dt_1 \int_{t_1}^{T\wedge t_2} dB_\tau \frac{B_\tau - B_{t_1}}{(\epsilon + t_2 - \tau)^{\frac{d}{2}+1}} e^{\frac{-(B_\tau - B_{t_1})^2}{2(\epsilon + t_2 - \tau)}}$$

$$R^\epsilon_{T,c} = -\frac{1}{(2\pi)^{d/2}} \int_T^{+\infty} dt_2 \int_0^{T} dt_1 \int_{t_1}^{T} dB_\tau \frac{B_\tau - B_{t_1}}{(\epsilon + t_2 - \tau)^{\frac{d}{2}+1}} e^{\frac{-(B_\tau - B_{t_1})^2}{2(\epsilon + t_2 - \tau)}}.$$

3. Proof of the results

Proof of Theorem 2.1. Let $(\mathcal{F}_T)_{T\geq 0}$ be the Brownian filtration. Recall that if $\Gamma(\theta_T)\Phi \in (L^2)$ $\forall T > 0$, then $\mathbb{E}(\Phi|\mathcal{F}_T) = \Gamma(\theta_T)\Phi$ is a martingale with respect to the filtration $(\mathcal{F}_T)_{T\geq 0}$, thus

$$M^\epsilon_{T,c} = \int_0^{+\infty} dt_2 \int_0^{t_2} dt_1 \Gamma(\theta_T)\delta_{\epsilon,c}(B_{t_2} - B_{t_1})$$

is a martingale with respect to the Brownian filtration. Let us decompose $M^{\epsilon}_{T,c}$ as:

$$M^{\epsilon}_{T,c} = \int_0^T dt_2 \int_0^{t_2} dt_1 \Gamma(\theta_T)\delta_{\epsilon,c}(B_{t_2} - B_{t_1}) + \int_T^{+\infty} dt_2 \int_0^{t_2} dt_1 \Gamma(\theta_T)\delta_{\epsilon,c}(B_{t_2} - B_{t_1}).$$

Note that if $\Phi \in (L^2)$ we have

$$S\Gamma(\theta_T)\Phi(\cdot) = S\Phi(\theta_T \cdot). \tag{3.1}$$

To compute the S-transform of $\delta_\epsilon(B_{t_2} - B_{t_1})$ we note that

$$\delta_\epsilon(B_{t_2} - B_{t_1}) = (2\pi\epsilon)^{-d/2} \prod_{i=1}^{d} e^{-\frac{1}{2}\langle \omega_i, K\omega_i \rangle}$$

with $K = \epsilon^{-1}|t_2 - t_1|P$, where P is the projector onto the indicator function of the interval $[t_1, t_2]$. It is well known [9] that the S-transform of such a Gauss kernels is given by

$$S\delta_\epsilon(B_{t_2} - B_{t_1})(\xi) = (2\pi\epsilon \det(1+K))^{-d/2} e^{-\frac{1}{2}(\xi, \frac{K}{1+K}\xi)} \tag{3.2}$$

$$= \frac{1}{(2\pi(\epsilon + t_2 - t_1))^{d/2}} \exp\left\{ -\frac{\sum_{i=1}^d \left(\int_{t_1}^{t_2} \xi_i(u)du \right)^2}{2(\epsilon + t_2 - t_1)} \right\}$$

$$= \frac{1}{(2\pi(\epsilon + t_2 - t_1))^{d/2}} \exp\left\{ -\frac{\left(\int_{t_1}^{t_2} \xi(u)du \right)^2}{2(\epsilon + t_2 - t_1)} \right\}. \tag{3.3}$$

As a result we obtain

$$\Gamma(\theta_T)\delta_{\epsilon,c}(B_{t_2} - B_{t_1}) = \begin{cases} \delta_{\epsilon,c}(B_{t_2} - B_{t_1}) & T > t_2 \\ 0 & T < t_1 \\ \delta_{t_2 - T + \epsilon, c}(B_T - B_{t_1}) & t_1 < T < t_2. \end{cases}$$

So

$$\begin{aligned} L^{\epsilon}_{T,c} &= \int_0^T dt_2 \int_0^{t_2} dt_1 \delta_{\epsilon,c}(B_{t_2} - B_{t_1}) \\ &= \int_0^T dt_2 \int_0^{t_2} dt_1 \Gamma(\theta_T)\delta_{\epsilon,c}(B_{t_2} - B_{t_1}) \\ &= \int_0^{+\infty} dt_2 \int_0^{t_2} dt_1 \Gamma(\theta_T)\delta_{\epsilon,c}(B_{t_2} - B_{t_1}) \\ &\quad - \int_T^{+\infty} dt_2 \int_0^T dt_1 \delta_{t_2 - T + \epsilon, c}(B_T - B_{t_1}) \end{aligned}$$

and theorem 2 is proved. □

Proof of theorem 2.2. Using the formula (3.3) we obtain the following expansion in powers of ξ

$$
\begin{aligned}
&S\delta_{\epsilon,c}(B_{t_2}-B_{t_1})(\xi)\\
&=\frac{1}{(2\pi(\epsilon+t_2-t_1))^{\frac{d}{2}}}\sum_{\vec{n}\in\mathbb{N}^d,n\neq 0}\frac{1}{\vec{n}!}\left(\frac{-1}{2(\epsilon+t_2-t_1)}\right)^n\langle\xi^{\otimes 2\vec{n}};\mathbf{1}_{[t_1,t_2]}^{\otimes 2n}\rangle
\end{aligned}
$$

then in view of (2.1) and (3.1)

$$
\begin{aligned}
&\mathcal{M}^{\epsilon}_{2\vec{n}}=\\
&=(2\pi)^{-\frac{d}{2}}\frac{1}{\vec{n}!}\Big(\frac{-1}{2}\Big)^n\theta_T^{\otimes 2n}\int_0^{+\infty}dt_2\int_0^{t_2}dt_1(\epsilon+t_2-t_1)^{-n-d/2}\mathbf{1}_{[t_1,t_2]}^{\otimes 2n}\\
&=(2\pi)^{-\frac{d}{2}}\frac{1}{\vec{n}!}\Big(\frac{-1}{2}\Big)^n\theta_T^{\otimes 2n}\int_v^{+\infty}dt_2\int_0^{u}dt_1(\epsilon+t_2-t_1)^{-n-d/2}\mathbf{1}_{[t_1,t_2]}^{\otimes 2n}
\end{aligned}
$$

So we obtain

$$
\begin{aligned}
\mathcal{M}^{\epsilon}_{2\vec{n}}=(2\pi)^{-d/2}\frac{1}{\vec{n}!}(\frac{-1}{2})^n&\frac{1}{(n+d/2-1)(n+d/2-2)}\\
\cdot\,\theta_T^{\otimes 2n}&\left\{(\epsilon+v-u)^{-(n+d/2-2)}-(\epsilon+v)^{-(n+d/2-2)}\right\}
\end{aligned}
$$

where $v=v(s_1,\ldots,s_{2n})=\max_i s_i$ and $u=\min_i s_i$. □

An analogous argument produces the kernels of R^{ϵ}_T.

Proof of theorem 2.3. We first find a representation as in (2.2) for $\Gamma(\theta_T)\delta_\epsilon$

$$\Gamma(\theta_T)\delta_\epsilon(B_{t_2}-B_{t_1})=\mathbb{E}\Big(\Gamma(\theta_T)\delta_\epsilon(B_{t_2}-B_{t_1})\Big)+I(m^\epsilon)$$

note that $m^\epsilon=(m_i^\epsilon)_{1\leq i\leq d}$ is in $\mathbb{R}^d\otimes L^2(\mathbb{R})\otimes(L^2)$ and that of course it depends of t_1,t_2.

By theorem 4.1 in [2] we have

$$m_i^\epsilon(\tau)=S^{-1}\frac{\delta}{\delta\xi_i(\tau)}S\Big(\Gamma(\theta_T)\delta_\epsilon(B_{t_2}-B_{t_1})\Big)(\theta_\tau\xi)$$

or

$$
\begin{aligned}
&S\Big(\Gamma(\theta_T)\delta_\epsilon(B_{t_2}-B_{t_1})\Big)(\xi)=\\
&=[2\pi(t_2-t_1+\epsilon)]^{-d/2}\exp\Big\{-\frac{\mathbf{1}_{[0,T]}(t_1)\sum_{i=1}^d(\int_{t_1}^{t_2\wedge T}\xi_i du)^2}{2(t_2-t_1+\epsilon)}\Big\}\\
&=[2\pi(t_2-t_1+\epsilon)]^{-d/2}\exp\Big\{-\frac{\mathbf{1}_{[0,T]}(t_1)(\int_{t_1}^{t_2\wedge T}\xi du)^2}{2(t_2-t_1+\epsilon)}\Big\}.
\end{aligned}
$$

It is known from [2] that if $g(x_1,\dots,x_d)$ is a smooth function $g\colon \mathbb{R}^d \to \mathbb{R}$, $(h_i)_{1\le i\le d} \in \mathbb{R}^d \otimes L^2(\mathbb{R})$ and $E(\xi) = g(\langle \xi_1, h_1\rangle,\dots,\langle \xi_d, h_d\rangle)$ then

$$\frac{\delta E}{\delta \xi_i(\tau)}(\xi) = \frac{\partial g}{\partial x_i}(\langle \xi_1, h_1\rangle,\dots,\langle \xi_d, h_d\rangle) h_i(\tau).$$

Using this formula we obtain for fixed $i = 1,\dots,d$

$$\frac{\delta}{\delta \xi_i(\tau)}\Big(\Gamma(\theta_T)\delta_\epsilon(B_{t_2} - B_{t_1})\Big)(\theta_\tau \xi)$$
$$= \frac{-\mathbf{1}_{[0,T]}(t_1)\mathbf{1}_{[t_1,t_2\wedge T]}(\tau)}{(2\pi)^{d/2}(t_2 - t_1 + \epsilon)^{d/2+1}} \int_{t_1}^{\tau} \xi_i du \ \exp\Big\{-\frac{(\int_{t_1}^{\tau} \xi du)^2}{2(t_2 - t_1 + \epsilon)}\Big\}$$

we have now to compute the inverse of the S-transform of the last expression. For each $1 \le i \le d$, denote by

$$G_i(\xi) = \int_{t_1}^{\tau} \xi_i du \ \exp\Big\{-\frac{(\int_{t_1}^{\tau} \xi du)^2}{2(t_2 - t_1 + \epsilon)}\Big\}$$

so

$$G_i(\xi) = \sqrt{\tau - t_1}\langle h, \xi_i\rangle \exp\Big\{-\frac{(\tau - t_1)\langle h, \xi_i\rangle^2}{2(t_2 - t_1 + \epsilon)}\Big\} \prod_{j\neq i} \exp\Big\{-\frac{(\tau - t_1)\langle h, \xi_j\rangle^2}{2(t_2 - t_1 + \epsilon)}\Big\}$$

where $h = \frac{\mathbf{1}_{[t_1,\tau]}}{\sqrt{\tau - t_1}}$, then $S^{-1}G_i$ is also a product of functions, each of them depend only of $\langle \omega_j, h\rangle$.

Lemma 3.1. *Let $\|h\|_{L^2(\mathbb{R})} = 1$, $\eta \in S(\mathbb{R})$ and $0 < c < 1/2$. Let E and F be two functions of η defined by*

$$E(\eta) = \exp\Big(-c\langle h, \eta\rangle^2\Big)$$

and

$$F(\eta) = \langle h, \eta\rangle \exp\Big(-c\langle h, \eta\rangle^2\Big)$$

where $\langle h, \eta\rangle = \int h(t)\eta(t)dt$. Then

$$S^{-1}E(\omega) = \frac{1}{\sqrt{1-2c}} \exp\Big(\frac{c}{2c-1}\langle \omega, h\rangle^2\Big)$$

and

$$S^{-1}F(\omega) = \frac{\langle \omega, h\rangle}{(1-2c)^{3/2}} \exp\Big(\frac{c}{2c-1}\langle \omega, h\rangle^2\Big). \qquad \square$$

Using this lemma, we obtain for $i = 1, \ldots, d$

$$m_i^\epsilon(\tau) = \frac{\mathbf{1}_{[0,T]}(t_1)\mathbf{1}_{[t_1,t_2\wedge T]}(\tau)}{(2\pi)^{d/2}(\epsilon+t_2-\tau)^{d/2+1}}\left(B_\tau^i - B_{t_1}^i\right)e^{-\frac{(B_\tau-B_{t_1})^2}{2(\epsilon+t_2-\tau)}}$$

and finally we obtain the desired formula for $M_{T,c}^\epsilon$

$$M_{T,c}^\epsilon = \frac{1}{(2\pi)^{d/2}}\int_0^{+\infty} dt_2 \int_0^{T\wedge t_2} dt_1 \int_{t_1}^{T\wedge t_2} dB_\tau$$
$$\frac{\left(B_\tau - B_{t_1}\right)}{(\epsilon+t_2-\tau)^{d/2+1}}\exp\left\{-\frac{(B_\tau - B_{t_1})^2}{2(\epsilon+t_2-\tau)}\right\}.$$

We proceed analogously for R_T^ϵ, and the theorem is proved. □

References

[1] De Faria M., Hida T., Streit L. and Watanabe H.: Intersection local times as generalized white noise functionals, *Acta Appl. Math.* **46** (1997) 351–362.

[2] De Faria M., Oliveira M. J., Streit L.: Generalized Clark-Ocone formula, *Random Oper. Stochastic Equations* **8** (2000) 163–174.

[3] Deck Th., Potthoff J., Vage G.: A Review of White Noise Analysis from a Probabilistic Standpoint. *Acta Appl. Math.* **48** (1997) 91–112.

[4] Drumond C., de Faria M., Streit L.: The renormalization of self intersection local times I: The chaos expansion, *Infin. Dimens. Anal. Quantum Probab. Relat. Top.* **3** (2000) 223–236.

[5] Drumond C., de Faria M., Streit L.: The square of self intersection local time of Brownian motion. Stochastic processes, physics and geometry: new interplays, I (F. Gesztesy et al., eds.). CMS Conference Proceedings 28 (2000) 115–122.

[6] Dynkin E. B.: Polynomials of the occupation field and related random fields, *J. Funct. Anal.* **58** (1984) 20–52.

[7] Dynkin E. B.: Regularized self-intersection local times of planar Brownian motion, *Ann. Prob.* **16** (1988) 58–74.

[8] Hida T.: *Brownian Motion*, Springer, Berlin-Heidelberg- New York. 1980.

[9] Hida T., Kuo H-H., Potthoff J., Streit L.: *White Noise – An Infinite Dimensional Calculus*, Kluwer-Academic, 1993.

[10] Imkeller P., Yan J-A.: Multiple intersection local time of planar Brownian motion as a particular Hida distribution, *J. Funct. Anal.* **140** (1996) 256–273.

[11] Jenane Gannoun R.: PhD thesis, Tunis, 2001.

[12] Kuo H-H.: Donker's delta function as a generalized Brownian functional and its application, *Lecture Notes in Control and Information Sciences* **49** (1983) 167–178.

[13] Le Gall J.-F.: Sur le temps local d'intersection du mouvement Brownian plan et la méthode de renormalisation de Varadhan, Springer L.N.M. **1123** (1985) 314–331.

[14] Mendonça S., Streit L.: Multiple Intersection Local Time in Terms of White Noise, *Infin. Dimens. Anal. Quantum Probab. Relat. Top.* **4** (2001) 533–543.

[15] Nualart D., Vives J.: Chaos expansion and local times, *Publ. Mat.* **36** (1992) 827–836.

[16] Potthoff J., Timpel M.: On dual pair of spaces of smooth and generalized random variables, *Potential Analysis* **4** (1995) 637–654.

[17] Rosen J.: A local time approach to the self-intersections of Brownian paths in space, *Comm. Math. Phys.* **88** (1983) 327–338.

[18] Rosen J.: Tanaka's formula and renormalization for intersections of planar Brownian motion. Ann. Prob. 14 (1986) 1245–1251.

[19] Yor M.: Renormalisation et convergence en loi pour les temps locaux d'intersections du mouvement brownien dans $\mathbb{R}^3$. Springer L.N.M. **1123** (1985) 350–365.

[20] Varadhan S.R.S.: Appendix to "Euclidean quantum field theory" by Szymanzik K., in: R. Jost ed., *Local Quantum Theory*, Academic Press, New York. 1969.

[21] Watanabe H.: The local time of self-intersections of Brownian motions as generalized Brownian functionals. *Lett. Math. Phys.* **23** (1991) 1–9.

ITÔ FORMULA FOR GENERALIZED FUNCTIONALS OF BROWNIAN BRIDGE

Yuh-Jia Lee
Department of Applied Mathematics, National University of Kaohsiung, Kaohsiung 811, Taiwan
yjlee@nuk.edu.tw

Chen-Chih Huang
Department of Mathematics, National Cheng-Kung University, Tainan, Taiwan
l1688101@dec4000.cc.ncku.edu.tw

Abstract Employing the calculus on the classical Wiener space $(\mathcal{C}', \mathcal{C})$ we represent the Brownian motion $\{B(t)\}$ by $B(t,x) = (x, \alpha_t)$ for $x \in \mathcal{C}$, where $(\cdot,\cdot)$ is the C–C^* pairing and α_t is a function in $\mathcal{C}^*$ such that $\alpha_t(s) = \min\{t, s\}$ for $t \in [0,1]$ and for $s < t$. It follows that Brownian bridge is represented by $X(t,x) = (x, \beta_t)$ for $x \in \mathcal{C}$, where $\beta_t = \alpha_t - t\alpha_1$. Using such a representation, we define and study the generalized functionals associated with the Brownian bridge. It is shown that Itô formula for Brownian bridge may be derived without using the classical stochastic integration theory. In order to compare the Itô formula of Brownian bridge with the formula under the scheme of semimartingale theory we also consider the semimartingale version of the Brownian bridge represented by $\widehat{X}(t,x) = (x, \widehat{\beta}_t)$, for $x \in \mathcal{C}$, where $\widehat{\beta}_t(s) = -(1-t)ln(1 - s \wedge t)$ for $t < 1$ and $\widehat{\beta}_1 \equiv 0$. Its is shown that the Itô formula depends only on the variance parameter $t(1-t)$ of the Brownian bridge.

Keywords: Itô formula, white noise, Brownian motion, generalized functions AMS-classification (2000): Primary 60J65, 60H40; Secondary 46G12, 26E15

S. Albeverio et al. (eds.),
Proceedings of the International Conference on Stochastic Analysis and Applications, 107–127.

1. Introduction

A Brownian bridge is a Gaussian process $\{X(t) : 0 \leq t \leq 1\}$ with mean zero and the covariance function given by

$$\mathrm{Cov}(X(t)X(s)) = \ s \wedge t - s\, t.$$

Employing the formulation of the white noise calculus on the abstract Wiener space $(\mathcal{C}', \mathcal{C})$ [1], where $\mathcal{C}$ denotes the space of continuous functions which are defined on [0,1] and vanish at 0 and $\mathcal{C}'$ is the space of Cameron-Martin functions, we define and study the generalized functionals of Brownian bridge. Let $(\cdot, \cdot)$ denote the C–C^* pairing and represent the Brownian motion $\{B(t)\}$ by $B(t, x) = (x, \alpha_t)$ for $x \in \mathcal{C}$, where and α_t is a function in $\mathcal{C}^*$ defined by $\alpha_t(s) = t \wedge s = \min\{t, s\}$ for $s < t$ and $t \in [0, 1]$. Then we represent the Brownian bridge by $X(t, x) = (x, \beta_t)$ for $x \in \mathcal{C}$, where $\beta_t = \alpha_t - t\alpha_1$. Apply this representation, we develop the calculus of functionals associated with the Brownian bridge. In order to compare the Itô formula derived by the scheme of semimartingale theory, we also employ the semimartingale version of Brownian bridge which may be represented as a Brownian functional in the following form

$$\widehat{X}(t, x) = (x, \widehat{\beta}_t),$$

where $\widehat{\beta}_t(s) := 1_{[t,1]} - ln(1 - s \wedge t)$ and $\widehat{\beta}_1 \equiv 0$. (see section 2). When the former representation is used, the noise derivative with respect to Brownian bridge is the Fréchet derivative in the direction of $\dot{\widehat{\beta}}_t$; when the later representation is used the noise derivative with respect to Brownian bridge is the Fréchet derivative in the direction of $\dot{\widehat{\beta}}_t$. Since $\dot{\beta}_t$, $\dot{\widehat{\beta}}_t \in L_2[0, 1] \setminus \mathcal{C}$, the differential calculus of functionals associated the Brownian bridge can not be performed on the classical Wiener space. To overcome this difficulty, we replace the abstract Wiener pair $(\mathcal{C}', \mathcal{C})$ by a sequence of abstract Wiener pairs $(\mathcal{C}', E_p)$, where

$$\begin{aligned} E = \cap_{p=1}^{\infty} E_p \subset E_p \subset E_q \subset E_1 = L_2^* \subset \mathcal{C}^* \subset \mathcal{C}' \\ \subset \mathcal{C} \subset L_2 = E_{-1} \subset E_{-q} \subset E_{-p} \subset E^* \end{aligned}$$

for $p \geq q \geq 1$ and where L_2 denotes the space $L^2[0, 1]$. Let $\mathcal{E}_p$ be the space of exponential type entire functions on $E_{-p,c}$, the complexification of E_{-p} and the set $\mathcal{E}_\infty = \cap_1^\infty \mathcal{E}_p$ as test functionals. Then the calculus of generalized Wiener functionals is performed as linear functionals on $\mathcal{E}_\infty$.

In this paper, we first describe the definition of Brownian bridge on probability space in section 2, then in section 3 we construct a Gel'fand triple on the alternative classical Wiener space $(\mathcal{C}', L_2[0,1])$ which serves as the underlying probability space in our investigation, there the representation of Brownian bridge as a Wiener functional is given. In section 4 we briefly introduce the theory of generalized white noise functionals developed from [9]. Our main results of this paper is given in section 5 discussing the generalized Itô formula for Brownian bridge and its connection with the stochastic integration.

Notations

1 For a real linear space V, V_c denotes the complexification of V.

2 For each continuous function x on $[0,1]$, $\dot{x}(t)$, $\ddot{x}(t)$, $x^{(p)}(t)$ are respectively the first, second, and p-th order derivatives of x with respect to $t \in [0,1]$, if they exist.

3 For a n-linear operator T on a complex normed space X, $Tz^n := T(z,\ldots,z)$ (n copies) for each $z \in X$.

4 For any Hilbert space H, $\|\cdot\|_{HS^n(H)}$ denotes the n-linear Hilbert-Schmidt operator norm and $\|\cdot\|_{\mathcal{L}^n(H)}$ the n-linear operator norm over H.

2. The Brownian Bridge

We first review briefly the background concerning the Brownian bridge. For the detailed description, we refer the reader to [2, 3, 4].

2.1 Two versions of Brownian Bridge

Let $(\Omega, \Im, \mathcal{P})$ be a complete probability space. A Brownian bridge $X = \{X(t) : 0 \le t \le 1\}$ on $(\Omega, \Im, \mathcal{P})$ is a Gaussian process with mean zero and having the covariance function $\mathrm{Cov}(X(t)X(s))$ given by

$$\mathrm{Cov}(X(t)X(s)) = s \wedge t - st.$$

Let $B = \{B(t) : 0 \le t \le 1\}$ be the standard one-dimensional Brownian motion on $(\Omega, \Im, \mathcal{P})$. Then it is easy to see that $X(t) = B(t) - tB(1)$, $0 \le t \le 1$, is a representation of the Brownian bridge on $(\Omega, \Im, \mathcal{P})$. $X = \{X(t)\}$ is not adapted, not to say that it is a semimartingale. However it can be shown that a semimartingale version $\widehat{X}(t)$ of Brownian bridge may be defined by

$$\widehat{X}(t) = \int_0^t \frac{1-t}{1-s}\, dB(s) \quad \text{if } t < 1 \text{ and } \quad \widehat{X}(1) = 0$$

(see for example [17]). $\{\widehat{X}(t) : 0 \le t \le 1\}$ is a canonical representation of the Brownian bridge X relative to the Brownian motion B. Sometimes, one call $\widehat{X}$ the adapted Brownian bridge. Apply the semimartingales theory we obtain the Itô formula

$$f(\widehat{X}(b)) - f(\widehat{X}(a)) = \int_a^b f'(\widehat{X}(t))\, d\widehat{X}(t) + \frac{1}{2}\int_a^b f''(\widehat{X}(t))\, dt$$

The above formula will be derived by white noise calculus as shown in section 5.

2.2 The Brownian bridge as a Wiener functional

In this section, we shall define the functional representation of the Brownian bridge on the classical abstract Wiener space. For the related discussion of analysis on the abstract Wiener space, we refer the reader to [6, 7, 8, 9, 18].

Let $\mathcal{C}$ be the collection of real-valued continuous functions x which is defined on [0,1] and satisfies $x(0)$=0 and $\mathcal{C}'$ the subclass of $\mathcal{C}$ consisting of absolutely continuous function x whose derivative $\dot{x}$ satisfies

$$\int_0^1 |\dot{x}(t)|^2\, dt < \infty.$$

Then $\mathcal{C}$ is a Banach space with the sup-norm $|\cdot|_\infty$ and $\mathcal{C}'$ is a Hilbert space with norm $|\cdot|_0 = \sqrt{\langle\cdot,\cdot\rangle_0}$ and the inner product $\langle\cdot,\cdot\rangle_0$ defined by

$$\langle x, y\rangle_0 := \int_0^1 \dot{x}(t)\dot{y}(t)\, dt$$

for $x, y \in \mathcal{C}'$. The space $\mathcal{C}'$ is usually called the Cameron-Martin space and it is well known that $(\mathcal{C}', \mathcal{C})$ forms an abstract Wiener space (AWS, for abbreviation). The Wiener measure μ is then realized as the abstract Wiener measure with variance parameter $t = 1$. It is easy to see that the pair $(\mathcal{C}', L_2)$ is also an abstract Wiener space, where $L_2 = L^2[0,1]$ with the $|\cdot|_2$-norm. The Wiener measure μ is then extended to a measure, still denoted by μ, in such way that for any Borel subset E of L_2, $\mu(E) = \mu(E \cap \mathcal{C})$.

The dual space $\mathcal{C}^*$ of $\mathcal{C}$ may be identified as a dense subspace of $\mathcal{C}'$ given as follows:

$$\mathcal{C}^* = \{x \in \mathcal{C}' : \dot{x} \text{ of bounded variation, right continuous, } \dot{x}(1) = 0\}.$$

Under this identification, the $\mathcal{C}$–$\mathcal{C}^*$ pairing is given by

$$(x, y) = -\int_0^1 x(t)\, d\dot{y}(t), \qquad x \in \mathcal{C},\ y \in \mathcal{C}^*. \tag{2.1}$$

To define the functional representation of the Brownian bridge on $(\mathcal{C}, \mathcal{B}(C), \mu)$, first we note that there exist an function $\alpha_t \in \mathcal{C}^*$ such that $B(t; \cdot) = (\cdot, \alpha_t)$, $0 \le t \le 1$, represents the standard Brownian motion on $(\mathcal{C}, \mathcal{B}(\mathcal{C}), \mu)$, where $\alpha_t(s) = s \wedge t$ for $s < t$, $t \in [0,1]$ and $(\cdot, \cdot)$ is the $\mathcal{C}$–$\mathcal{C}^*$ pairing (see [9]), then the Brownian bridge $\{X(t) : t \in [0,1]\}$ on $(\mathcal{C}, \mathcal{B}(\mathcal{C}), \mu)$ can be represented by

$$X(t, x) = (x, \beta_t) \tag{2.2}$$

for every $x \in \mathcal{C}$, where $\beta_t = \alpha_t - t\alpha_1$, with $\dot{\alpha}_1 = \mathbf{1}_{[0,1)}$. Moreover we have

Lemma 2.1. *$d/dt\, \beta_t = h_t$ in L_2, where $h_t(s) = 1_{[t,1]}(s) - s$ for $s \in [0,1]$.*

The representation of Brownian bridge in (2.2) is not adapted with respect to the filtration $\sigma\{B(u) : 0 \le u \le t\}$, not to say that it is a semimartingale. As the stochastic integral with respect to the Brownian bridge is concerned. It is desirable to replace the above representation by an adapted one. A desired representation is given in the following

Lemma 2.2. *Let $\widehat{\beta}_t(s) = (t-1)\ln(1 - s \wedge t)$, for $s \in [0,1]$ and $0 \le t < 1$; and let $\widehat{\beta}_1 \equiv 0$. Then we have*

(a) *$\widehat{\beta}_t \in \mathcal{C}^*$ for $t \in [0,1]$; and the adapted Brownian bridge $\{X(t) : t \in [0,1]\}$ can be represented by*

$$X(t, x) = (x, \widehat{\beta}_t), \quad \text{for every } x \in \mathcal{C}. \tag{2.3}$$

(b) *$d/dt\, \widehat{\beta}_t = h_t$ in L_2, where $h_t(s) = 1_{[t,1]}(s) + \ln(1 - s \wedge t)$ for $s \in [0,1]$, and for $0 \le t < 1$.*

Since $h_t \in L_2 \backslash C$ and since the derivative in the direction of h_t will play an essential role in the study of stochastic integrals with respect to the Brownian bridge, we are led to consider the functionals defined on the alternative AWS $(\mathcal{C}', L_2)$. Observe that the identify mapping is a continuous embedding from $\mathcal{C}$ into L_2, one can identify the dual space ${L_2}^*$ of L_2 as a subspace of $\mathcal{C}^*$ in the sense that, for $y \in {L_2}^*$ and $x \in C$, then $(x, y)_2 = (x, y)$, where $(\cdot, \cdot)_2$ denotes the pairing of L_2–${L_2}^*$. Furthermore, ${L_2}^*$ can be identified as a dense subspace of $\mathcal{C}^*$ as follows:

$$\begin{aligned} {L_2}^* = \{x \in \mathcal{C}^* : \dot{x} \text{ is absolutely continuous with} \\ \ddot{x} \in L_2,\ \dot{x}(1) = 0, \text{ and } \dot{x}(0) = 0\}. \end{aligned}$$

By the above identification, we have

$$(x, y)_2 = -\int_0^1 x(t)\, \ddot{y}(t)\, dt$$

for $x \in L_2$ and $y \in L_2{}^*$.

To conclude this section, we summary the above results into a Lemma for future applications.

Lemma 2.3. (a) *If we identify the dual space of $\mathcal{C}'$ by itself, then $\mathcal{C}^*$ and $L_2{}^*$ can be identified as subspace of $\mathcal{C}'$. Then we have the following inclusive relations in which the smaller space are densely embedded in the larger space.*

$$L_2{}^* \subset \mathcal{C}^* \equiv \mathcal{C}' \subset \mathcal{C} \subset L_2.$$

(b) *Let $\gamma_t = \beta_t$ or $\widehat{\beta}_t$. Then the Brownian bridge can be represented by*

$$X(t,x) = (x,\, \gamma_t), \qquad \text{for all } x \in \mathcal{C}$$

or

$$X(t,x) = \langle x,\, \gamma_t\rangle, \qquad \text{for } w\text{-almost all } x \in L_2,$$

where β_t is defined as in either (2.2) *or* (2.3).
(c) *For $x \in \mathcal{C}'$ and $y \in L_2{}^*$, the following identity holds:*

$$(x,\, y) \;=\; (x,\, y)_2 = \;\langle x,\, y\rangle_0.$$

3. A Gel'fand triple associated with the classical Wiener space

Let K be a bounded linear operator on $L^2[0,1]$ defined by

$$Kx(t) = \int_0^1 (s \wedge t)\, x(s)\, ds, \quad x \in L^2[0,1].$$

Then K is a positive self-adjoint compact operator having a complete orthonormal basis (CONS, for abbreviation) consisting of functions $\{e_n(t) = \sqrt{2}\,\sin(n - 1/2)\pi t \,:\, n = 1,2,\dots\}$ with the corresponding eigenvalues given by $\{\,1/((n-1/2)\pi)^2 : n = 1,2,\dots\}$. For $n \in \mathbb{N}$, let

$$f_n(t) = \sqrt{K}\, e_n(t), \quad t \in [0,1]. \tag{3.1}$$

Observe that, for each $n \in \mathbb{N}$ and for $t \in [0,1]$, $e_n(1-t) = (-1)^{n+1}\, \dot{f}_n(t)$, then $\{f_n : n \in \mathbb{N}\} \subset \mathcal{C}^*$. Moreover, $\{f_n : n \in \mathbb{N}\}$ is also a CONS of $L^2[0,1]$. Thus, $\{f_n : n \in \mathbb{N}\}$ is a CONS of $\mathcal{C}'$. Let A be the inverse operator of $\sqrt{K}$ on $\mathcal{C}'$. Then A is self-adjoint densely defined on $\mathcal{C}'$ and $A\, f_n = \lambda_n\, f_n$ for each $n \in \mathbb{N}$, where $\lambda_n = (n - 1/2)\,\pi$

For any $p \geq 0$, let E_p be the domain of A^p. Then E_p is a real Hilbert space with the norm $|x|_p = |A^p x|_0$ ($\mathcal{C}' = E_0$) and $\{(1/\lambda_n^p)\, f_n : n \in \mathbb{N}\}$

forms a CONS of E_p. The increasing family $\{|\cdot|_p : p \geq 0\}$ of norms are compatible and comparable; and the embedding from $E_{p+\alpha}$ into E_p is of Hilbert-Schmidt type whenever $\alpha > 1/2$. Next, let E_{-p} be the completion of $\mathcal{C}'$ with respect to the norm $|x|_{-p} = |A^{-p}\, x|_0$. Then E_{-p} is a Hilbert spaces with a CONS $\{\lambda_n^p\, f_n : n \in \mathbb{N}\}$. Identify $x \in E_p^*$ with the element $\sum_{n=1}^{\infty} (x,\, f_n)\, f_n$ in E_{-p}, where $(\cdot,\cdot)$ is the E_p^*–E_p pairing, then E_{-p} becomes the dual space of E_p. Set $E = \cap_{p\geq 0} E_p$ and endow on E with the projective limit topology induced by E_p's. Then E is a nuclear space with the dual $E^* = \cup_{p>0} E_{-p}$ and $E \subset \mathcal{C}' \subset E^*$ forms a Gel'fand triple.

Observe that, for $x \in \mathcal{C}'$ and $p \geq 1$,

$$|x|_{-p}^2 = \sum_{n=1}^{\infty} \lambda_n^{-2p}\, \langle x,\, f_n\rangle_0^2 = \int_0^1 x(t)^2\, dt \leq |x|_{\infty}^2.$$

It follows that $\mathcal{C} \subset E_{-p}$ for all $p \geq 1$. Further more, by the denseness of $\mathcal{C}'$ in $\mathcal{C}$ and in E_{-p}, we have the following chain of continuous inclusion:

$$\begin{aligned} E \subset E_p \subset E_q \subset E_1 = L_2^* \subset \mathcal{C}^* \subset \mathcal{C}' \\ \subset \mathcal{C} \subset L_2 = E_{-1} \subset E_{-q} \subset E_{-p} \subset E^*, \end{aligned}$$

where $p \geq q \geq 1$ and L_2 denotes the space $L^2[0,1]$. For notational convenience, we will use the notation $(\cdot,\cdot)$ to stand for all the dual pairings of E^*–E, E_{-p}–E_p $(p \geq 1)$, and $\mathcal{C}$–$\mathcal{C}^*$.

For $p \geq 1$, $(\mathcal{C}', E_{-p})$ is an AWS. Let μ and μ_{-p}, $p \geq 1$, be the abstract Wiener measures of $\mathcal{C}$ and E_{-p} respectively. Then the measurable support of μ_{-p} is contained in $\mathcal{C}$ and , for any integrable complex-valued function φ on (E_{-p}, μ_{-p}), the restriction of φ on $\mathcal{C}$ is $\mathcal{B}(\mathcal{C})$-measurable and

$$\int_{E_{-p}} \varphi(x)\, \mu_{-p}(dx) = \int_{\mathcal{C}} \varphi(x)\, \mu(dx).$$

E_p may be identified as a subspace of $\mathcal{C}'$ given as follows.

Proposition 3.1. *For $p \in \mathbb{N}$, E_p is the class consisting of all functions $x \in \mathcal{C}'$ satisfying: (i) $\dot{x}, \ddot{x}, \ldots, x^{(p)}$ are absolutely continuous with $x^{(p+1)} \in L^2[0,1]$ and (ii) $x^{(2k)}(0) = x^{(2k+1)}(1) = 0$ for $k = 0, 1, \ldots, [p/2]$. Moreover,*

$$(x,\, y) = -\int_0^1 x(t)\, d\dot{y}(t) \quad \text{for } x \in L_2 \text{ and } y \in E_p. \tag{3.2}$$

Corollary 3.2. *The nuclear space E consists of all real-valued almost everywhere differentiable functions x defined on $[0,1]$ such that $x^{(2k)}(0) = x^{(2k+1)}(1) = 0$ for each $k \in \mathbb{N} \cup \{0\}$.*

4. The Spaces of test and generalized functions

If V is a real normed space with the $|\cdot|_V$-norm, then the $|\cdot|_{V_c}$-norm of V_c is given by $|x+iy|_{V_c} = \sup\{\|e^{i\theta}(x+i\,y)\|_{V_c} : \theta \in [0,2\pi]\}$ for $x, y \in V$, where $\|x+i\,y\|_{V_c}^2 = |x|_V^2 + |y|_V^2$. In general, $\|\cdot\|_{V_c}$ is a quasi-norm and for any $x, y \in V$, we have the inequality:

$$\|x+i\,y\|_{V_c} \le |x+i\,y|_{V_c} \le \sqrt{2}\,\|x+i\,y\|_{V_c}.$$

Note that when V is a Hilbert space, the quasi-norm $\|\cdot\|_{V_c}$ coincides with the $|\cdot|_{V_c}$-norm.

In this section, we construct test functionals on the AWS $(\mathcal{C}', \mathcal{C})$ and the AWS $(\mathcal{C}', E_{-p})$ for $p \ge 1$. For notational simplicity, we use the symbols $|\cdot|_\infty$ and $|\cdot|_{-p}$ to stand also for $|\cdot|_{\mathcal{C}_c}$ and $|\cdot|_{E_{-p,c}}$, respectively. For a fixed Banach space B which is either $\mathcal{C}$ or $\mathcal{E}_{-p}$, let $\mathcal{E}(B)$ be the class of those functions φ defined on $\mathcal{C}$ so that φ has an analytic extension $\widetilde{\varphi}(z)$ to B_c and satisfies the exponential growth condition: $|\varphi(z)| \le c\exp\{c'|z|_{B_c}\}$ for some constants $c, c' > 0$. It is clear that $\mathcal{E}(B) \subset L^2(\mathcal{C}, \mu)$ by the Fernique theorem (see [6]). For $m \in \mathbb{N}$ and $\varphi \in \mathcal{E}(B)$, define

$$\|\varphi\|_{\mathcal{E}_m(B)} = \sup\{|\widetilde{\varphi}(z)|\, e^{-m\,|z|_{B_c}} : z \in B_c\}.$$

Let $\mathcal{E}_m(B) = \{\varphi \in \mathcal{E}(B) : \|\varphi\|_{\mathcal{E}_m(B)} < +\infty\}$. Then $\{(\mathcal{E}_m(B), \|\cdot\|_{\mathcal{E}_m(B)})\}$ is an increasing sequence of Banach spaces and $\mathcal{E}(B) = \cup_{m\in\mathbb{N}}\, \mathcal{E}_m(B)$. Endow $\mathcal{E}(B)$ with the inductive limit topology induced by the family $\{\mathcal{E}_m(B)\}$. Then $\mathcal{E}(B)$ becomes a locally convex topological algebra. For notational simplicity, let $\mathcal{E}_{m,p} = \mathcal{E}_m(E_p), \mathcal{E}_{m,0} = \mathcal{E}(\mathcal{C}), \mathcal{E}_p = \mathcal{E}(E_{-p})$ and $\mathcal{E}_0 = \mathcal{E}(\mathcal{C})$. Then we have the following chain of continuous inclusion: as $p \ge q \ge 1$

$$\mathcal{E}_\infty = \cap_{p\ge 1}\, \mathcal{E}_p \subset \mathcal{E}_p \subset \mathcal{E}_q \subset \mathcal{E}_0 \subset L^2(\mathcal{C}, \mu),$$

where $\mathcal{E}_\infty$ is topologized as the projective limit of $\{\mathcal{E}_p\}$. Thus a sequence $\{\varphi_n\}$ in $\mathcal{E}_\infty$ converges to φ in $\mathcal{E}_\infty$ iff for any $p \ge 1$, φ_n converges to φ in $\mathcal{E}_p$ as $n \to \infty$. $\mathcal{E}_\infty$ will serve as the space of test functions for the generalized functions in our investigation.

For $f \in L^2(\mathcal{C}, \mu)$, the Wiener-Itô decomposition theorem assures that f can be decomposed into an orthogonal direct sum of multiple Wiener integrals $I_n(f)$ of order n, $n \in \mathbb{N}$. Let $\mu * f$ be the convolution of f and μ defined on $\mathcal{C}'$, i.e., $\mu * f(h) = \int_{\mathcal{C}} f(h+y)\,\mu(dy)$ for $h \in \mathcal{C}'$. Then $\mu * f$ is infinitely Fréchet-differentiable in the directions of $\mathcal{C}'$ and $I_n(f)$ can be represented by

$$I_n(f) = L^2(\mathcal{C}, \mu)\text{-}\lim_{k\to\infty} \frac{1}{n!} \int_{\mathcal{C}} D^n \mu * f(0)(P_k(\cdot) + i\,P_k(y))^n\, \mu(dy)$$

(see [9, 10]), where D^n denotes the n-th Fréchet derivative in the directions of $\mathcal{C}'$ and $P_k(z) = \sum_{j=1}^k (z, f_j)\, f_j$ for $z \in E_c^*$. Moreover,

$$\int_{\mathcal{C}} |f(x)|^2 \, \mu(dx) = \sum_{n=0}^{\infty} \frac{1}{n!} \, \|D^n \mu * f(0)\|^2_{HS^n(\mathcal{C}')}.$$

Next, we briefly describe the space (E) of test functionals, which was introduced by Meyer and Yan (see [14, 16]). For $m \in \mathbb{N}$ and $p \geq 1$, define the $\|\cdot\|_{m,p}$-norm on $L^2(\mathcal{C}, \mu)$ by

$$\|f\|^2_{m,p} = \sum_{n=0}^{\infty} \frac{1}{m^{2n}} \, \|D^n \mu * f(0)\|^2_{HS^n(E_{-p})}. \tag{4.1}$$

Denote by $(E)_{m,p}$ the class of functions f in $L^2(\mathcal{C}, \mu)$ such that $\|f\|_{m,p} < +\infty$. Then $\{(E)_{m,p} : m \in \mathbb{N}\}_m$ is an increasing sequence of Hilbert spaces of with the inner products induced by the $\|\cdot\|_{m,p}$-norms. Let $(E)_p$ be the inductive limit of $\{(E)_{m,p}\}$ and let $(E) = \bigcap_{p \geq 1} (E)_p$ be the projective limit of $(E)_p$. Then the following chain of continuous inclusions hold:

$$(E) \subset (E)_p \subset (E)_q \subset L^2(\mathcal{C}, \mu) \quad \text{whenever } p \geq q \geq 1.$$

Obviously, each member f of (E) satisfies

$$\sum_{n=0}^{\infty} \frac{1}{n!} \, \|D^n \mu * f(0)\|^2_{HS^n(E_{-p})} < +\infty \quad \forall p \geq 1.$$

According to the work by Lee [11, 14], there exists a unique analytic function $\widetilde{f}$ defined on E_c^* such that $\widetilde{f} = f$ μ-almost everywhere on $\mathcal{C}_c$ and

$$\widetilde{f}(z) = \sum_{n=0}^{\infty} \frac{1}{n!} \int_{\mathcal{C}} D^n \mu * \widetilde{f}(z + i\, y)^n \, \mu(dy), \quad z \in E_c^*,$$

where the series converges absolutely and uniformly on bounded subsets of E_c^*. $\widetilde{f} \in \mathcal{E}_\infty$ and the space $\mathcal{E}_\infty$ is exactly the collection of analytic version of members of (E). We reformulate the related results and growth estimates as follows.

Theorem 4.1 ([14, 15]). (i) *Let f be in (E). For any $p \geq 1$, let $m_p \in \mathbb{N}$ so that $f \in (E)_{m_p,p}$. Then*

$$\|\widetilde{f}\|_{\mathcal{E}_{m_p,p}} \leq c_{m_p} \, \|f\|_{\mathcal{E}_{m_p,p}}, \tag{4.2}$$

where $c_m = \int_{\mathcal{C}} e^{m|y|_\infty} \, \mu(dy)$ for any $m \in \mathbb{N}$.

(ii) *Let φ be in $\mathcal{E}_\infty$. For any $p \geq 1$, let $m_p \in \mathbb{N}$ so that $\varphi \in \mathcal{E}_{m_p}(E_{-p})$. Let r be a real number such that $r > \ln 2/(2\ln\pi - 2\ln 2)$, and choose $\widetilde{m}_p$ be the least integer greater than $\sqrt{m_p e}$. Then*

$$\|\varphi\|_{\widetilde{m}_p, p-r} \leq \beta_{m_p,p} \|\varphi\|_{\mathcal{E}_{m_p,p}}, \tag{4.3}$$

where $\beta_{m_p,p}$ is a constant depending only on p and m_p.

(iii) *The mapping $\varphi \to \mu * \varphi$ is a homeomorphism from $\mathcal{E}_\infty$ onto $\mathcal{E}_\infty$. In fact, let φ be assumed as above. Then*

$$(1/c_k)\, \|\varphi\|_{\mathcal{E}_{\widetilde{m_p,p}}} \leq \|\mu * \varphi\|_{\mathcal{E}_{m_p,p}} \leq c_{m_p}\, \|\varphi\|_{\mathcal{E}_{m_p,p}}, \tag{4.4}$$

where k is the least integer greater than e^{m_p}.

(iv) *$\mathcal{E}_\infty \subset (E)$ and $\mathcal{E}_\infty = \widetilde{(E)}$ as vector spaces, where $\widetilde{(E)} = \{\widetilde{f} : f \in (E)\}$.*

Proof. The statements (i), (ii), and (iv) follow from [14, Theorem 4.1 and Theorem 4.4]. For the statement (iii), it is obvious that $\|\mu * \varphi\|_{\mathcal{E}_{m_p,p}} \leq c_{m_p}\, \|\varphi\|_{\mathcal{E}_{m_p,p}}$. On the other hand, take an arbitrary $\psi \in \mathcal{E}_\infty$ and define

$$f(z) = \sum_{n=0}^{\infty} \frac{1}{n!} \int_{\mathcal{C}} D^n\psi(0)(z + i\,y)^n\, \mu(dy), \quad z \in E_c^*.$$

Then $\mu * f = \psi$. For any $p \geq 1$, let $m_p \in \mathbb{N}$ so that $\psi \in \mathcal{E}_{m_p,p}$. Then, by the Cauchy integral formula, $|D^n\psi(0)z^n| \leq e^{m_p n}\, \|\psi\|_{\mathcal{E}_{m_p,p}}\, |z|_{-p}^n$ for each $n \in \mathbb{N}$. Thus, for any $z \in E_{-p,c}$,

$$\begin{aligned} |f(z)| &\leq \sum_{n=0}^{\infty} \frac{1}{n!} \int_{\mathcal{C}} |D^n\psi(0)(z + i\,y)^n|\, \mu(dy) \\ &\leq \|\psi\|_{\mathcal{E}_{m_p,p}} \int_{\mathcal{C}} \exp\{e^{m_p}\,(|z|_{-p} + |y|_{-p})\}\, \mu(dy). \end{aligned}$$

Dividing both sides of the above estimation by $e^{k|z|_{-p}}$, we obtain (4.4). □

Example 4.2. (1) For $f_1, f_2 \ldots, f_n \in \mathcal{C}'$, define

$$: \widetilde{f}_1 \ldots \widetilde{f}_n : (x) = \int_{\mathcal{C}} \prod_{j=1}^{n} (x + i\,y,\, f_j)\, \mu(dy).$$

Then $: \widetilde{f}_1 \ldots \widetilde{f}_n : \;\in L^2(\mathcal{C}, \mu)$. It is worth to note that

$$I_n(f_1 \otimes \ldots f_n) =: \widetilde{f}_1 \ldots \widetilde{f}_n : .$$

Moreover, for $\eta_1, \ldots, \eta_n \in E_c$, we have $: \widetilde{\eta}_1, \ldots, \widetilde{\eta}_n : \in \mathcal{E}_\infty$.

(2) The exponential vector functional $\varepsilon(\eta)$ associated with $\eta \in E_c$ which is given by

$$\varepsilon(\eta) = \exp\Big\{(\cdot, \eta) - \frac{1}{2}\int_0^1 \dot{\eta}(t)^2\, dt\Big\}.$$

Then $\varepsilon(\eta) \in \mathcal{E}_\infty$.

For more examples, we refer the reader to [9].

Remark 4.3. **(a)** Identify (E) with $\mathcal{E}_\infty$, then Theorem 4.1 implies that two families of norms $\{\|\cdot\|_{m,p}\}$ and $\{\|\cdot\|_{\mathcal{E}_{m,p}}\}$ are equivalent. In other words, the space $\mathcal{E}_\infty$ is equivalent to the Yan-Meyer space (E).

(b) Let f_{j_i}'s be functions as given in (3.1). Then for any $\varphi \in \mathcal{E}_\infty$, the series

$$\sum_{n=0}^{\infty} \frac{1}{n!}\Big\{\sum_{j_1,\dots,j_n=1}^{\infty} D^n\mu * \varphi(0)(f_{j_1},\dots,f_{j_n}) : \widetilde{f}_{j_1}\dots\widetilde{f}_{j_n} : (z)\Big\},$$

converges to φ in $\mathcal{E}_\infty$ and, for any $p \geq 1$, the series also converges absolutely and uniformly on each bounded set of $E_{-p,c}$. As a consequence, the linear space $\mathcal{P}$ spanned by all cylinder polynomials of the form $\phi((\cdot,\eta_1),\dots,(\cdot,\eta_n))$ for any $n \in \mathbb{N}\cup\{0\}$ is dense in $\mathcal{E}_\infty$ and in $\mathcal{E}_p$ for $p \geq 1$, where $\phi(x_1,\dots,x_n)$ is a n-variable polynomial with complex coefficients. However, it is not clear to us if $\mathcal{P}$ is dense in $\mathcal{E}_0$.

Denote the dual spaces of $\mathcal{E}_0$, $\mathcal{E}_p$ $(p \geq 1)$, and $\mathcal{E}_\infty$ respectively by $\mathcal{E}_0^*$, $\mathcal{E}_p^*$, and $\mathcal{E}_\infty^*$ which are topolozied by the weak*-topologies. Then we have the following chain of continuous inclusions:

$$\mathcal{E}_\infty \subset \mathcal{E}_p \subset \mathcal{E}_q \subset \mathcal{E}_0 \subset L^2(\mathcal{C},\mu) \subset \mathcal{E}_0^* \subset \mathcal{E}_q^* \subset \mathcal{E}_p^* \subset \mathcal{E}_\infty^*, \text{ for } p \geq q \geq 1.$$

Members of $\mathcal{E}_\infty^*$ will be referred as the generalized Wiener functionals .

Next, let $\langle\langle\cdot,\cdot\rangle\rangle_0$, $\langle\langle\cdot,\cdot\rangle\rangle_p$, and $\langle\langle\cdot,\cdot\rangle\rangle_\infty$ stand for the dual pairing of $\mathcal{E}_0^*$–$\mathcal{E}_0$, $\mathcal{E}_p^*$–$\mathcal{E}_p$, and $\mathcal{E}_\infty^*$–$\mathcal{E}_\infty$, respectively.

Example 4.4 ([8, 9]). Let $(\mathcal{C}', B)$ be an AWS with $B = \mathcal{C}$ or $B = E_{-p}$, $p = 0, 1, 2, \dots$ and let $|\cdot|_B$ denote the corresponding norm.

(a) Denote by $L^1_{\exp}(B,\mu)$ the space of all measurable functions f defining on $(B, \mathcal{B}(B))$ such that

$$\int_B f(x)\, e^{m\,|x|_B}\, \mu(dx) \quad \text{for all } m \in \mathbb{N}.$$

By the Fernique theorem, $L^1_{\exp}(B,\mu)$ can be regarded as a subspace of $\mathcal{E}(B)^*$ by identifying each $f \in L^1_{\exp}(B,\mu)$ with the functional G_f defined by

$$\langle\langle G_f, \varphi \rangle\rangle = \int_B f(x)\,\varphi(x)\,\mu(dx),$$

for $\varphi \in \mathcal{E}(B)$. Members of $L^1_{\exp}(B,\mu)$ will be called regular generalized Wiener functionals on B. For more examples we refer the reader to [9].

(b) Let ν be a Borel measure defined on B so that, for $m \in \mathbb{N}$,

$$\int_B e^{m|x|_B}\nu(dx) < \infty.$$

Then ν generates a generalized Wiener functional, still denoted by ν, defined by

$$\langle\langle \nu, \psi \rangle\rangle = \int_B \psi(x)\,\nu(dx)$$

for all $\psi \in \mathcal{E}(B)$

(c) Let $h \in \mathcal{C}'$. Then $\widetilde{h} \in L^2(\mu) \subset \mathcal{E}^*_\infty$ and

$$\langle\langle \widetilde{h}, \psi \rangle\rangle = \int_B \langle h, x\rangle\,\psi(x)\,\mu(dx) = (D\mu\psi(0), h),$$

where $D\mu\psi(0)$ denotes the Fréchet derivative of $\mu\psi$ at 0. For a proof see [8]. The above identity still make sense even for $h \in B$, this define the functional $\widetilde{h}$ as a generalized functional for $h \in B$.

(d) Let $h \in \mathcal{C}'$ and $\alpha \in \mathbb{C}$, then we have

$$\int_B e^{\alpha\langle x,\,h\rangle}\varphi(x)\,\mu(dx) = e^{\frac{1}{2}\alpha^2|h|_{\mathcal{C}'}{}^2}\mu\varphi(\alpha h).$$

Note that the term $\mu\varphi(\alpha h)$ in the above identity still makes sense even for $h \in B$ and the mapping $\varphi \to \mu\varphi(\alpha h)$ is continuous. The term

$$e^{-\frac{1}{2}\alpha^2|h|_{\mathcal{C}'}{}^2}e^{\alpha\langle x,\,h\rangle}$$

defines a generalized functional, denoted customarily by $: e^{\alpha\widetilde{h}} :$. This functional is called a multiplicative renormalization of $e^{\alpha\widetilde{h}}$. This lead us to the following definition

$$\langle\langle : e^{\alpha\widetilde{h}} :, \varphi \rangle\rangle := \mu\varphi(\alpha h).$$

Denote the dual space of (E) by $(E)^*$ which is endowed with the weak*-topology and let $\langle\langle \cdot, \cdot \rangle\rangle$ denote the dual pairing of $(E)^*$ and (E).

Definition 4.5. The S-transform SF of $F \in \mathcal{E}_\infty^*$ is defined as a complex-valued functional on E_c by

$$SF(\eta) = \langle\!\langle F, \varepsilon(\eta)\rangle\!\rangle_\infty, \qquad \eta \in E_c.$$

We remark that as $F \in L^2(\mathcal{C}, \mu)$, $SF = \mu * F$.

Since (E) is dense in $(E)_p$ for any $p \geq 1$, by [5, Theorem 6, pp 290], $(E)^* = \cup_{p\geq 1} (E)_p^*$ which is topologized by the inductive limit topology. This implies that if $\widetilde{F} \in (E)^*$, then there exists $p \geq 1$ so that $\widetilde{F} \in (E)_p^*$, the dual of $(E)_p$. Since $(E)_p^* = \cap_{m\geq 1}(E)_{m,p}^*$, we have the following

Proposition 4.6 ([14]). *Let $\widetilde{F}$ be in $(E)^*$. Then there exists $p \geq 1$ so that for any $m \in \mathbb{N}$,*

$$\sum_{n=0}^{\infty} \frac{m^{2n}}{(n!)^2} \|D^n S\widetilde{F}(0)\|_{HS^n(E_p)}^2 < +\infty. \tag{4.5}$$

According to Theorem 4.1, the embedding $j : \mathcal{E}_\infty \to (E)$ is continuous. So that, for any $G \in (E)^*$, $G \circ j \in \mathcal{E}_\infty^*$. Thus $(E)^*$ can be identified as a subspace of $\mathcal{E}_\infty^*$. Conversely, for a fixed $F \in \mathcal{E}_\infty^*$, define a functional $\widetilde{F}$ on (E) by

$$\langle\!\langle \widetilde{F}, f\rangle\!\rangle := \langle\!\langle F, \widetilde{f}\rangle\!\rangle_\infty, \quad f \in (E),$$

where $\widetilde{f}$ is the analytic version of f on E_c^*. It follows from Theorem 4.1 that $\widetilde{F} \in (E)^*$.

Recall that a functional G on E_c is analytic if it satisfies the following two conditions:

(A-1): for all $\eta, \phi \in E_c$, the mapping $\mathbb{C} \ni \lambda \mapsto G(\eta + \lambda\phi)$ is entire on $\mathbb{C}$.

(A-2): there exists $p \geq 1$ such that for any $c > 0$,

$$\sup\{|G(\eta)| : \eta \in E_c, |\eta|_p \leq c\} < +\infty.$$

The following theorem characterizes the generalized functionals in $\mathcal{E}_\infty^*$ in terms of its S-transforms.

Theorem 4.7 ([15]). *Let $F \in \mathcal{E}_\infty^*$ be fixed. Then the S-transform SF of F is an analytic function on E_c such that the number*

$$n_{m,-p}(F) := \sum_{n=0}^{\infty} \frac{m^{2n}}{(n!)^2} \|D^n SF(0)\|_{HS^n(E_p)}^2. \tag{4.6}$$

is finite.

Conversely, suppose that G is an analytic function defined on E_c *and satifies the condition* (A-2) *for some* $p \geq 1$*. Let* q *be sufficiently large so that* $e^2 \cdot \sum_{j=1}^{\infty} \lambda_j^{-2(q-p)} < 1$.

Then there exists a unique $F \in \mathcal{E}_\infty^*$ *such that* $n_{m,-q}(F) < +\infty$ *and* $SF = G$*, where* $\lambda_j = (j - (1/2))\pi$ *for* $j \in \mathbb{N}$.

Proposition 4.8. (i) $SF \equiv 0$ *iff* $F \equiv 0$ *for* $F \in \mathcal{E}_\infty^*$.

(ii) $\mathcal{E}_\infty$ *is a dense subspace of* $\mathcal{E}_\infty^*$.

Proof. The first assertion (i) follows from Theorem 4.7. For the second assertion (ii), let $F \in \mathcal{E}_\infty^*$. By Theorem 4.7, there exists $p \geq 1$ so that $n_{m,-p}(F)$ is finite for each $m \in \mathbb{N}$. Let

$$F_k(z) = \sum_{n=0}^{k} \frac{1}{n!} \int_C D^n SF(0)(P_k(z) + i\, P_k(y))^n \,\mu(dy), \quad z \in E_{p,c}, \tag{4.7}$$

for any $k \in \mathbb{N}$. Then F_k's are all in $\mathcal{E}_\infty$. Take an arbitrary $\varphi \in \mathcal{E}_\infty$. Then $\varphi \in \mathcal{E}_m(E_{-p-r})$ for some $m \in \mathbb{N}$; and by Theorem 4.1, $\|\varphi\|_{\widetilde{m},p} < +\infty$, where r and $\widetilde{m}$ are defined as in (4.3). Then, by Theorem 4.7 and the Cauchy-Schwarz inequality, we obtain

$$|\langle\!\langle F_k - F, \varphi \rangle\!\rangle_\infty| = |\langle\!\langle F_k - F, \varphi \rangle\!\rangle| \leq \|\varphi\|_{\widetilde{m},p} \cdot n_{\widetilde{m},-p}(F_k - F) \to 0.$$

Thus, $F_k \to F$ in $\mathcal{E}_\infty^*$ as $k \to \infty$. It implies that $\mathcal{E}_\infty$ is dense in $\mathcal{E}_\infty^*$. □

4.1 Basic operations for generalized Brownian functionals

In this subsection, let B denote the space $\mathcal{C}$ or E_{-p} in order to simplify the notations.

Multiplication by functions. Let ψ be a fixed but arbitrary function in $\mathcal{E}(B)$. Then the mapping $\varphi \to \varphi\psi$ is continuous on $\mathcal{E}(B)$. This fact leads to define ψF as follows.

Definition 4.9. Let $F \in \mathcal{E}$ and $\varphi \in \mathcal{E}$. Define

$$\langle\!\langle \psi F, \varphi \rangle\!\rangle := \langle\!\langle F, \psi\varphi \rangle\!\rangle,$$

for all $\varphi \in \mathcal{E}(B)$.

Fourier-Wiener Transform. For a measurable function on B and for $\alpha, \beta \in \mathbb{C}$, define

$$\mathcal{F}_{\alpha,\beta} f(y) = \int_B f(\alpha x + \beta y)\, \mu(dx) \tag{4.8}$$

provided that the integral on the right hand side of (4.8) exists (see [8]).

Proposition 4.10 ([8]). **(a)** *For* $f \in \mathcal{E}^m(B)$, *we have*

$$\|\mathcal{F}_{\alpha,\beta} f\|_{m\beta^*} \leq \Big(\int_B e^{m|\alpha||x|_B}\, p_1(dx)\Big)\|f\|_m,$$

where

$$\beta^* = \min\{k : \; k \textit{ an integer and } k \geq \beta\}.$$

(b) $\mathcal{F}_{\alpha,\beta}(\mathcal{E}^m(B)) \subset \mathcal{E}^{m\beta^*}(B)$ *and the mapping* $f \to \mathcal{F}_{\alpha,\beta} f$ *is continuous on* $\mathcal{E}(B)$.

Differentiation.

Definition 4.11. Let $F \in \mathcal{E}(B)^*$. Then

(a) For any $x \in B$, define

$$\langle\!\langle D_x^* F,\, \varphi\rangle\!\rangle := \langle\!\langle F,\, D_x\varphi\rangle\!\rangle$$

for all $\varphi \in \mathcal{E}(B)$, where

$$D_x\varphi(y) = (x,\, D\varphi(y)) = D\varphi(y)x \quad (x, y \in B).$$

(b) For any $z \in B^*$ and for any $\varphi \in \mathcal{E}(B)$, define

$$\langle\!\langle D_z F, \varphi\rangle\!\rangle := \langle\!\langle F,\, \widetilde{z}\varphi\rangle\!\rangle - \langle\!\langle F,\, D_z\varphi\rangle\!\rangle$$

for all $\varphi \in \mathcal{E}(B)$.

Let $h_t = \frac{d}{dt}\,\beta_t$ or $h_t = \frac{d}{dt}\,\widehat{\beta}_t$. Then $h_t \in L_2$ (hence in E_{-p} for $p \in \mathbb{N}$). Define $\partial_t = D_{h_t}$. For $F \in \mathcal{E}_p^*$ and for $\varphi \in \mathcal{E}_p$, define

$$\langle\!\langle \partial_t^* F,\, \varphi\rangle\!\rangle := \langle\!\langle F,\, \partial_t\varphi\rangle\!\rangle.$$

Then $D_x^* F \in \mathcal{E}^*, D_z F \in \mathcal{E}^*$ and $\partial_t^* F$ are members of $\mathcal{E}_p^*$.

5. Generalized Itô Formula for the Brownian Bridge

For $f \in \mathcal{S}$, the Schwartz space of rapidly decreasing functions on $\mathbb{R}$, and a nonzero $h \in \mathcal{C}'$, we have

$$f(\langle x,\, h\rangle) = \frac{1}{\sqrt{2\pi}} \int_{-\infty}^{\infty} e^{i\langle x,\, h\rangle y}\, \widehat{f}(y)\, dy,$$

where $\widehat{f}(y) = \frac{1}{\sqrt{2\pi}} \int_{-\infty}^{+\infty} f(u)\, e^{-iuy}\, du$ for $y \in \mathbb{R}$. Moreover, for $\psi \in \mathcal{E}_p$, we see that

$$
\begin{aligned}
\langle\!\langle f(\langle x,\, h\rangle),\, \psi\rangle\!\rangle &= \int_C f(\langle x,\, h\rangle)\, \psi(x)\, p_1(dx) \\
&= \frac{1}{\sqrt{2\pi}} \int_C \int_{-\infty}^{\infty} e^{i\langle x,\, h\rangle y}\, \widehat{f}(y)\, \psi(x)\, p_1(dx) dy \\
&= \frac{1}{\sqrt{2\pi}} \int_{-\infty}^{\infty} e^{-\frac{1}{2}y^2 |h|_0^2} \mathcal{F}_{1,i}\, \psi(y\, h)\, \widehat{f}(y)\, dy \\
&= (G_{h,\psi},\, \widehat{f}) \\
&= (\widehat{G_{h,\psi}},\, f).
\end{aligned} \tag{5.1}
$$

where $G_{h,\psi}(y) = \frac{1}{\sqrt{2\pi}} \mathcal{F}_{1,i}\, \psi(y\, h) e^{-\frac{1}{2}y^2|h|^2}$ and $(\cdot, \cdot)$ denotes the $\mathcal{S}$–$\mathcal{S}'$ pairing. It is easy to see that $G_{h,\psi} \in \mathcal{S}$. It follows from a routine argument that the mapping $\psi \mapsto \widehat{G_{\beta_t,\psi}}$ is continuous on $\mathcal{E}_p$.

Let $\{X(t) : 0 \leq t \leq 1\}$ be the Brownian bridge on $(\mathcal{C}, \mathcal{B}(C), \mu)$. We express $X(t, x) = (x, \beta_t)$ and let $h_t = \frac{d}{dt}\beta_t$ in L_2 for $0 < t < 1$. Then, for $f \in \mathcal{S}'$, the dual space of $\mathcal{S}$, $f(X(t))$ can be regarded as a generalized function by the following

Definition 5.1. For $f \in \mathcal{S}'$, we define $\langle\!\langle F(X(t)),\, \psi\rangle\!\rangle := (\widehat{G}_{\beta_t,\psi},\, F)$ for $\psi \in \mathcal{E}$, where $\widehat{G_{\beta_t,\psi}}$ is defined as in (5.1) and $0 < t < 1$.

Theorem 5.2 (Generalized Itô formula). *Let $\partial_t = D_{h_t}$ and $\partial_t{}^* = D^*_{h_t}$. Then, for $f \in \mathcal{S}'$, we have*

$$
f(X(b)) = f(X(a)) + \int_a^b \partial_t{}^* f'(X(t)) dt + \frac{1}{2} \int_a^b (1-2t) f''(X(t)) dt, \tag{5.2}
$$

which exists in the generalized sense, where $0 < a < b < 1$.

Proof. To prove the identity (5.2), it is equivalent to prove that the identity

$$
\frac{d}{dt} f(X(t)) = \partial_t{}^* f'(X(t)) + \frac{1}{2}(1 - 2t) f''(X(t)) \tag{5.3}
$$

holds in $\mathcal{E}^*$ for $0 < t < 1$. To verify the identity (5.3), we first note that

$$
\varepsilon^{-1}(\beta_{t+\varepsilon} - \beta_t) \to h_t \quad \text{in } \mathcal{S}'.
$$

Then, for $\psi \in \mathcal{E}_\infty$, we have

$$\begin{aligned}
&\frac{d}{dt}\langle\!\langle f(X(t)), \psi\rangle\!\rangle \\
&= \lim_{\epsilon\to 0}\left\langle\!\!\left\langle \frac{f(X(t+\epsilon) - f(X(t)))}{\epsilon}, \psi \right\rangle\!\!\right\rangle \\
&= \lim_{\epsilon\to 0}\left(\frac{G_{\beta_{t+\epsilon},\psi} - G_{\beta_t,\psi}}{\epsilon}, f\right) \\
&= \lim_{\epsilon\to 0}\frac{1}{\epsilon}\int_{t+0\wedge\epsilon}^{t+0\vee\epsilon}\left(\frac{d}{ds}G_{\beta_s,\psi}, \widehat{f}\right) ds \\
&= \left(\frac{d}{dt}\widehat{G}_{\beta_t,\psi}, f\right) \\
&= \left(\frac{d}{dt}\Big(\frac{1}{2\pi}\int_{-\infty}^{\infty}\mathcal{F}_{1,i}\psi(y\beta_t)e^{-\frac{1}{2}y^2t(1-t)}e^{-iy\xi}dy\Big), f_{[\xi]}\right) \\
&= \left(\frac{1}{2\pi}\int_{-\infty}^{\infty} -\frac{1}{2}(1-2t)y^2e^{-\frac{1}{2}y^2t(1-t)}\mathcal{F}_{1,i}\psi(y\beta_t)e^{-iy\xi}dy, f_{[\xi]}\right) \\
&\quad + \left(\frac{1}{2\pi}\int_{-\infty}^{\infty} e^{-\frac{1}{2}y^2t(1-t)}(iy)[\mathcal{F}_{1,i}D_{h_t}\psi(y\beta_t)]e^{-iy\xi}dy, f_{[\xi]}\right) \\
&= \frac{1}{2}(1-2t)\left(\frac{1}{2\pi}\int_{-\infty}^{\infty} e^{-\frac{1}{2}y^2t(1-t)}\mathcal{F}_{1,i}\psi(y\beta_t)e^{-iy\xi}dy, {f_{[\xi]}}''\right) \\
&\quad + \left(\frac{1}{2\pi}\int_{-\infty}^{\infty} e^{-\frac{1}{2}y^2t(1-t)}[\mathcal{F}_{1,i}D_{h_t}\psi(y\beta_t)]e^{-iy\xi}dy, {f_{[\xi]}}'\right) \\
&= \frac{1}{2}(1-2t)\langle\!\langle f''(X(t)), \psi\rangle\!\rangle + \langle\!\langle f'(X(t)), \partial_t\psi\rangle\!\rangle \\
&= \langle\!\langle {\partial_t}^* f'(X(t)), \psi\rangle\!\rangle + \frac{1}{2}(1-2t)\langle\!\langle f''(X(t)), \psi\rangle\!\rangle.
\end{aligned}$$

where $(\cdot,\cdot)$ denotes the $\mathcal{S}$–$\mathcal{S}'$ pairing. This proves the identity (5.3). □

Remark 5.3. In view of the argument of the above proof, the identity (5.2) depends only on the property that $|h_t|_2^2 = t(1-t)$, and is independent of the choice of β_t.

Let $\{Y_t : a \le t \le b\}, 0 < a < b < 1$ be a continuous $\mathcal{E}(C)^*$-valued process. We define

$$\int_a^b Y_t\, dX(t+) := \lim_{|\Gamma|\to 0}\sum_{j=1}^{n}(\widetilde{\beta}_{t_j} - \widetilde{\beta}_{t_{j-1}})Y_{t_{j-1}} \text{ in } \mathcal{E}_\infty^*$$

provided that the limit exists, where $\Gamma = \{a = t_0 < t_1 < \cdots < t_n = b\}$.

Proposition 5.4. *Let $f \in C^2(\mathbb{R})$ such that $f' \in \mathcal{S}'$. Then*

$$\int_a^b \partial_t{}^* f'(X(t))dt = \int_a^b f'(X(t))dX(t+) + \int_a^b tf''(X(t))dt$$

in $\mathcal{E}(L_2)^*$, where $0 < a < b < 1$.

Proof. Let $f \in C^2(R)$ and $\Gamma = \{a = t_0 < t_1 < ... < t_n = b\}$ be any partition. Then

$$\begin{aligned}
&\sum_{j=1}^n f'(X(t_{j-1}))(X(t_j) - X(t_{j-1})) \\
&= \sum_{j=1}^n (\widetilde{\beta}_{t_j} - \widetilde{\beta}_{t_{j-1}}) f'(\widetilde{\beta}_{t_{j-1}}) \\
&= \sum_{j=1}^n D_{\beta_{t_j} - \beta_{t_{j-1}}} f'(\widetilde{\beta}_{t_{j-1}}) + \sum_{j=1}^n D^*_{\beta_{t_j} - \beta_{t_{j-1}}} f'(\widetilde{\beta}_{t_{j-1}}) \text{ in } \mathcal{E}^*. \qquad (5.4)
\end{aligned}$$

Observe that

$$\begin{aligned}
&\sum_{j=1}^n D_{\beta_{t_j} - \beta_{t_{j-1}}} f'(\widetilde{\beta}_{t_{j-1}}) \\
&= \sum_{j=1}^n D_{\beta_{t_j}} f'(\widetilde{\beta}_{t_{j-1}}) - D_{\beta_{t_{j-1}}} f'(\widetilde{\beta}_{t_{j-1}}) \\
&= \sum_{j=1}^n \left(f''(\widetilde{\beta}_{t_{j-1}}) \langle \beta_{t_j}, \beta_{t_{j-1}} \rangle_0 - f''(\widetilde{\beta}_{t_{j-1}}) \langle \beta_{t_{j-1}}, \beta_{t_{j-1}} \rangle_0 \right) \\
&= \sum_{j=1}^n f''(\widetilde{\beta}_{t_{j-1}})(t_{j-1} - t_j t_{j-1} - t_{j-1} + t_{j-1}^2) \\
&= -\sum_{j=1}^n f''(\widetilde{\beta}_{t_{j-1}}) t_{j-1}(t_j - t_{j-1}) \qquad (5.5)
\end{aligned}$$

Then, for all $x \in \mathcal{C}$:

$$\lim_{|\Gamma| \to 0} \sum_{j=1}^n D_{\beta_{t_j} - \beta_{t_{j-1}}} f'(\widetilde{\beta}_{t_{j-1}}(x)) = -\int_a^b tf''(X(t;x))dt \qquad (5.6)$$

On the other hand, let $\varphi \in \mathcal{E}_m$ for some $m \in \mathbb{N}$,

$$\begin{aligned}
&\sum_{j=1}^n \langle\langle D^*_{\beta_{t_j}-\beta_{t_{j-1}}} f'(\widetilde{\beta}_{t_{j-1}}), \varphi\rangle\rangle \\
&= \sum_{j=1}^n \langle\langle f'(\widetilde{\beta}_{t_{j-1}}), D_{\beta_{t_j}-\beta_{t_{j-1}}} \varphi\rangle\rangle \\
&= \sum_{j=1}^n \left\langle\!\!\left\langle f'(\widetilde{\beta}_{t_{j-1}}), \int_{t_{j-1}}^{t_j} \partial_t \varphi \, dt \right\rangle\!\!\right\rangle \\
&= \sum_{j=1}^n \langle\langle f'(\widetilde{\beta}_{t_{j-1}}), (t_{j-1} - t_j)\partial_{t_{j-1}}\varphi\rangle\rangle && (5.7) \\
&\quad + \sum_{j=1}^n \left\langle\!\!\left\langle f'(\widetilde{\beta}_{t_{j-1}}), \int_{t_j-1}^{t_j} (\partial_t \varphi - \partial_{t_{j-1}}\varphi) \, dt \right\rangle\!\!\right\rangle. && (5.8)
\end{aligned}$$

Since the process $\{f'(\widetilde{\beta}_t) : a \leq t \leq b\}$ is a continuous $\mathcal{E}(\mathcal{C})^*$-valued process, for any $m \geq 1$ there exists an $M > 0$ such that

$$\sup_{t\in[a,b]} \|f'(\widetilde{\beta}_t)\|_{\mathcal{E}_m(\mathcal{C})^*} = M < +\infty.$$

Let $\phi(t) = \langle\langle f'(\widetilde{\beta}_t), \partial_t \varphi\rangle\rangle$, $t \in [a, b]$. Then we have

$$\begin{aligned}
|\phi(t) - \phi(t')| &\leq |\langle\langle f'(\widetilde{\beta}_t) - f'(\widetilde{\beta}'_t), \partial_t\varphi\rangle\rangle| \\
&\quad + |\langle\langle f'(\widetilde{\beta}_t) - f'(\widetilde{\beta}_{t'}), \partial_t\varphi - \partial_{t'}\varphi\rangle\rangle| \\
&\quad + |\langle\langle f'(\widetilde{\beta}_t), \partial_t\varphi - \partial_{t'}\varphi\rangle\rangle| \\
&\leq |\langle\langle f'(\widetilde{\beta}_t) - f'(\widetilde{\beta}'_t), \partial_t\varphi\rangle\rangle| + 3Me^m \|\varphi\|_m |h_t - h'_t|_2 \\
&\quad \to 0, \text{ as } t' \to t.
\end{aligned}$$

Thus ϕ is a continuous functions on $[a, b]$. It implies that the sum in (5.7) tends to $\int_a^b \langle\langle \partial_t^* f'(\widetilde{\beta}_t), \varphi\rangle\rangle \, dt$ as $|\Gamma| \to 0$. For the sum in (5.8),

$$\begin{aligned}
&\left| \sum_{j=1}^n \left\langle\!\!\left\langle f'(\widetilde{\beta}_{t_{j-1}}), \int_{t_j-1}^{t_j} (\partial_t\varphi - \partial_{t_{j-1}}\varphi) \, dt \right\rangle\!\!\right\rangle \right| \\
&\leq M\|\varphi\|_m \, e^m \sum_{j=1}^n \int_{t_{j-1}}^{t_j} |h_t - h_{t_{j-1}}|_2 \, dt \\
&= M\|\varphi\|_m \, e^m \sum_{j=1}^n \int_{t_{j-1}}^{t_j} \sqrt{(t - t_{j-1}) - (t - t_{j-1})^2} \, dt \to 0 \text{ as } |\Gamma| \to 0.
\end{aligned}$$

Therefore

$$\lim_{|\Gamma|\to 0}\sum_{j=1}^{n}\langle\langle D^*_{\beta_{t_j}-\beta_{t_{j-1}}} f'(\widetilde{\beta}_{t_{j-1}}),\varphi\rangle\rangle = \int_a^b \langle\langle \partial_t^* f'(\widetilde{\beta}_t),\varphi\rangle\rangle\, dt. \qquad (5.9)$$

Combining (5.3) with (5.6), (5.9), and letting $|\Gamma|\to 0$ in (5.4), we obtain that

$$\int_a^b f'(X(t))dX(t+) = \int_a^b \partial_t{}^* f'(X(t))dt - \int_a^b tf''(X(t))dt \quad \text{in } \mathcal{E}^*.$$

We complete the proof. □

Remark 5.5. If $X(t)=\langle\cdot,\widehat{\beta}_t\rangle$ is a adapted representation of the Brownian bridge, the $\int_a^b f'(X(t))\,dX(t+)$ is exactly the stochastic integral of the integrand $f'(X(t))$ with the semimartingale $\{X(t) : a\le t\le b\}$.

By Theorem 5.2 and Proposition 5.4, we see that for $f\in\mathcal{C}^2(\mathbb{R})$ with $f'\in\mathcal{S}'$,

$$f(X(b)) = f(X(a)) + \int_a^b f'(X(t))dX(t+) + \frac{1}{2}\int_a^b f''(X(t))dt$$

which exists in the generalized sense. It is exactly the classical Itô formula as shown in section 2.

Example 5.6. Let $f(x)=x^2/2$. Then, by Theorem 5.2,

$$\frac{1}{2}X^2(b) - \frac{1}{2}X^2(a) = \int_a^b \partial_t{}^* X(t)dt + \frac{1}{2}\int_a^b (1-2t)dt \text{ in } \mathcal{E}(L_2)^*.$$

or

$$\int_a^b \partial_t{}^* X(t)dt = \frac{1}{2}X^2(b) - \frac{1}{2}X^2(a) - \frac{1}{2}(b-a) + \frac{1}{2}(b^2-a^2).$$

In fact, we can directly compute $\int_a^b \partial_t{}^* X(t)dt$ as follows.

$$\begin{aligned}
\left\langle\!\!\left\langle \int_a^b \partial_t{}^*\widetilde{\beta}_t dt,\ \psi\right\rangle\!\!\right\rangle &= \int_a^b \langle\langle \partial_t{}^*\widetilde{\beta}_t,\ \psi\rangle\rangle\, dt\\
&= \int_a^b \langle\langle \widetilde{\beta}_t,\ \partial_t{}^*\psi\rangle\rangle\, dt = \int_a^b\int_C \langle D^2\psi(x)\beta_t, h_t\rangle\, w(dx)dt\\
&= \int_C\int_a^b \frac{d}{dt}\frac{1}{2}(D^2\psi(x)\beta_t{}^2)dt\ w(dx)\\
&= \frac{1}{2}\int_C (D^2\psi(x)\beta_b{}^2 - D^2\psi(x)\beta_a{}^2)\, w(dx)\\
&= \frac{1}{2}\int_C [\langle x,\beta_b\rangle^2 - \langle\beta_b,\beta_b\rangle_0]\psi(x) - [\langle x,\beta_a\rangle^2 - \langle\beta_a,\beta_a\rangle_0]\psi(x) w(dx)\\
&= \left\langle\!\!\left\langle \frac{1}{2}(X^2(b) - b(1-b) - X^2(a) + a(1-a)),\ \psi\right\rangle\!\!\right\rangle
\end{aligned}$$

Acknowledgments

The first author was supported by the National Science of Taiwan, ROC

References

[1] Gross L.: Abstract Wiener Space, In "*Proceedings 5th Berkeley Symp. Math. Stat. Probab.*", vol.2, 31–42, 1965

[2] Hida T.: *Brownian Motion*, Springer-Verlag, 1980

[3] Kallianpur G.: *Stochastic Filtering Theory*, Springer-Verlag, 1980

[4] Klebaner F. C. : *Introduction to stochastic calculus with applications*, Imperial College Press, 1998

[5] Köthe G.: *Topological Vector Space I*, Springer-Verlag, New York/ Heidelberg/ Berlin, 1966

[6] Kuo H.-H.: *Gaussian Measures in Banach Spaces*, Lecture Notes in Math. Vol.463, Springer-Verlag, 1975

[7] Kuo H.-H.: *White Noise Distribution Theory*, CRC press, 1996.

[8] Lee Y.-J.: Generalized Functions on Infinite Dimensional Spaces and its Application to White Noise Calculus, *J. Funct. Anal.* **82** (1989) 429–464

[9] Lee Y.-J.: Generalized White Noise Functionals on Classical Wiener Space, *J. Korean Math. Soc.* **35**, No.3 (1998) 613–635

[10] Lee Y.-J.: On the convergence of Wiener-Itô decomposition, *Bull. Inst. Math. Acad. Sinica* **17** (1989) 305–312

[11] Lee Y.-J.: Analytic version of test functionals, Fourier transform and a characterization of measures in white noise calculus, *J. Funct. Anal.* **100** (1991) 359–380

[12] Lee Y.-J.: Positive generalized functions on infinite dimensional spaces, In "*Stochastic Process, a Festschrift in Honour of Gopinath Kallianpur*", Springer-Verlag, 1993, 225–234

[13] Lee Y.-J.: Convergence of Fock expansion and transformations of Brownian functionals, in "*Functional Analysis and Global Analysis*", Edited by T. Sunada and P.-W. Sy, Springer-Verlag, 1997, 142–156

[14] Lee Y.-J.: Integral representation of second quantization and its application to white noise analysis, *J. Funct. Anal.* **133** (1995) 253–276

[15] Lee Y.-J., Shih H.-H.: Generalized Clark formula on the classical Wiener space, preprint.

[16] Meyer P. -A., Yan J.-A. : Les "fonctions caractéristiques" des distributions sur l'espace de Wiener, in "*Sém. Probab. XXV*", Lecture Notes in Math., Vol. 1485, 1991, 61–78

[17] Revuz D., Yor M.: *Continuous Martingales and Brownian Motion*, Springer-Verlag, 1992

[18] Watanabe S.: *Lectures on Stochastic Differential Equations and Malliavin Calculus*, Tata Inst. of Fundamental Research, Bombay, 1984

FOCK SPACE OPERATOR VALUED MARTINGALE CONVERGENCE

Un Cig Ji
Department of Mathematics, Research Institute of Mathematical Finance, Chungbuk National University, Cheongju, 361-763 Korea
uncigji@chungbuk.ac.kr

Kyung Pil Lim
Department of Mathematics, Sogang University, Seoul, 121-742 Korea

Abstract In this paper we discuss the strong convergence of quantum martingale on (Boson) Fock space. In particular, the strong convergence of basic martingales are established.

1. Introduction

A non–commutative generalization of classical conditional expectation initiated by Umegaki is a norm one projection from a von Neumann algebra onto a von Neumann subalgebra and the notion of (non–commutative) martingale was also generalized. Several types of convergence (e.g., pointwise convergence, strong convergence, norm convergence, etc.) of martingales have been extensively studied ([6, 7, 8, 31], etc.). The notion of conditional expectation in the sense of Umegaki has been generalized by Accardi and Cecchini [1] by using the Tomita–Takesaki theory. The strong convergence of martingale of (generalized) conditional expectation in the sense of Accardi and Cecchini was proved by Hiai and Tsukada [11]. In [2], Accardi and Cecchini proved the martingale convergence of generalized conditional expectation in the weak topology.

Meanwhile, a non–commutative stochastic calculus of Itô type based on quantum Brownian motion has been formulated by Hudson and Parthasarathy [13]. Since then it has been extensively developed in [21, 27] and the references cited therein. In particular, the stochastic integral representation of (bounded) regular quantum martingale which

S. Albeverio et al. (eds.),
Proceedings of the International Conference on Stochastic Analysis and Applications, 129–143.

is quantum analogue of the classical Kunita and Watanabe theorem has been proved by Parthasarathy and Sinha [29]. In the recent paper [14], the notion of bounded regular martingale has been extended to unbounded martingale and then the stochastic integral representation of unbounded regular quantum martingale was proved based on the Fock riggings. The (nuclear) Fock riggings have widely been used in white noise theory [12, 18, 22] which give a new approach to quantum stochastic calculus [5, 23, 24, 25]. One of the benefit using Fock rigging in quantum stochastic calculus is that a (unbounded) quantum stochastic process can be considered as a process of bounded operators on Fock rigging.

In this paper we study the strong convergence of quantum martingale on Fock space which is motivated by a result in [28]. For the study of quantum martingales we adapt the Fock rigging used in [14] and for the limit of a quantum martingale we need more bigger space than the space in which the martingale lies, see Section 5.

This paper is organized as follows: In Section 2 we review the basic construction of riggings of Fock space. In Section 3 we recall the definition of thc Bargmann–Segal space after [10] (see, also, [15, 16]). In Section 4 we prove characterization of a continuous operator from Fock space into another Fock space (Theorem 4.1) which is a slight generalization of Theorem 5.2 in [15]. In Section 5 we study the strong convergence of quantum martingale on Fock space (Theorem 5.2). As an example, the strong convergence of basic martingales are established.

2. Construction of Riggings of Fock Space

Let $\mathcal{H}_{\mathbb{R}}$ be a real separable Hilbert space with norm $|\cdot|_0$ induced by the inner product $\langle\cdot,\cdot\rangle$. The complexification is denoted by $\mathcal{H}$ whose norm is denoted by the same symbol. The (Boson) Fock space $\Gamma(\mathcal{H})$ over $\mathcal{H}$ is the space of all sequence $\phi = (f_n)_{n=0}^{\infty}$, where f_n is an element of the n-fold symmetric tensor power $\mathcal{H}^{\widehat{\otimes} n}$ and

$$\|\phi\|_0^2 = \sum_{n=0}^{\infty} n!\,|\,f_n\,|_0^2 < \infty.$$

The canonical $\mathbb{C}$-bilinear form on $\Gamma(\mathcal{H})$ is denoted by $\langle\!\langle\cdot,\cdot\rangle\!\rangle$.

Let K be a selfadjoint operator in $\mathcal{H}$ with domain $\mathrm{Dom}(K)$ and assume that $K \geq 1$. For each $p \geq 0$ the dense subspace $\mathrm{Dom}(K^p) \subset \mathcal{H}$ becomes a Hilbert space, denoted by $\mathcal{D}_p \equiv \mathcal{D}_p(K)$, equipped with the norm

$$|\,\xi\,|_p = |\,K^p\xi\,|_0, \qquad \xi \in \mathrm{Dom}(K^p).$$

Since $K \geq 1$, we see that $|\xi|_p \leq |\xi|_q$ for $0 \leq p \leq q$ and thus we have the following riggings:

$$\begin{aligned} \mathcal{D}(K) \equiv \operatorname*{proj\,lim}_{p\to\infty} \mathcal{D}_p \subset \cdots \subset \mathcal{D}_q \subset \cdots \subset \mathcal{D}_p \subset \cdots \\ \cdots \subset \mathcal{D}_0 = \mathcal{H} \subset \cdots \subset \mathcal{D}_{-p} \subset \cdots \subset \mathcal{D}_{-q} \subset \cdots \subset \mathcal{D}(K)^*, \end{aligned} \tag{2.1}$$

where $q \geq p \geq 0$, $\mathcal{D}_{-p} \equiv \mathcal{D}_{-p}(K)$ is the completion of $\mathcal{H}$ with respect to the norm $|\xi|_{-p} = |K^{-p}\xi|_0$ and $\mathcal{D}^* \equiv \mathcal{D}(K)^*$ is the strong dual space of $\mathcal{D} \equiv \mathcal{D}(K)$. Then $\mathcal{D}$ becomes a countable Hilbert space. It is known from general theory that $\mathcal{D}^* = \operatorname{ind\,lim}_{p\to\infty} \mathcal{D}_{-p}$. With these riggings, for any $r \geq 0$ the operator K^r is naturally considered as an isometry from $\mathcal{D}_p$ onto $\mathcal{D}_{p-r}$, $p \in \mathbb{R}$.

We next construct a riggings of Fock space associated with the rigged Hilbert spaces (2.1). We set

$$\mathfrak{G}_{K;p} = \Gamma(\mathcal{D}_p), \qquad p \in \mathbb{R}.$$

By definition, the norm of $\mathfrak{G}_{K;p}$ is given as

$$\begin{aligned} \|\phi\|_p^2 &= \|\Gamma(K^p)\phi\|_0^2 \\ &= \sum_{n=0}^{\infty} n!\,|(K^{\otimes n})^p f_n|_0^2 = \sum_{n=0}^{\infty} n!\,|f_n|_p^2, \quad \phi = (f_n), \end{aligned} \tag{2.2}$$

where $\Gamma(K^p)$ is the second quantized operator of K. Then we have

$$\begin{aligned} \mathfrak{G}_K \equiv \operatorname*{proj\,lim}_{p\to\infty} \mathfrak{G}_{K;p} \subset \cdots \subset \mathfrak{G}_{K;q} \subset \cdots \subset \mathfrak{G}_{K;p} \subset \cdots \\ \cdots \subset \mathfrak{G}_{K;0} = \Gamma(\mathcal{H}) \subset \cdots \subset \mathfrak{G}_{K;-p} \subset \cdots \subset \mathfrak{G}_{K;-q} \subset \cdots \subset \mathfrak{G}_K^*, \end{aligned} \tag{2.3}$$

where $\mathfrak{G}_K$ becomes a countable Hilbert space equipped with the Hilbertian norms defined in (2.2), and $\mathfrak{G}_K^*$ is the strong dual space of $\mathfrak{G}_K$, or equivalently $\mathfrak{G}_K^* = \operatorname{ind\,lim}_{p\to\infty} \mathfrak{G}_{K;-p}$. The canonical $\mathbb{C}$-bilinear form on $\mathfrak{G}_K^* \times \mathfrak{G}_K$ which is also denoted by $\langle\!\langle \cdot\,,\ \cdot \rangle\!\rangle$, we have

$$\langle\!\langle \Phi, \phi \rangle\!\rangle = \sum_{n=0}^{\infty} n!\,\langle F_n, f_n \rangle, \qquad \Phi = (F_n) \in \mathfrak{G}_K^*, \quad \phi = (f_n) \in \mathfrak{G}_K,$$

where $\langle \cdot\,, \cdot \rangle$ is the canonical $\mathbb{C}$–bilinear form on $(\mathcal{D}^{\otimes n})^* \times \mathcal{D}^{\otimes n}$ for any n. Moreover, the Schwarz inequality takes the form:

$$|\langle\!\langle \Phi, \phi \rangle\!\rangle| \leq \|\Phi\|_{-p}\,\|\phi\|_p\,.$$

There are several interesting versions of the riggings of Fock space which are suitable to discuss special problems.

If $K = -\mathrm{d}^2/\mathrm{d}t^2 + t^2 + 1$, the Fock rigging (2.3) associated to K is equivalent to Hida–Kuo–Takenaka space [17] in white noise analysis [12, 18, 22]. The rigging (2.3) with $K = cI$ $(c > 1)$ is widely used to discuss several problems [3, 10, 30]; also [4, 19, 20], particularly, to discuss quantum martingales [14]. In this case, for notational convenience, we use $\mathfrak{G}$ and $\mathfrak{G}_p$ for $\mathfrak{G}_{cI}$ and $\mathfrak{G}_{cI;p}$, respectively for any $p \in \mathbb{R}$.

For each $\xi \in \mathcal{H}$ the associated vector

$$\phi_\xi = \left(1, \xi, \frac{\xi^{\otimes 2}}{2!}, \cdots, \frac{\xi^{\otimes n}}{n!}, \cdots\right)$$

in $\Gamma(\mathcal{H})$ is called an *exponential vector* or a *coherent vector*. Moreover, we have

$$\|\phi_\xi\|_p^2 = \sum_{n=0}^{\infty} \frac{1}{n!} \, |\, \xi \,|_p^{2n} = \mathrm{e}^{|\,\xi\,|_p^2}, \qquad p \in \mathbb{R}.$$

This implies that for any $p \in \mathbb{R}$ an exponential vector ϕ_ξ belongs to $\mathfrak{G}_{K;p}$ if and only if ξ belongs to $\mathcal{D}_p$. Therefore, ϕ_ξ belongs to $\mathfrak{G}_K$ (resp. $\mathfrak{G}_K^*$) if and only if ξ belongs to $\mathcal{D}$ (resp. $\mathcal{D}^*$). In fact, for any dense subset $\mathbf{D}$ of $\mathcal{D}$ the exponential vectors $\{\phi_\xi\,;\, \xi \in \mathbf{D}\}$ span a dense subspace of $\mathfrak{G}_K$, hence of $\mathfrak{G}_{K;p}$ for all $p \in \mathbb{R}$ and of $\mathfrak{G}_K^*$. Thus each $\Phi \in \mathfrak{G}_K^*$ is uniquely determined by its values for the exponential vectors. For $\Phi \in \mathfrak{G}_K^*$, the $\mathbb{C}$-valued function defined on $\mathcal{D}$ by

$$S\Phi(\xi) = \langle\!\langle \Phi, \phi_\xi \rangle\!\rangle\,, \qquad \xi \in \mathbf{D},$$

is called the *S-transform* of Φ. For $\Phi = (F_n)$ we have

$$S\Phi(\xi) = \sum_{n=0}^{\infty} n! \left\langle F_n, \frac{\xi^{\otimes n}}{n!} \right\rangle = \sum_{n=0}^{\infty} \langle F_n, \xi^{\otimes n} \rangle\,, \qquad \xi \in \mathbf{D}.$$

Let $\mathcal{L}(\mathfrak{X}, \mathfrak{Y})$ be the space of all bounded linear operators from a locally convex $\mathfrak{X}$ into another locally convex space $\mathfrak{Y}$. Then a continuous linear operator $\Xi \in \mathcal{L}(\mathfrak{G}_K, \mathfrak{G}_K^*)$ is uniquely determined by its matrix elements with respect to the exponential vectors. For $\Xi \in \mathcal{L}(\mathfrak{G}_K, \mathfrak{G}_K^*)$, we set

$$\widehat{\Xi}(\xi, \eta) = \langle\!\langle \Xi\phi_\xi, \phi_\eta \rangle\!\rangle\,, \qquad \xi, \eta \in \mathbf{D}.$$

This $\mathbb{C}$-valued function $\widehat{\Xi}$ defined on $\mathbf{D} \times \mathbf{D}$ is called the *symbol* of Ξ. There is a obvious relation between the symbol and the S-transform in the following way:

$$\widehat{\Xi}(\xi, \eta) = S(\Xi\phi_\xi)(\eta) = S(\Xi^*\phi_\eta)(\xi), \qquad \xi, \eta \in \mathbf{D}.$$

Then for each $r, s \in \mathbb{R}$ the S-transform of $\Phi \in \mathfrak{G}_{K;r}$ is uniquely extended to an entire function on $\mathcal{D}_{-r}$ and the symbol of $\Xi \in \mathcal{L}(\mathfrak{G}_{K;r}, \mathfrak{G}_{K;s})$ is uniquely extended to an entire function on $\mathcal{D}_r \times \mathcal{D}_{-s}$.

3. Bargmann–Segal Space

From now on, to recall the Bargmann–Segal space ([10, 16]), we assume that the operator K satisfies that $K(\mathrm{Dom}(K) \cap \mathcal{H}_{\mathbb{R}}) \subset \mathcal{H}_{\mathbb{R}}$ and there is a real nuclear space $\mathcal{N}_{\mathbb{R}}$ which is densely and continuously imbedded in $\mathcal{D} \cap \mathcal{H}_{\mathbb{R}}$ and is kept invariant under K. Then we have a real Gelfand triple:

$$\mathcal{N}_{\mathbb{R}} \subset \mathcal{H}_{\mathbb{R}} \subset \mathcal{N}_{\mathbb{R}}^*. \tag{3.1}$$

Let $\mu_{1/2}$ be the Gaussian measure on $\mathcal{N}_{\mathbb{R}}^*$ of which the characteristic function is given by

$$\mathrm{e}^{-\frac{1}{4}\langle \xi, \xi \rangle} = \int_{\mathcal{N}_{\mathbb{R}}^*} \mathrm{e}^{i\langle x, \xi \rangle} \mu_{1/2}(\mathrm{d}x). \qquad \xi \in \mathcal{N}_{\mathbb{R}},$$

where the canonical $\mathbb{C}$–bilinear form on $\mathcal{N}_{\mathbb{R}}^* \times \mathcal{N}_{\mathbb{R}}$ is denoted by $\langle \cdot, \cdot \rangle$ again. A probability measure ν on $\mathcal{N}^* = \mathcal{N}_{\mathbb{R}}^* + \mathrm{i}\mathcal{N}_{\mathbb{R}}^*$ is defined by

$$\nu(\mathrm{d}z) = \mu_{1/2}(\mathrm{d}x) \times \mu_{1/2}(\mathrm{d}y), \qquad z = x + \mathrm{i}\,y, \quad x, y \in \mathcal{N}_{\mathbb{R}}^*.$$

Then the probability space $(\mathcal{N}^*, \nu)$ is called the (standard) *complex Gaussian space* associated with (3.1).

The *Bargmann–Segal space*, denoted by $E^2(\nu)$, is defined as the space of entire functions $g \colon \mathcal{H} \to \mathbb{C}$ such that

$$\|g\|_{E^2(\nu)}^2 \equiv \sup_{P \in \mathcal{P}} \int_{\mathcal{N}^*} |g(Pz)|^2 \nu(\mathrm{d}z) < \infty,$$

where $\mathcal{P}$ is the set of all finite rank projections on $\mathcal{H}_{\mathbb{R}}$ with range contained in $\mathcal{N}_{\mathbb{R}}$. Note that $P \in \mathcal{P}$ is naturally extended to a continuous operator from $\mathcal{N}^*$ into $\mathcal{H}$ which is denoted by the same symbol. The Bargmann–Segal space $E^2(\nu)$ is a Hilbert space with norm $\|\cdot\|_{E^2(\nu)}$. For $\phi = (f_n)_{n=0}^{\infty} \in \Gamma(\mathcal{H})$ the following series:

$$\sum_{n=0}^{\infty} \langle z^{\otimes n}, f_n \rangle, \quad z \in \mathcal{H}$$

converges uniformly on each bounded subset of $\mathcal{H}$ and the limit is denoted by $J\phi(z)$ for each $z \in \mathcal{H}$. Hence $J\phi$ becomes an entire function on $\mathcal{H}$. Moreover, it is easily checked by definition that $J\phi \in E^2(\nu)$ and J is a unitary isomorphism from $\Gamma(\mathcal{H})$ onto $E^2(\nu)$ ([9, 10]). In particular, $\|J\phi\|_{E^2(\nu)} = \|\phi\|_0$, or equivalently,

$$\sup_{P \in \mathcal{P}} \int_{\mathcal{N}^*} |\langle\!\langle \Gamma(P)\phi, \phi_z \rangle\!\rangle|^2 \, \nu(\mathrm{d}z) = \|\phi\|_0^2, \qquad \phi \in \Gamma(\mathcal{H}).$$

The map J is called the *duality transform* and is related with the S-transform:

$$J\phi|_{\mathcal{D}} = S\phi, \quad \phi \in \Gamma(\mathcal{H}).$$

For a more detailed study of relations between $\Gamma(\mathcal{H})$, $E^2(\nu)$ and $L^2(E^*, \mu)$, we refer to [26].

Theorem 3.1 ([10, 16]). *Let $p \in \mathbb{R}$. Then a $\mathbb{C}$-valued function g on $\mathcal{D}$ is the S-transform of some $\Phi \in \mathfrak{G}_{K;p}$ if and only if g can be extended to a continuous function on $\mathcal{D}_{-p}$ and $g \circ K^p \in E^2(\nu)$.*

4. Operators on Fock Space

Let $\mathbf{D}$ be a dense subset of $\mathcal{N}$. Now, we study a characterization of symbols of continuous operators from $\mathfrak{G}_{K;p}$ into $\mathfrak{G}_{K;q}$ in terms of the Bargmann–Segal space.

Theorem 4.1. *Let p, $q \in \mathbb{R}$. A $\mathbb{C}$-valued function Θ on $\mathbf{D} \times \mathbf{D}$ is the symbol of some $\Xi \in \mathcal{L}(\mathfrak{G}_{K;p}, \mathfrak{G}_{K;q})$ if and only if*

(i) *for each $\xi \in \mathbf{D}$, $\Theta(\xi, \cdot)$ can be extended to entire function on $\mathcal{D}_{-q}$.*

(ii) *there exists a constant $C \geq 0$ such that*

$$\left\| \sum_{i=1}^{k} a_i \Theta(\xi_i, K^q \cdot) \right\|_{E^2(\nu)}^2 \leq C \left\| \sum_{i=1}^{k} a_i \phi_{\xi_i} \right\|_p^2 \tag{4.1}$$

for any $k \geq 1$ and any choice of $\xi_i \in \mathbf{D}$ and $a_i \in \mathbb{C}$, $i = 1, \cdots, k$.

In this case $\|\Xi\|_{K;p,q} \leq C$, where $\|\cdot\|_{K;p,q}$ is the operator norm on $\mathcal{L}(\mathfrak{G}_{K;p}, \mathfrak{G}_{K;q})$.

Proof. The proof is a simple modification of the proof of Theorem 5.2 in [16]. Suppose that $\Theta = \widehat{\Xi}$ with $\Xi \in \mathcal{L}(\mathfrak{G}_{K;p}, \mathfrak{G}_{K;q})$. Then condition (i) is obviously satisfied since $\Theta(\xi, \eta) = \langle\langle \Xi\phi_\xi, \phi_\eta \rangle\rangle$ is well-defined for $\xi \in \mathbf{D} \subset \mathcal{D}_p$ and $\eta \in \mathcal{D}_{-q}$. As for (ii), we take $C \geq 0$ such that $\|\Xi\phi\|_q \leq C\|\phi\|_p$ since $\Xi \in \mathcal{L}(\mathfrak{G}_{K;p}, \mathfrak{G}_{K;q})$. On the other hand,

$$\left\| \sum_{i=1}^{k} a_i \Theta(\xi_i, K^q \cdot) \right\|_{E^2(\nu)}^2 = \sup_{P \in \mathcal{P}} \int_{\mathcal{N}_{\mathbb{C}}^*} |\langle\langle \Xi\psi, \phi_{K^q P z} \rangle\rangle|^2 \, \nu(\mathrm{d}z) = \|\Xi\psi\|_q^2,$$

where $\psi = \sum_{i=1}^{k} a_i \phi_{\xi_i}$. Hence (4.1) follows.

We next prove the converse. For each $\xi \in \mathbf{D}$ we define a function $F_\xi \colon \mathcal{D}_{-q} \to \mathbb{C}$ by

$$F_\xi(\eta) = \Theta(\xi, \eta), \qquad \eta \in \mathcal{D}_{-q}.$$

Then by condition (i) the function F_ξ is entire on $\mathcal{D}_{-q}$. Moreover, $F_\xi \circ K^q \in E^2(\nu)$; in fact, by (ii) we have

$$\|F_\xi \circ K^q\|_{E^2(\nu)} = \|\Theta(\xi, K^q\cdot)\|_{E^2(\nu)} \le C\,\|\phi_\xi\|_p < \infty.$$

Then by Theorem 3.1, there exists a unique $\Phi_\xi \in \mathfrak{G}_{K;q}$ such that $F_\xi = S\Phi_\xi$, i.e.,

$$S\Phi_\xi(\eta) = F_\xi(\eta) = \Theta(\xi, \eta). \tag{4.2}$$

Since the exponential vectors are linearly independent, a linear operator Ξ is uniquely specified by

$$\Xi\phi_\xi = \Phi_\xi, \qquad \xi \in \mathbf{D}. \tag{4.3}$$

Then we see from (4.2) that $\Theta = \widehat{\Xi}$. Hence, to our goal, we need to show that Ξ is extended to a bounded operator in $\mathcal{L}(\mathfrak{G}_{K;p}, \mathfrak{G}_{K;q})$. Let $k \ge 1$ and take $\xi_i \in \mathbf{D}$ and $a_i \in \mathbb{C}$, $i = 1, 2, \ldots, k$. Then by direct computation we have

$$\begin{aligned}\left\|\Xi\Big(\sum_{i=1}^k a_i\phi_{\xi_i}\Big)\right\|_q^2 &= \sup_{P\in\mathcal{P}}\int_{\mathcal{N}_{\mathbb{C}}^*}\left|\left\langle\!\!\left\langle \Gamma(K^q)\Big(\sum_{i=1}^k a_i\Phi_{\xi_i}\Big), \phi_{Pz}\right\rangle\!\!\right\rangle\right|^2 \nu(\mathrm{d}z)\\ &= \sup_{P\in\mathcal{P}}\int_{\mathcal{N}_{\mathbb{C}}^*}\left|\sum_{i=1}^k a_i\Theta(\xi_i, K^qPz)\right|^2 \nu(\mathrm{d}z)\\ &= \left\|\sum_{i=1}^k a_i\Theta(\xi_i, K^q\cdot)\right\|_{E^2(\nu)}^2.\end{aligned}$$

Consequently, by (4.1) we have

$$\left\|\Xi\Big(\sum_{i=1}^k a_i\phi_{\xi_i}\Big)\right\|_q \le C\left\|\sum_{i=1}^k a_i\phi_{\xi_i}\right\|_p,$$

which proves that there exists $\Xi \in \mathcal{L}(\mathfrak{G}_{K;p}, \mathfrak{G}_{K;q})$ characterized by (4.3) since the exponential vectors $\{\phi_\xi\,;\, \xi \in \mathbf{D}\}$ span a dense subspace of $\mathfrak{G}_{K;p}$. □

Let $\mathcal{H}_1$, $\mathcal{H}_2$ be Hilbert spaces and let $\{\Xi_n\}_{n=1}^\infty$ be a sequence of bounded operators in $\mathcal{L}(\mathcal{H}_1, \mathcal{H}_2)$. Then by the uniform boundedness principle we can easily see that Ξ_n converges strongly to a bounded linear operator in $\mathcal{L}(\mathcal{H}_1, \mathcal{H}_2)$ if and only if $\{\Xi_n\}_{n=1}^\infty$ is bounded in $\mathcal{L}(\mathcal{H}_1, \mathcal{H}_2)$ and $\Xi_n(\xi)$ converges in $\mathcal{H}_2$ for each ξ in a dense subset of $\mathcal{H}_1$. Therefore, by the fact that for any $\Xi \in \mathcal{L}(\mathfrak{G}_{K;p}, \mathfrak{G}_{K;q})$ and $\xi \in \mathbf{D}$

$$\|\widehat{\Xi}(\xi, K^q\cdot)\|_{E^2(\nu)}^2 = \sup_{P\in\mathcal{P}}\int_{\mathcal{N}^*}\left|\widehat{\Xi}(\xi, K^qPz)\right|^2 \nu(\mathrm{d}z) = \|\Xi\phi_\xi\|_q^2,$$

we can easily see the following theorem.

Theorem 4.2. *Let* $\{\Xi_n\}_{n=1}^{\infty}$ *be a sequence in* $\mathcal{L}(\mathfrak{G}_{K;p}, \mathfrak{G}_{K;q})$. *Then* Ξ_n *converges strongly to a bounded linear operator in* $\mathcal{L}(\mathfrak{G}_{K;p}, \mathfrak{G}_{K;q})$ *if and only if* $\{\Xi_n\}_{n=1}^{\infty}$ *is bounded in* $\mathcal{L}(\mathfrak{G}_{K;p}, \mathfrak{G}_{K;q})$ *and* $\widehat{\Xi}_n(\xi, K^q\cdot)$ *converges in* $E^2(\nu)$ *for each* $\xi \in \mathbf{D}$.

From now on we assume that $\|K^{-1}\|_{\mathrm{OP}} < 1$. Let p and q be real numbers. Then, for each $K_{l,m} \in \mathcal{L}(\mathcal{D}_p^{\widehat{\otimes} m}, \mathcal{D}_q^{\widehat{\otimes} l})$ and $\phi = (f_n)$, we define $\Xi_{l,m}(K_{l,m})\phi = (g_n)$ by

$$g_n = 0, \quad 0 \le n < l; \qquad g_{l+n} = \frac{(n+m)!}{n!} S_{l+n}(K_{l,m} \otimes I^{\otimes n}) f_{n+m}, \quad n \ge 0,$$

where S_{l+m} is the symmetrizing operator. By the same arguments of those used in [22, Section 4.3] we have

$$\|\Xi_{l,m}(K_{l,m})\phi\|_q \le \|K_{l,m}\|_{m,l;p,q} M_{l,m} \|\phi\|_{(p\vee q)+r}, \tag{4.4}$$

where $p \vee q = \max\{p, q\}$, $r > 0$ is arbitrary and $\|\cdot\|_{m,l;p,q}$ is the operator norm in $\mathcal{L}(\mathcal{D}_p^{\widehat{\otimes} m}, \mathcal{D}_q^{\widehat{\otimes} l})$ and

$$M_{l,m} = \rho^{r(m-1/2)}(l^l m^m)^{1/2} \left(\frac{\rho^{-r/2}}{-re \log \rho} \right)^{(l+m)/2}.$$

Hence, $\Xi_{l,m}(K_{l,m}) \in \mathcal{L}(\mathfrak{G}_{K;(p\vee q)+r}, \mathfrak{G}_{K;q})$. Such an operator $\Xi_{l,m}(K_{l,m})$ is called an *integral kernel operator* with the kernel $K_{l,m}$. The symbol is given by

$$\Xi_{l,m}(K_{l,m})\widehat{\ }(\xi, \eta) = \left\langle K_{l,m}\xi^{\otimes m}, \eta^{\otimes l} \right\rangle e^{\langle \xi, \eta \rangle}.$$

Let $\eta \in \mathcal{H}$ and let $K_\eta \in \mathcal{L}(\mathcal{H}, \mathbb{C})$ be defined by $K_\eta(\xi) = \langle \eta, \xi \rangle$ for any $\xi \in \mathcal{H}$. For simple notation, we identify $\eta = K_\eta = K_\eta^*$, where K_η^* is the adjoint operator of K_η, i.e., $K_\eta^*(a) = a\eta$ for all $a \in \mathbb{C}$.

Let $\mathcal{H} = L^2_{\mathbb{C}}(\mathbb{R}_+, \mathrm{d}t)$, where $\mathrm{d}t$ is the Lebesgue measure on $\mathbb{R}_+ = [0, \infty)$. Then for each $t \ge 0$, we easily see that for any $p, q \ge 0$

$$\begin{aligned} &\|\chi_{[0,t]}\|_{1,0;p,q} \le |\chi_{[0,t]}|_{-p}, \\ &\|\chi_{[0,t]}\|_{0,1;p,-q} \le |\chi_{[0,t]}|_{-q}, \\ &\|\chi_{[0,t]}\|_{1,1;p,-p} \le 1, \end{aligned} \tag{4.5}$$

where χ_B is the indicator function on $B \subset \mathbb{R}_+$. In the last inequality of (4.5), the indicator function is considered as the multiplication operator. Therefore, for each $t \ge 0$ the following operators

$$A_t = \Xi_{0,1}(\chi_{[0,t]}), \quad A_t^* = \Xi_{1,0}(\chi_{[0,t]}), \quad \Lambda_t = \Xi_{1,1}(\chi_{[0,t]}) \tag{4.6}$$

are well-defined as operators in $\mathcal{L}(\mathfrak{G}_{K;(p\vee q)+r}, \mathfrak{G}_{K;q})$, $\mathcal{L}(\mathfrak{G}_{K;p+r}, \mathfrak{G}_{K;-q})$ and $\mathcal{L}(\mathfrak{G}_{K;p+r}, \mathfrak{G}_{K;-p})$, respectively, for any $p, q \geq 0$ and $r > 0$. For each $t \geq 0$, A_t and A_t^* are called the *annihilation operator* and the *creation operator*, respectively.

5. Quantum Martingale Convergence

In this section, let $\mathcal{H} = L^2_{\mathbb{C}}(\mathbb{R}_+, \mathrm{d}t)$. For each $\xi \in \mathcal{H}$, we write $\xi_B = \xi\chi_B$, where $B \subset \mathbb{R}_+$. For notational convenience, we write $\xi_{t]} = \xi_{[0,t]}$ and $\xi_{[t} = \xi_{[t,\infty)}$ for any $t > 0$. Then we have the decomposition

$$\mathcal{H} = \mathcal{H}_{s]} \oplus \mathcal{H}_{[s,t]} \oplus \mathcal{H}_{[t}, \qquad 0 < s < t < \infty,$$

where $\mathcal{H}_{s]} = \{\xi_{s]} | \, \xi \in \mathcal{H}\}$, $\mathcal{H}_{[s,t]} = \{\xi_{[s,t]} | \, \xi \in \mathcal{H}\}$ and $\mathcal{H}_{[t} = \{\xi_{[t} | \, \xi \in \mathcal{H}\}$. Put

$$\mathfrak{H}_{s]} = \Gamma(\mathcal{H}_{s]}), \quad \mathfrak{H}_{[s,t]} = \Gamma(\mathcal{H}_{[s,t]}) \quad \text{and} \quad \mathfrak{H}_{[t} = \Gamma(\mathcal{H}_{[t}).$$

Then we have the identification

$$\mathfrak{H} = \mathfrak{H}_{s]} \otimes \mathfrak{H}_{[s,t]} \otimes \mathfrak{H}_{[t}$$

via the following decomposition:

$$\phi_\xi = \phi_{\xi_{s]}} \otimes \phi_{\xi_{[s,t]}} \otimes \phi_{\xi_{[t}}, \qquad \xi \in \mathcal{H}.$$

Moreover, for any $p \in \mathbb{R}$ and $0 < s < t < \infty$, we have

$$\mathfrak{G}_p = \mathfrak{G}_{p;s]} \otimes \mathfrak{G}_{p;[s,t]} \otimes \mathfrak{G}_{p;[t},$$

where $\mathfrak{G}_{p;s]} = \mathfrak{G}_p \cap \mathfrak{H}_{s]}$, $\mathfrak{G}_{p;[s,t]} = \mathfrak{G}_p \cap \mathfrak{H}_{[s,t]}$, $\mathfrak{G}_{p;[t} = \mathfrak{G}_p \cap \mathfrak{H}_{[t}$ and their completion for $p \leq 0$.

A family of operators $\{\Xi_t\}_{t\geq 0} \subset \mathcal{L}(\mathfrak{G}, \mathfrak{G}^*)$ is called a *quantum stochastic process* if there exist $p, q \in \mathbb{R}$ such that $\Xi_t \in \mathcal{L}(\mathfrak{G}_p, \mathfrak{G}_q)$ for all $t \geq 0$ and for each $\phi \in \mathfrak{G}_p$ the map $t \mapsto \Xi_t\phi$ is strongly measurable. A quantum stochastic process $\{\Xi_t\}_{t\geq 0}$ is said to be *adapted* if for each $t \geq 0$ there exists $\Xi_{t]} \in \mathcal{L}(\mathfrak{G}_{p;t]}, \mathfrak{G}_{q;t]})$ (here $p \geq q$ are independent of t) such that $\Xi_t = \Xi_{t]} \otimes I_{[t}$, where $I_{[t} : \mathfrak{G}_{p;[t} \hookrightarrow \mathfrak{G}_{q;[t}$ is the inclusion map.

For each $t \in \mathbb{R}_+$, the *conditional expectation* $\mathbb{E}_t$ (see [3, 14, 24]) is defined by the second quantization operator $\Gamma(\chi_{[0,t]})$ of $\chi_{[0,t]}$, i.e., for each $t \in \mathbb{R}_+$

$$\mathbb{E}_t\Phi = (\chi_{[0,t]}^{\otimes n} f_n), \qquad \Phi = (f_n) \in \mathfrak{G}^*.$$

Then $\mathbb{E}_t \in \mathcal{L}(\mathfrak{G}_p, \mathfrak{G}_p)$ and $\mathbb{E}_t$ is an orthogonal projection on $\mathfrak{G}_p$ for any $p \in \mathbb{R}$ and $t \in \mathbb{R}_+$. Moreover, $\mathbb{E}_t \in \mathcal{L}(\mathfrak{G}, \mathfrak{G})$ and $\mathbb{E}_t \in \mathcal{L}(\mathfrak{G}^*, \mathfrak{G}^*)$.

An adapted process of operators $\{\Xi_t\}_{t\geq 0}$ is called a *(quantum) martingale* if for any $0 \leq s \leq t$

$$\mathbb{E}_s \Xi_t \mathbb{E}_s = \mathbb{E}_s \Xi_s \mathbb{E}_s.$$

The processes $\{A_t\}_{t\geq 0}$, $\{A_t^*\}_{t\geq 0}$ and $\{\Lambda_t\}_{t\geq 0}$ defined in (4.6) are quantum martingales and called the *annihilation*, *creation* and *number (or gauge)* processes, respectively. These martingales are called the *basic martingales.* The quantum stochastic process $Q_t = A_t + A_t^*$ is called the *quantum Brownian motion* or the *position process.*

From now on we assume that $\mathcal{N}$ contains a dense subset $\mathbf{D}$ consisting of functions with compact support in $\mathbb{R}_+$ and $K \geq cI$, and K preserves the support of functions in $\mathbf{D}$. For example, the harmonic oscillator $K = -\mathrm{d}^2/\mathrm{d}t^2 + t^2 + 1$ satisfies these assumptions with condition $1 < c \leq 2$.

Lemma 5.1. *Let* $\{\Xi_t\}_{t\geq 0}$ *be a martingale such that*

$$\{\Xi_t\}_{t\geq 0} \subset \mathcal{L}(\mathfrak{G}_{K;p}, \mathfrak{G}_{K;q}) \text{ for some } p \geq q.$$

Then for each $\xi \in \mathbf{D}$ *with* $\mathrm{supp}(\xi) \subset [0,a]$, $\{\Gamma(K^q)\Xi_t\phi_\xi\}_{t\geq a}$ *is a classical martingale in the standard Gaussian space.*

Proof. It is obvious that $\{\Gamma(K^q)\Xi_t\phi_\xi\}_{t\geq a} \subset \Gamma(\mathcal{H})$ for each $\xi \in \mathbf{D}$ with $\mathrm{supp}(\xi) \subset [0,a]$. On the other hand, for any $a \leq s \leq t$ and $\eta \in \mathcal{H}$ there exists a sequence $\{\eta_n\}_{n=1}^\infty \subset \mathbf{D}$ with $\mathrm{supp}(\eta_n) \subset [0,s]$ such that η_n converges to $\eta_{s]}$ in $\mathcal{H}$ and then we have

$$\begin{aligned}
\langle\langle \mathbb{E}_s\Gamma(K^q)\Xi_t\phi_\xi, \phi_\eta\rangle\rangle &= \lim_{n\to\infty} \langle\langle \Xi_t\phi_\xi, \Gamma(K^q)\phi_{\eta_n}\rangle\rangle \\
&= \lim_{n\to\infty} \langle\langle \Xi_t\phi_\xi, \mathbb{E}_s\Gamma(K^q)\phi_{\eta_n}\rangle\rangle \\
&= \lim_{n\to\infty} \langle\langle \Gamma(K^q)\mathbb{E}_s\Xi_t\phi_\xi, \phi_{\eta_n}\rangle\rangle \\
&= \langle\langle \Gamma(K^q)\Xi_s\phi_\xi, \phi_\eta\rangle\rangle ,
\end{aligned}$$

where we used that the operator K preserves the support of η_n for any n. Therefore, we have

$$\mathbb{E}_s\Gamma(K^q)\Xi_t\phi_\xi = \Gamma(K^q)\Xi_s\phi_\xi$$

which desires the proof. □

Theorem 5.2. *Let* $\{\Xi_t\}_{t\geq 0} \subset \mathcal{L}(\mathfrak{G}_{K;p}, \mathfrak{G}_{K;q})$ *be a martingale for some* $p \geq q$. *If there exists* $M \geq 0$ *such that*

$$\sup_{t\geq 0} \|\Xi_t\|_{K;p,q} \leq M, \tag{5.1}$$

then there exists an operator Ξ *in* $\mathcal{L}(\mathfrak{G}_{K;p}, \mathfrak{G}_{K;q})$ *such that* Ξ_t *converges strongly to* Ξ *in* $\mathcal{L}(\mathfrak{G}_{K;p}, \mathfrak{G}_{K;q})$ *as* $t \to \infty$.

Proof. For each given $\xi \in \mathbf{D}$ with $\mathrm{supp}(\xi) \subset [0, a]$, by Lemma 5.1, $\{\Gamma(K^q)\Xi_t\phi_\xi\}_{t \geq a}$ is a classical martingale in the standard Gaussian space with finite mean square norm. Therefore, by the classical martingale convergence theorem, there exists $\psi_\xi \in \Gamma(\mathcal{H})$ such that

$$\begin{aligned} &\lim_{t\to\infty} \|\Gamma(K^q)\Xi_t\phi_\xi - \psi_\xi\|_0 \\ &= \lim_{t\to\infty} \|\widehat{\Xi}_t(\xi, K^q\cdot) - \langle\langle \Gamma(K^{-q})\psi_\xi, \phi_{K^q\cdot}\rangle\rangle\|_{E^2(\nu)} = 0. \end{aligned} \tag{5.2}$$

Define a linear operator Ξ on the linear span of $\{\phi_\xi; \xi \in \mathbf{D}\}$ by

$$\Xi\phi_\xi = \Gamma(K^{-q})\psi_\xi, \qquad \xi \in \mathbf{D}$$

and define a $\mathbb{C}$-valued function Θ on $\mathbf{D} \times \mathbf{D}$ by

$$\Theta(\xi, \eta) = \langle\langle \Xi\phi_\xi, \phi_\eta\rangle\rangle, \qquad \xi, \eta \in \mathbf{D}.$$

Then for each $\xi \in \mathbf{D}$, $\Theta(\xi, \cdot)$ can be extended to an entire function on $\mathcal{D}_{-q}$. On the other hand, by (5.1), we have

$$\left\| \sum_{i=1}^k a_i \Theta(\xi_i, K^q\cdot) \right\|_{E^2(\nu)}^2 \leq M \left\| \sum_{i=1}^k a_i \phi_{\xi_i} \right\|_p^2$$

for any $k \geq 1$ and any choice of $\xi_i \in \mathbf{D}$ and $a_i \in \mathbb{C}$, $i = 1, \cdots, k$. Hence by Theorem 4.1, Ξ is a continuous linear operator from $\mathfrak{G}_{K;p}$ into $\mathfrak{G}_{K;q}$. Moreover, by (5.2) and Theorem 4.2, Ξ_t converges strongly to Ξ in $\mathcal{L}(\mathfrak{G}_{K;p}, \mathfrak{G}_{K;q})$. □

Remark 5.3. In this section we assume that $K \geq cI$ to discuss convergence of martingale. Then for any $p, q \geq 0$ we have the following continuous inclusions:

$$\mathfrak{G}_{K;p} \subset \mathfrak{G}_p \subset \Gamma(\mathcal{H}) \subset \mathfrak{G}_{-q} \subset \mathfrak{G}_{K;-q}$$

and

$$\mathcal{L}(\mathfrak{G}_p, \mathfrak{G}_{-q}) \subset \mathcal{L}(\mathfrak{G}_{K;p}, \mathfrak{G}_{K;-q}).$$

In general, a martingale $\{\Xi_t\}_{t\geq 0}$ in our sense lives in $\mathcal{L}(\mathfrak{G}_p, \mathfrak{G}_{-q})$ for some $p, q \geq 0$. According to Theorem 5.2, if we consider the convergence of $\{\Xi_t\}_{t\geq 0}$, we need more bigger space $\mathcal{L}(\mathfrak{G}_{K;p}, \mathcal{G}_{K;-q})$.

Corollary 5.4. *Let* $\{\Xi_{l,m}(K_{l,m}(t))\}_{t\geq 0} \subset \mathcal{L}(\mathfrak{G}_p, \mathfrak{G}_q)$ *be a martingale for some* $p \geq q$. *If for each* $t \geq 0$, $K_{l,m}(t) \in \mathcal{L}(\mathcal{D}_p^{\widehat{\otimes} m}, \mathcal{D}_q^{\widehat{\otimes} l})$ *and if*

$$\sup_{t\geq 0} \|K_{l,m}(t)\|_{m,l;p,q} < \infty,$$

then there exists an operator Ξ *in* $\mathcal{L}(\mathfrak{G}_{K;p\vee q+r}, \mathfrak{G}_{K;q})$ *such that*

$$\Xi_{l,m}(K_{l,m}(t)) \text{ converges strongly to } \Xi$$

in $\mathcal{L}(\mathfrak{G}_{K;p\vee q+r}, \mathfrak{G}_{K;q})$ *as* $t \to \infty$ *for any* $r > 0$.

Proof. By (4.4) we have

$$\| \Xi_{l,m}(K_{l,m}(t))\phi\|_q \leq \|K_{l,m}(t)\|_{m,l;p,q} M_{l,m} \|\phi\|_{(p\vee q)+r},$$

where $r > 0$ is arbitrary. Therefore,

$$\sup_{t\geq 0} \| \Xi_{l,m}(K_{l,m}(t))\|_{K;(p\vee q)+r,q} \leq \sup_{t\geq 0} \|K_{l,m}(t)\|_{m,l;p,q} M_{l,m}.$$

Hence the proof follows from Theorem 5.2. □

Theorem 5.5. *Let* $\Xi \in \mathcal{L}(\mathfrak{G}, \mathfrak{G}^*)$. *Then* $\mathbb{E}_t\Xi\mathbb{E}_t \otimes I_{[t}$ *converges strongly to* Ξ *in* $\mathcal{L}(\mathfrak{G}, \mathfrak{G}^*)$ *as* $t \to \infty$.

Proof. For simple notation, for each $t \geq 0$ we put

$$\Xi_t = \mathbb{E}_t\Xi\mathbb{E}_t \otimes I_{[t}.$$

Then $\{\Xi_t\}_{t\geq 0}$ is an martingale. On the other hand, there exists $p \geq 0$ such that $\Xi \in \mathcal{L}(\mathfrak{G}_p, \mathfrak{G}_{-p})$. Therefore, we have

$$\|\Xi_t\|_{cI;p,-p} = \sup_{\phi\in\mathfrak{G}_p, \|\phi\|_p\leq 1} \|\Xi_t\phi\|_{-p} \leq \|\Xi\|_{cI;p,-p}.$$

Hence by Theorem 5.2, there exists an operator X in $\mathcal{L}(\mathfrak{G}_p, \mathfrak{G}_{-p})$ such that Ξ_t converges strongly to X. Moreover, for any $\xi, \eta \in \mathcal{H}$ we have

$$\widehat{X}(\xi, \eta) = \lim_{t\to\infty} \left\langle\!\left\langle \Xi\phi_{\xi_{t]}}, \phi_{\eta_{t]}} \right\rangle\!\right\rangle \left\langle\!\left\langle \phi_{\xi_{[t}}, \phi_{\eta_{[t}} \right\rangle\!\right\rangle = \widehat{\Xi}(\xi, \eta).$$

Therefore by Theorem 4.1, $X = \Xi$. We complete the proof. □

Theorem 5.6. *Let* $\{\Xi_t\}_{t\geq 0} \subset \mathcal{L}(\mathfrak{G}_p, \mathfrak{G}_q)$ $(p \geq q)$ *be a martingale. If there exists* $\Xi \in \mathcal{L}(\mathfrak{G}, \mathfrak{G}^*)$ *such that* Ξ_t *converges weakly to* Ξ *as* $t \to \infty$, *then*

$$\Xi_t = \mathbb{E}_t\Xi\mathbb{E}_t \otimes I_{[t}, \qquad t \geq 0.$$

Proof. Since $\{\Xi_t\}_{t\geq 0}$ is an adapted process, for each $t \geq 0$ there exists $\Xi_{t]} \in \mathcal{L}(\mathfrak{G}_{p;t]}, \mathfrak{G}_{q;t]})$ such that $\Xi_t = \Xi_{t]} \otimes I_{[t}$. On the other hand, for each $t \geq 0$ and any $\xi, \eta \in \mathcal{H}$ we have

$$\begin{aligned}(\mathbb{E}_t\Xi\mathbb{E}_t \otimes I_{[t})\widehat{\ }(\xi,\eta) &= \left\langle\!\left\langle \mathbb{E}_t\Xi\mathbb{E}_t\phi_{\xi_{t]}}, \phi_{\eta_{t]}} \right\rangle\!\right\rangle \left\langle\!\left\langle \phi_{\xi_{[t}}, \phi_{\eta_{[t}} \right\rangle\!\right\rangle \\ &= \lim_{s\to\infty} \left\langle\!\left\langle \mathbb{E}_t\Xi_s\mathbb{E}_t\phi_{\xi_{t]}}, \phi_{\eta_{t]}} \right\rangle\!\right\rangle \left\langle\!\left\langle \phi_{\xi_{[t}}, \phi_{\eta_{[t}} \right\rangle\!\right\rangle \\ &= \left\langle\!\left\langle \Xi_{t]}\phi_{\xi_{t]}}, \phi_{\eta_{t]}} \right\rangle\!\right\rangle \left\langle\!\left\langle \phi_{\xi_{[t}}, \phi_{\eta_{[t}} \right\rangle\!\right\rangle \\ &= \widehat{\Xi}_t(\xi,\eta).\end{aligned}$$

Thus the proof follows from Theorem 4.1. □

Example 5.7 (Basic martingales). Let K be the harmonic oscillator, i.e., $K = \mathrm{d}^2/\mathrm{d}t^2 - t^2 + 1$ and $\mathcal{N}$ is the Schwartz space, i.e., $\mathcal{N} = \mathcal{S}(\mathbb{R})$. Then we have the following martingale convergence theorem for the basic quantum martingales:

Note that there exist $M \geq 0$ and $p, q \geq 0$

$$\sup_{t\geq 0}\left\{\|A_t\|_{K;(p\vee q)+r,q}, \|A_t^*\|_{K;p+r,-q}, \|\Lambda_t\|_{K;p+r,-p}\right\} \leq M$$

for arbitrary $r > 0$, where constants M and p, q can be chosen differently for the basic martingales. Hence, by Theorem 5.2, there exist operators Ξ_A, Ξ_{A^*} and Ξ_Λ in $\mathcal{L}(\mathfrak{G}_{K;(p\vee q)+r}, \mathfrak{G}_{K;q})$, $\mathcal{L}(\mathfrak{G}_{K;p+r}, \mathfrak{G}_{K;-q})$, and $\mathcal{L}(\mathfrak{G}_{K;p+r}, \mathfrak{G}_{K;-p})$, respectively, such that A_t, A_t^* and Λ_t converge strongly to Ξ_A, Ξ_{A^*} and Ξ_Λ, respectively, as $t \to \infty$. In fact, $\Xi_A = \Xi_{0,1}(\chi_{\mathbb{R}_+})$, $\Xi_{A^*} = \Xi_{1,0}(\chi_{\mathbb{R}_+})$ and $\Xi_\Lambda = \Xi_{1,1}(\chi_{\mathbb{R}_+})$. Therefore, the quantum Brownian motion $Q_t = A_t + A_t^*$ converges strongly to $\Xi_{0,1}(\chi_{\mathbb{R}_+}) + \Xi_{1,0}(\chi_{\mathbb{R}_+})$ in $\mathcal{L}(\mathfrak{G}_{K;(p\vee q)+r}, \mathfrak{G}_{K;-q})$ as $t \to \infty$.

Acknowledgments

The first author is most grateful to Professors D.M. Chung and N. Obata for helpful comments and valuable discussions that improved this paper. This work was supported by the Brain Korea 21 Project. The authors thank the referee for several comments that improved this paper.

References

[1] Accardi L., Cecchini C.: Conditional expectation in von Neumann algebras and a Theorem of Takesaki, *J. Funct. Anal.* **45** (1982) 245–273.

[2] Accardi L., Longo R.: Martingale convergence of generalized conditional expectations, *J. Funct. Anal.* **118** (1993) 119–130.

[3] Benth F.E., Potthoff J.: On the martingale property for generalized stochastic processes, *Stochastics Stochastics Rep.* **58** (1996) 349–367.

[4] Belavkin V.P.: A quantum nonadapted Ito formula and stochastic analysis in Fock scale, *J. Funct. Anal.* **102** (1991) 414–447.

[5] Chung D.M., Ji U.C., Obata N.: Quantum stochastic analysis via white noise operators in weighted Fock space, *Rev. Math. Phys.* **14** (2002) 241–272.

[6] Cuculescu I.: Martingales on von Neumann algebras, *J. Multivariate Anal.* **1** (1971) 17–27.

[7] Dang-Ngoc N.: Pointwise convergence of martingales in von Neumann algebras, *Israel J. Math.* **34** (1979) 273–280.

[8] Goldstein S.: Convergence of martingales in von Neumann algebras, *Bull. Acad. Polon. Sci. Sér. Sci. Math.* **27** (1979) 853–859.

[9] Gross L., Malliavin P.: Hall's transform and the Segal–Bargmann map, in "Itô's Stochastic Calculus and Probability Theory (N. Ikeda, S. Watanabe, M. Fukushima and H. Kunita, Eds.)," pp. 73–116, Springer-Verlag, 1996.

[10] Grothaus M., Kondratiev Yu.G., Streit L.: Complex Gaussian analysis and the Bargmann–Segal space, *Methods Funct. Anal. Topology* **3** (1997) 46–64.

[11] Hiai F., Tsukada M.: Strong martingale convergence of generalized conditional expectations on von Neumann algebras, *Trans. Amer. Math. Soc.* **282** (1984) 791–798.

[12] Hida T.: "Analysis of Brownian Functionals," Carleton Math. Lect. Notes, no. 13, Carleton University, Ottawa, 1975.

[13] Hudson R.L., Parthasarathy K.R.: Quantum Itô's formula and stochastic evolutions, *Commun. Math. Phys.* **93** (1984) 301–323.

[14] Ji U.C.: Stochastic integral representation theorem for quantum semimartingales, *J. Funct. Anal.* **201** (2003) 1–29.

[15] Ji U.C., Obata N.: Segal–Bargmann transform of white noise operators and white noise differential equations, RIMS Kokyuroku **1266** (2002) 59–81.

[16] Ji U.C., Obata N.: A role of Bargmann-Segal spaces in characterization and expansion of operators on Fock space, preprint, 2000.

[17] Kubo I., Takenaka S.: Calculus on Gaussian white noise I–IV, *Proc. Japan Acad. Ser. A Math. Sci.* **56A** (1980) 376–380; 411–416; **57A** (1981) 433–437; **58A** (1982) 186–189.

[18] Kuo H.-H.: *White Noise Distribution Theory*, CRC Press, 1996.

[19] Lindsay J.M., Maassen H.: An integral kernel approach to noise, in "Quantum Probability and Applications III (L. Accardi and W. von Waldenfels Eds.)." Lecture Notes in Math. Vol. 1303, pp. 192–208, Springer-Verlag, 1988.

[20] Lindsay J.M., Parthasarathy K.R.: Cohomology of power sets with applications in quantum probability, *Commun. Math. Phys.* **124** (1989) 337–364.

[21] Meyer P.-A.: "Quantum Probability for Probabilists," Lect. Notes in Math., Vol. 1538, Springer-Verlag, 1993.

[22] Obata N.: "White Noise Calculus and Fock Space," Lect. Notes in Math., Vol. 1577, Springer–Verlag, 1994.

[23] Obata N.: Generalized quantum stochastic processes on Fock space, *Publ. RIMS, Kyoto Univ.*, **31** (1995) 667–702.

[24] Obata N.: White noise approach to quantum martingales, in "Probability Theory and Mathematical Statistics", pp. 379–386, World Sci. Publishing, 1995.

[25] Obata N.: Conditional expectation in classical and quantum white noise calculi, *RIMS Kokyuroku* **923** (1995) 154–190.

[26] Obata N.: Inverse S-transform, Wick product and overcompleteness of exponential vectors, in "Quantum Information IV (T. Hida and K. Saitô, Eds.)," pp. 147–176, World Sci. Publishing, 2002.

[27] Parthasarathy K.R.: "An introduction to quantum stochastic calculus", Birkhäuser, 1992.

[28] Parthasarathy K.R.: Some additional remarks on Fock space stochastic calculus, in Lect. Notes in Math., Vol. 1204, pp. 331–333, Springer–Verlag, 1986.

[29] Parthasarathy K.R., Sinha K.B.: Stochastic integral representation of bounded quantum martingales in Fock space, *J. Funct. Anal.* **67** (1986) 126–151.

[30] Potthoff J., Timpel M.: On a dual pair of spaces of smooth and generalized random variables, *Potential Analysis* **4** (1995) 637–654.

[31] Tsukada M.: Strong convergence of martingales in von Neumann algebras, *Proc. Amer. Math. Soc.* **88** (1983) 537–540.

STOCHASTIC INTEGRATION FOR COMPENSATED POISSON MEASURES AND THE LÉVY-ITÔ FORMULA

Barbara Rüdiger
Institut für Angewandte Mathematik, Abteilung Stochastik, Universität Bonn, Wegelerstr. 6, D-53115 Bonn, Germany
ruediger@wiener.iam.uni-bonn.de

Abstract This is a review paper which presents in a unified way part of the results obtained in [38] and [2], where stochastic integrals of Banach valued random (resp. deterministic) functions w.r.t. compensated Poisson random measures are studied. As a consequence, the Lévy-Itô decomposition theorem for additive processes on Banach spaces is presented here in its stronger formulation (than [17], [8]), proposed in [2], for the special case where the additive processes are Lévy processes.

Keywords: Stochastic integrals, martingales measures, Lévy-Itô decomposition, càdlàg processes on Banach spaces, independent increments, Lévy measures, type 2 spaces.
AMS-classification (2000): 60G51, 60H05, 47G30, 46B09

1. Introduction

Processes with jumps are used for the description of interesting phenomena in mathematical physics. We have described some of them in [4]. We also know that for the description of critical phenomena not only diffusion processes are useful, but more generally additive processes (see e.g. [19, 13, 45] for physical motivations). Such kinds of phenomena are described by stochastic differential equations (SDEs) with (non Gaussian) additive-noise on infinite dimensional state spaces, as e.g. separable Hilbert spaces and separable Banach spaces. The analysis of strong solutions to this kind of SDEs requires a previous research on stochastic integrals of random or deterministic Hilbert valued (resp. Banach valued) functions with respect to additive noises, given e.g. by compensated Poisson random measures associated to additive processes. In [2] we give a direct definition of stochastic integrals for deterministic Banach val-

S. Albeverio et al. (eds.),
Proceedings of the International Conference on Stochastic Analysis and Applications, 145–167.

ued functions $f\colon E \to F$ with respect to compensated Poisson random measures of Lévy processes $(L_t)_{t\geq 0}$ and find sufficient conditions for the existence. In [38] we give a direct definition of stochastic integrals for *random Banach valued functions* with respect to compensated Poisson random measures of additive processes, and find sufficient conditions for the existence of these integrals. In [39] we establish the correspondent Ito formula. In [28] we use the results presented in Section [38] to prove existence and uniqueness of initial value problems for Banach valued non linear SDEs with non Gaussian additive noise. All these results are new also on Hilbert spaces and in [29] we use them to describe the dynamics of an infinite interacting particle system in statistical mechanics. (In [3] other kinds of SDEs with non Gaussian additive noise on infinite-dimensional spaces are obtained by subordination of diffusion processes.) The results of [38] have also applications in financial mathematics: according to the discussion and concluding remarks of Section 7 in [7], the stochastic integrals introduced in [38], correspond, for the case where the Banach space F is the space of continuous real valued functions defined on a compact subset of $[0, \infty]$, with the uniform norm, to the mathematical tool demanded in [7], to generalize the theory of [7] of the zero coupon bond markets.

In Section 2 we introduce the compensated Poisson random measures associated to additive Banach valued processes, as done in [38] (see also [2, 14]). In Section 3, and Section 4 we present (part of) the results concerning the stochastic integration w.r.t. compensated Poisson random measure of [2] and [38] in a unified way, presenting a generalization of the results of [2], to the case where the compensated Poisson random measure is associated to additive processes (and not only Lévy processes, like done in [2]).

We remark that the stochastic integrals introduced and analyzed in [2] and [38] are defined as a natural generalization to Hilbert and Banach valued spaces of the real valued stochastic integrals w.r.t. general martingale measures introduced in [6, 41] and [16], for the case where the martingale measures are compensated Poisson random measures. This is stressed in [38] (and mentioned also in Section 3), where the relation w.r.t. these two different presentations of real valued stochastic integrals is also understood and discussed. We refer also to [7, 8, 9, 10, 11, 22, 30, 31, 32, 34, 35, 36, 37, 42, 43] where different stochastic integrals of infinite dimensional valued functions w.r.t. general random measures are defined and studied.

The motivation in [2], to analyze stochastic integrals (of deterministic functions) w.r.t. compensated Poisson random measures associated to Lévy processes, was the analysis of the martingale part in the Lévy-Itô decomposition theorem on separable Banach spaces. This decomposition theorem, which states that any additive process is decomposed in a pure jump semimartingale and a continuous martingale driven by a centered Brownian motion, was conceived by Lévy [23, 24], and formulated and proved by Ito [17], for the case where the state space is the real line. It was proved in [42] on (co-)nuclear spaces and in [8] on general Banach spaces. In [2] we prove that under suitable integrability conditions of the function $f(x) = x$ (which are always satisfied on separable Hilbert spaces and hence also on $\mathbb{R}^d$) the pure jump martingale part in the decomposition is a stochastic integral (of $f(x) = x$) in a stronger sense than the stochastic integral obtained in the proof of [17] or [8]. In our terminology it is a "strong p-integral", $p = 1, 2$ (Definition 4.3 in Section 4) w.r.t. the associated compensated Poisson random measure. In Section 5 we present the Lévy-Itô decomposition theorem for Banach valued additive processes in its strong formulation, as done in [2], however for the case of Banach valued additive processes. The proof is less straight than the proof proposed for Lévy processes in [2], and we need, for this more general case, to refer to part of the proof done in [8]. This is a rather sketchy proof, but in [2] we give, in a final remark, precise references to this proof proposed by E. Dettweiler (we refer to Section 5).

2. Poisson and Lévy measures of additive processes on separable Banach spaces

We assume that a filtered probability space $(\Omega, \mathcal{F}, (\mathcal{F}_t)_{0 \leq t \leq +\infty}, \mathsf{P})$, satisfying the "usual hypothesis", is given:

i) $\mathcal{F}_0$ contains all null sets of $\mathcal{F}$

ii) $\mathcal{F}_t = \mathcal{F}_t^+$, where $\mathcal{F}_t^+ = \cap_{u>t} \mathcal{F}_u$ for all t such that $0 \leq t < +\infty$, i.e., the filtration is right continuous.

In this Section we introduce the compensated Poisson random measures associated to additive processes on $(\Omega, \mathcal{F}, (\mathcal{F}_t)_{0 \leq t \leq +\infty}, P)$ with values in $(E, \mathcal{B}(E))$, where in the whole paper we assume that E is a separable Banach space with norm $\| \cdot \|$ and $\mathcal{B}(E)$ is the corresponding Borel σ-algebra.

Definition 2.1. A process $(X_t)_{t \geq 0}$ with state space $(E, \mathcal{B}(E))$ is an $\mathcal{F}_t$-additive process on $(\Omega, \mathcal{F}, \mathsf{P})$ if

i) $(X_t)_{t \geq 0}$ is adapted (to $(\mathcal{F}_t)_{t \geq 0}$)

ii) $X_0 = 0$ a.s.

iii) $(X_t)_{t\geq 0}$ has increments independent of the past, i.e. $X_t - X_s$ is independent of $\mathcal{F}_s$ if $0 \leq s < t$

iv) $(X_t)_{t\geq 0}$ is stochastically continuous

v) $(X_t)_{t\geq 0}$ is càdlàg.

An additive process is a Lévy process if the following condition is satisfied

vi) $(X_t)_{t\geq 0}$ has stationary increments, that is $X_t - X_s$ has the same distribution as X_{t-s}, $0 \leq s < t$.

Remark 2.2. Any process satisfying i)-iv) has a càdlàg version. This follows from its being, after compensation, a martingale (see e.g. [12, 26, 32]).

Let $(X_t)_{t\geq 0}$ be an additive process on $(E, \mathcal{B}(E))$ (in the sense of Definition 2.1). Set $X_{t-} := \lim_{s\uparrow t} X_s$ and $\Delta X_s := X_s - X_{s-}$.

The following results, i.e. Theorem 2.3, Theorem 2.4, Corollary 2.5, Theorem 2.6, Corollary 2.7, Theorem 2.8 are known (see e.g. [14]). (The proofs of Theorem 2.4, Corollary 2.5, Theorem 2.6, Corollary 2.7, Theorem 2.8 can e.g. be done following [2]).

Theorem 2.3. *Let $\Lambda \in \mathcal{B}(E)$, $0 \in (\overline{\Lambda})^c$ (where as usual $\overline{\Lambda}$ denotes the closure of the set Λ and with N^c we denote the complementary of a set N), then*

$$N_t^\Lambda := \sum_{0<s\leq t} \mathbf{1}_\Lambda(\Delta X_s) = \sum_{n\geq 1} \mathbf{1}_{t\geq T_n^\Lambda} \tag{2.1}$$

where

$$T_1^\Lambda := \inf\{s > 0 : \Delta X_s \in \Lambda\} \tag{2.2}$$

$$T_{n+1}^\Lambda := \inf\{s > T_n^\Lambda : \Delta X_s \in \Lambda\}, \quad n \in \mathbb{N}. \tag{2.3}$$

N_t^Λ is an adapted counting process without explosions. Moreover:

$$\mathsf{P}(N_t^\Lambda = k) = \exp(-\nu_t(\Lambda)) \frac{(\nu_t(\Lambda))^k}{k!} \tag{2.4}$$

$$\nu_t(\Lambda) := E[N_t^\Lambda] \tag{2.5}$$

Theorem 2.4. *Let $\mathcal{B}(E \setminus \{0\})$ be the trace σ-algebra on $E \setminus \{0\}$ of the Borel σ-algebra $\mathcal{B}(E)$ on E , and let*

$$\mathcal{F}(E \setminus \{0\}) := \{\Lambda \in \mathcal{B}(E \setminus \{0\}) : 0 \in (\overline{\Lambda})^c\}, \tag{2.6}$$

then $\mathcal{F}((E\setminus\{0\})$ *is a ring and for all* $\omega\in\Omega$ *the set function*

$$\begin{aligned} N_t^{\cdot} := N_t(\omega,\cdot)\colon \mathcal{F}(E\setminus\{0\}) &\longrightarrow \mathbb{R}_+ \\ \Lambda &\longmapsto N_t^{\Lambda}(\omega) \end{aligned} \tag{2.7}$$

is a σ*-finite pre- measure (in the sense of e.g. [5]).*

Corollary 2.5. *For any* $\omega\in\Omega$ *there is a unique* σ*-finite measure on* $\mathcal{B}(E\setminus\{0\})$

$$\begin{aligned} N_t(\omega,\cdot)\colon \mathcal{B}(E\setminus\{0\}) &\longrightarrow \mathbb{R}_+ \\ A &\longmapsto N_t^{A}(\omega) \end{aligned} \tag{2.8}$$

which is the continuation of the σ*-finite pre-measure on* $\mathcal{F}(E\setminus\{0\})$ *given by Theorem 2.4.*

From Theorem 2.4, Corollary 2.5 it follows that $N_t\colon \Lambda\to N_t^{\Lambda}$ is a random measure on $(E\setminus\{0\},\mathcal{B}(E\setminus\{0\}))$.

Theorem 2.6. *The set function* $\nu_t(\Lambda) := E[N_t^{\Lambda}(\omega)]\in\mathbb{R}$, $\Lambda\in\mathcal{F}(E\setminus\{0\})$, $\omega\in\Omega$ *satisfies:*

$$\begin{aligned} \nu_t\colon \mathcal{F}(E\setminus\{0\}) &\longrightarrow \mathbb{R}_+ \\ \Lambda &\longmapsto E[N_t^{\Lambda}(\omega)] \end{aligned} \tag{2.9}$$

and is a σ*-finite pre-measure on* $((E\setminus\{0\}),\mathcal{F}(E\setminus\{0\}))$

Corollary 2.7. *There is a unique* σ*-finite measure on the* σ*-algebra* $\mathcal{B}(E\setminus\{0\})$

$$\begin{aligned} \nu_t\colon \mathcal{B}(E\setminus\{0\}) &\longrightarrow \mathbb{R}_+ \\ A &\longmapsto E[N_t^{A}(\omega)] \end{aligned} \tag{2.10}$$

which is the continuation to $\mathcal{B}(E\setminus\{0\})$ *of the* σ*-finite pre-measure* ν_t *on the ring* $(E\setminus\{0\},\mathcal{F}(E\setminus\{0\}))$*, given by Theorem 2.6.*

Theorem 2.8. *The* σ*-finite measure* ν_t *of Corollary 2.7 is a Lévy measure.*

We recall the definition of Lévy measures on separable Banach spaces (see e.g. [1, 25]).

Definition 2.9. A σ-finite positive measure ν on $(E\setminus\{0\},\mathcal{B}(E\setminus\{0\}))$ is a *"Lévy measure"*, if there is a probability measure μ on $(E,\mathcal{B}(E))$ such that the Fourier transform $\hat{\mu}(F)$, $F\in E'$ satisfies

$$\hat{\mu}(F) = \exp\int_{E\setminus\{0\}} \exp(iF(x)-1-iF(x)\mathbf{1}_{\|x\|\le 1})\nu(\mathrm{d}x) \tag{2.11}$$

Definition 2.10. We call $N_t(\mathrm{d}x)(\omega)$ the Poisson random measure at time t associated to the additive process $(X_t)_{t\geq 0}$ and $\nu_t(\mathrm{d}x)$ the correspondent Lévy-measure. We call $N_t(\mathrm{d}x)(\omega) - \nu_t(\mathrm{d}x)$ the compensated Poisson random measure associated to the additive process $(X_t)_{t\geq 0}$ at time t. (We omit sometimes to write the dependence on $\omega \in \Omega$.)

Let $\mathcal{S}(\mathbb{R}_+)$ be the semi-ring of sets $(t_1, t_2]$, $0 \leq t_1 < t_2$, and $\mathcal{S}(\mathbb{R}_+) \times \mathcal{B}(E \backslash \{0\})$ be the semi-ring of the product sets $(t_1, t_2] \times A$, $A \in \mathcal{B}(E \backslash \{0\})$. Let

$$N((t_1, t_2] \times A)(\omega) = N_{t_2}(A)(\omega) - N_{t_1}(A)(\omega) \quad \forall A \in \mathcal{B}(E \backslash \{0\}) \quad \forall \omega \in \Omega \tag{2.12}$$

For all $\omega \in \Omega$ fixed $N(\mathrm{d}t\,\mathrm{d}x)(\omega)$ is a σ-finite pre-measure on the product semi-ring $S(\mathbb{R}_+) \times \mathcal{B}(E \setminus \{0\})$.

Let us denote also with $N(\mathrm{d}t\,\mathrm{d}x)(\omega)$ the measure which is the unique extension of the pre-measure to the σ-algebra $\mathcal{B}(\mathbb{R}_+ \times (E \backslash \{0\}))$ generated by $\mathcal{S}(\mathbb{R}_+) \times \mathcal{B}(E \backslash \{0\})$ (see e.g. [5, Satz 5.7, Chap. I, §5], [20, Theorem 1, Chap. V, §2] for the existence of a unique minimal σ-algebra containing a product semi-ring).

Let

$$\nu((t_1, t_2] \times A) = \nu_{t_2}(A) - \nu_{t_1}(A) \quad \forall A \in \mathcal{B}(E \setminus \{0\}) \tag{2.13}$$

$\nu(\mathrm{d}t\,\mathrm{d}x)$ is a σ-finite pre-measure on $S(\mathbb{R}_+) \times \mathcal{B}(E \setminus \{0\})$. Let us denote also by $\nu(\mathrm{d}t\,\mathrm{d}x)$ the σ-finite measure, which is the unique extension of this pre-measure on $\mathcal{B}(\mathbb{R}_+ \times (E \setminus \{0\}))$.

Definition 2.11. We call $N(\mathrm{d}t\,\mathrm{d}x)(\omega)$ the Poisson random measure associated to the additive process $(X_t)_{t\geq 0}$ and $\nu(\mathrm{d}t\,\mathrm{d}x)$ its compensator. We call $N(\mathrm{d}t\,\mathrm{d}x)(\omega) - \nu(\mathrm{d}t\,\mathrm{d}x)$ the compensated Poisson random measure associated to the additive process $(X_t)_{t\geq 0}$. (We omit sometimes to write the dependence on $\omega \in \Omega$.)

3. Stochastic integration of Banach valued random functions w.r.t. compensated Poisson random measures

We describe in this Section the results in [38] where we define stochastic integration of Banach valued random functions $f(t, x, \omega)$ with respect to the compensated Poisson random measures

$$q(\mathrm{d}t\,\mathrm{d}x)(\omega) := N(\mathrm{d}t\,\mathrm{d}x)(\omega) - \nu(\mathrm{d}t\,\mathrm{d}x)$$

associated to additive processes $(X_t)_{t\geq 0}$, defined below.

Let F be a separable Banach space with norm $\|\cdot\|_F$. (When no misunderstanding is possible we write $\|\cdot\|$ instead of $\|\cdot\|_F$.) Let $F_t := \mathcal{B}(\mathbb{R}_+ \times (E \setminus \{0\})) \otimes \mathcal{F}_t$ be the product σ -algebra generated by the semi-ring $\mathcal{B}(\mathbb{R}_+ \times (E \setminus \{0\})) \times \mathcal{F}_t$ of the product sets $A \times F$, $A \in \mathcal{B}(\mathbb{R}_+ \times (E \setminus \{0\}))$, $F \in \mathcal{F}_t$ (where as usual $\mathcal{F}_t$ is the filtration of the additive process $(X_t)_{t\geq 0}$). Let $T > 0$, and

$$\begin{aligned} M^T(E/F) := \{f\colon \mathbb{R}_+ \times (E \setminus \{0\}) \times \Omega \to F, \text{ such that} \\ f \text{ is } F_T/\mathcal{B}(F) \text{ measurable and } f(t,x,\omega) \\ \text{is } \mathcal{F}_t\text{-adapted } \forall x \in E \setminus \{0\}, \forall t \in (0,T]\}. \end{aligned} \tag{3.1}$$

There is a "natural definition" of stochastic integral w.r.t. $q(\mathrm{d}t\,\mathrm{d}x)(\omega)$ on those sets $(0,T] \times \Lambda$ where the measures $N(\mathrm{d}t\,\mathrm{d}x)(\omega)$ (ω fixed) and $\nu(\mathrm{d}t\,\mathrm{d}x)$ are finite, i.e. $0 \notin \overline{\Lambda}$ (Definition 3.1).

Definition 3.1. Let $t \in (0,T]$, $\Lambda \in \mathcal{F}(E \setminus \{0\})$ (defined in (2.6)), $f \in M^T(E/F)$. Assume that $f(\cdot,\cdot,\omega)$ is Bochner integrable on $(0,T] \times \Lambda$ w.r.t. ν, for all $\omega \in \Omega$ fixed (see e.g. [46, Chapter V, §5] for the definition of Bochner integral). The *natural integral* of f on $(0,t] \times \Lambda$ w.r.t. the compensated Poisson random measure $q(\mathrm{d}t\,\mathrm{d}x) := N(\mathrm{d}t\,\mathrm{d}x)(\omega) - \nu(\mathrm{d}t\,\mathrm{d}x)$ is

$$\begin{aligned} &\int_0^t \int_\Lambda f(s,x,\omega)\,(N(\mathrm{d}s\,\mathrm{d}x)(\omega) - \nu(\mathrm{d}s\,\mathrm{d}x)) \\ &:= \sum_{0<s\leq t} f(s,(\Delta X_s)(\omega),\omega)\,\mathbf{1}_\Lambda(\Delta X_s(\omega)) \qquad (3.2) \\ &\quad - \int_0^t \int_\Lambda f(s,x,\omega)\nu(\mathrm{d}s\,\mathrm{d}x), \quad \omega \in \Omega \qquad (3.3) \end{aligned}$$

where the last term is understood as a Bochner integral, (for $\omega \in \Omega$ fixed) of $f(s,x,\omega)$ w.r.t. the measure ν.

Definition 3.2. A function f belongs to the set $\Sigma(E/F)$ of *simple functions*, if $f \in M^T(E/F)$, $T > 0$ and there exist $n \in \mathbb{N}$, $m \in \mathbb{N}$, such that

$$f(t,x,\omega) = \sum_{k=1}^{n-1} \sum_{l=1}^{m} \mathbf{1}_{A_{k,l}}(x)\mathbf{1}_{F_{k,l}}(\omega)\mathbf{1}_{(t_k,t_{k+1}]}(t)a_{k,l} \tag{3.4}$$

where $A_{k,l} \in \mathcal{F}(E\setminus\{0\})$ (i.e. $0 \notin \overline{A_{k,l}}$), $t_k \in (0,T]$, $t_k < t_{k+1}$, $F_{k,l} \in \mathcal{F}_{t_k}$, $a_{k,l} \in F$ and $A_{k_1,l_1} \cap A_{k_2,l_2} = \emptyset$, $F_{k_1,l_1} \cap F_{k_2,l_2} = \emptyset$, if $l_1 \neq l_2$,.

Proposition 3.3. *Let $f \in \Sigma(E/F)$ be of the form (3.4), then*

$$\begin{aligned}&\int_0^T \int_A f(t,x,\omega) q(\mathrm{d}t\,\mathrm{d}x)(\omega)\\&= \sum_{k=1}^{n-1}\sum_{l=1}^{m} a_{k,l}\mathbf{1}_{F_{k,l}}(\omega) q((t_k,t_{k+1}]\cap(0,T]\times A_{k,l}\cap A)(\omega).\end{aligned} \tag{3.5}$$

for all $A \in \mathcal{B}(E\setminus\{0\})$, $T>0$.

Remark 3.4. The random variables $\mathbf{1}_{F_{k,l}}$ in (3.5) are independent of $q((t_k,t_{k+1}]\cap(0,T]\times A_{k,l}\cap A)$ for all $k\in 1...n-1$, $l\in 1...m$ fixed.

Proof of Proposition 3.3. The proof is an easy consequence of the Definition 2.11 of the random measure $q(\mathrm{d}t\,\mathrm{d}x)$. □

It is more difficult to define the stochastic integral on those sets $(0,T]\times A$, $A\in\mathcal{B}(E\setminus\{0\})$, such that $\nu((0,T]\times A)=\infty$. For real valued functions this problem was already discussed e.g. in [41] and [6, 45, 44] (for general martingale measures). Different definitions of stochastic integrals were proposed. What we call the "simple-p-integral" (Definition 3.21) corresponds, when $p=2$, $f(t,x,\omega)$ is real valued and has a version which is left-continuous in time, to the stochastic integral introduced e.g. in [16] (for point processes). Our definition of "strong-p-integral" (Definition 3.10) corresponds, when $p=2$, f is real valued, and $\nu(\mathrm{d}t\,\mathrm{d}x)=\mathrm{d}t\beta(\mathrm{d}x)$ (dt being the Lesbegue measure on $(\mathbb{R}_+,\mathcal{B}(\mathbb{R}_+))$), to the stochastic integrals introduced e.g. in [41] (for $(X_t)_{t\geq 0}$ an α-stable Lévy process) and [6] (for martingale measures on $\mathbb{R}^d$). In [38] we introduced also the concept of strong $p\wedge q$ stochastic-integral, $1\leq q<p$, which is used in Theorem 3.17.

Let $\nu\times P$ be the product measure on the semi-ring $\mathcal{B}(\mathbb{R}_+\times(E\setminus\{0\}))\times\mathcal{F}_\infty$ of the product sets $A\times F$, $A\in\mathcal{B}(\mathbb{R}_+\times(E\setminus\{0\}))$, $F\in\mathcal{F}_\infty$. Let us also denote by $\nu\otimes P$ the unique extension of $\nu\times P$ on the product σ-algebra $F_\infty := \mathcal{B}(\mathbb{R}_+\times(E\setminus\{0\}))\otimes\mathcal{F}_\infty$ generated by $\mathcal{B}(\mathbb{R}_+\times(E\setminus\{0\}))\times\mathcal{F}_\infty$.

Let $p\geq 1$,

$$M_\nu^{T,p}(E/F) := \left\{ f\in M^T(E/F) : \int_0^T\int E[\|f(t,x)\|^p]\,\nu(\mathrm{d}t\,\mathrm{d}x)<\infty \right\} \tag{3.6}$$

where $E[f]$ we denotes the expectation with respect to the probability measure P.

In [38] we define the strong p-integral (Definition 3.10 below) and strong $p\wedge q$-integral (Definition 3.12 below) through approximation of the natural integrals of simple functions. Before we list some properties of the functions $f \in M_\nu^{T,p}(E/F)$, established in [38] (for the proofs we refer to [38]).

Definition 3.5. Let $f: \mathbb{R}_+ \times (E \setminus \{0\}) \times \Omega \to F$ be given. A sequence $\{f_n\}_{n\in\mathbb{N}}$ of $F_T/\mathcal{B}(F)$ measurable functions is L^p-approximating f on $(0,T] \times A \times \Omega$ w.r.t. $\nu \otimes P$, if f_n is $\nu \otimes P$-a.s. converging to f, when $n \to \infty$, and

$$\lim_{n\to\infty} \int_0^T \int_A E[\|f_n(t,x) - f(t,x)\|^p]\, d\nu = 0 \tag{3.7}$$

i.e. $\|f_n - f\|$ converges to zero in $L^p((0,T] \times A \times \Omega, \nu \otimes P)$, when $n \to \infty$.

Theorem 3.6. *Let $p \geq 1$. Suppose that the compensator $\nu(\mathrm{d}t\,\mathrm{d}x)$ of the Poisson random measure $N(\mathrm{d}t\,\mathrm{d}x)(\omega)$ satisfies the following hypothesis:*

Hypothesis A: *ν is a product measure $\nu = \alpha \otimes \beta$ on the σ-algebra generated by the semi-ring $\mathcal{S}(\mathbb{R}_+) \times \mathcal{B}(E \setminus \{0\})$, of a σ-finite measure α on $\mathcal{S}(\mathbb{R}_+)$, such that $\alpha([0,T]) < \infty$, $\forall T > 0$, and a σ-finite measure β on $\mathcal{B}(E \setminus \{0\})$.*

Let $T > 0$, then for all $f \in M_\nu^{T,p}(E/F)$ and all $A \in \mathcal{B}(E \setminus \{0\})$, there is a sequence of simple functions $\{f_n\}_{n\in\mathbb{N}}$ which satisfies the following

Property P: *$f_n \in \Sigma(E/F)\, \forall n \in \mathbb{N}$, and f_n is L^p-approximating f on $(0,T] \times A \times \Omega$ w.r.t. $\nu \otimes P$.*

Theorem 3.7. *Let $p \geq 1$. Let $T > 0$, $f \in M_\nu^{T,p}(E/F)$, and suppose that the following hypothesis is satisfied:*

Hypothesis B: *$f(t,x,\omega)$ is left-continuous for $t \in (0,T]$, for each $x \in E \setminus \{0\}$ and P-a.e. $\omega \in \Omega$ fixed.*

Then for all $A \in \mathcal{B}(E \setminus \{0\})$, there is a sequence of simple functions $\{f_n\}_{n\in\mathbb{N}}$, $f_n \in \Sigma(E/F)\, \forall n \in \mathbb{N}$, such that f_n is L^p-approximating f on $(0,T] \times A \times \Omega$ w.r.t. $\nu \otimes P$.

Remark 3.8. From the proofs of Theorem 3.6 and Theorem 3.7 it follows that if $f \in M_\nu^{T,p}(E/F) \cap M_\nu^{T,q}(E/F)$ with $p,q \geq 1$, $T > 0$, and hypothesis A in Theorem 3.6 or hypothesis B in Theorem 3.7 is satisfied, then, for all $A \in \mathcal{B}(E \setminus \{0\}$ there is a sequence of simple functions $\{f_n\}_{n\in\mathbb{N}}$, $f_n \in \Sigma(E/F)\, \forall n \in \mathbb{N}$, such that f_n is both L^p- and L^q-approximating f on $(0,T] \times A \times \Omega$ w.r.t. $\nu \otimes P$.

Definition 3.9. Let $p \geq 1$, $L_p^F(\Omega, \mathcal{F}, P)$ is the space of F-valued random variables, such that $E\|Y\|^p = \int \|Y\|^p dP < \infty$. We denote by $\|\cdot\|_p$ the quasi-norm ([25]) given by $\|Y\|_p = (E\|Y\|^p)^{1/p}$. Given $(Y_n)_{n\in\mathbb{N}}, Y \in L_p^F(\Omega, \mathcal{F}, P)$, we write

$$\lim\nolimits^p_{n\to\infty} Y_n = Y$$

if $\lim_{n\to\infty} \|Y_n - Y\|_p = 0$

Definition 3.10. Let $p \geq 1, T > 0$. We say that f is strong p-integrable on $(0,T] \times A$ w.r.t. $q(\mathrm{d}t\,\mathrm{d}x)(\omega)$, $A \in \mathcal{B}(E \setminus \{0\})$, if there is a sequence $\{f_n\}_{n\in\mathbb{N}} \in \Sigma(E/F)$, such that f_n is L^p-approximating f on $(0,T]\times A\times\Omega$ w.r.t. $\nu \otimes P$, and for any such sequence the limit of the natural integrals of f_n w.r.t. $q(\mathrm{d}t\,\mathrm{d}x)$ exists in $L_p^F(\Omega, \mathcal{F}, P)$ for $n \to \infty$, i.e.

$$\int_0^T \int_A f(t,x,\omega) q(\mathrm{d}t\,\mathrm{d}x)(\omega) := \lim_{n\to\infty}^p \int_0^T \int_A f_n(t,x,\omega) q(\mathrm{d}t\,\mathrm{d}x)(\omega) \tag{3.8}$$

exists. Moreover, the limit (3.8) does not depend on the sequence $\{f_n\}_{n\in\mathbb{N}} \in \Sigma(E/F)$, which is L^p-approximating f on $(0,T] \times A \times \Omega$ w.r.t. $\nu \otimes P$.

We call the limit in (3.8) the strong p-integral of f w.r.t. $q(\mathrm{d}t\,\mathrm{d}x)$ on $(0,T] \times A$

Remark 3.11. Let $f \in M_\nu^{T,r}(E/F) \cap M_\nu^{T,q}(E/F)$, $r, q \geq 1$. If f is both r- and q- strong integrable on $(0,T] \times A$, $r, q \geq 1$, then from Remark 3.8 it follows that the strong r-integral coincides with the strong q-integral. In fact from any sequence f_n, for which the limit in (3.8) holds with $p = r$ and $p = q$, it is possible to extract a subsequence for which the convergence (3.8) holds also P- a.s..

It is useful (see e.g. in Theorem 3.17) to introduce also the following definition of "strong $p \wedge q$-integral", with $1 \leq q < p$, which is approximated by the natural integrals of a smaller class of simple functions, than the strong p-integral.

Definition 3.12. Let $1 \leq q < p$, $A \in \mathcal{B}(E \setminus \{0\})$, $T > 0$. We say that f is strong $p \wedge q$-integrable on $(0,T] \times A$ w.r.t. $q(\mathrm{d}t\,\mathrm{d}x)(\omega)$, if there is a sequence $\{f_n\}_{n\in\mathbb{N}} \in \Sigma(E/F)$, such that f_n is both L^p- and L^q-approximating f on $(0,T] \times A \times \Omega$ w.r.t. $\nu \otimes P$, and for any such sequence the limit of the natural integrals of f_n w.r.t. $q(\mathrm{d}t\,\mathrm{d}x)$ exists in $L_p^F(\Omega, \mathcal{F}, P)$ for $n \to \infty$; i.e. (3.8) exists. Moreover, the limit (3.8) does not depend on the sequence $\{f_n\}_{n\in\mathbb{N}} \in \Sigma(E/F)$, such that f_n is L^p- and L^q-approximating f on $(0,T] \times A \times \Omega$ w.r.t. $\nu \otimes P$.

We call the limit in (3.8) the strong $p \wedge q$-integral of f w.r.t. $q(\mathrm{d}t\,\mathrm{d}x)$ on $(0,T] \times A$

Remark 3.13. If $f \in M_\nu^{T,p} \cap M_\nu^{T,q}$, and f is strong p -integrable and strong $p \wedge q$-integrable on $(0,T] \times A$, then the strong p-integral of f coincides with the strong $p \wedge q$-integral on $(0,T] \times A$.

Proposition 3.14. *Let $p, q \geq 1$. Let f be strong p (resp. strong $p \wedge q$)-integrable on $(0,T] \times A$, $A \in \mathcal{B}(E \setminus \{0\})$. Then the strong p (resp. strong $p \wedge q$)-integral $\int_0^t \int_A f(s,x) q(\mathrm{d}s\,\mathrm{d}x)$, $t \in [0,T]$, is an $\mathcal{F}_t$ -martingale with mean zero.*

(For the proof of Proposition 3.14 we refer to [38].) In [38] we find sufficient conditions for the existence of the strong p- (resp. strong $p \wedge q$) - integral. We report these results here, but only with a sketch of the proof. Starting from here, we assume in this Section that either hypothesis A or hypothesis B is satisfied.

Theorem 3.15. *Let $f \in M_\nu^{T,1}(E/F)$, then f is strong 1-integrable w.r.t. $q(\mathrm{d}t, \mathrm{d}x)$ on $(0,t] \times A$, for any $0 < t \leq T$, $A \in \mathcal{B}(E \setminus \{0\})$. Moreover*

$$E\Big[\Big\|\int_0^t \int_A f(s,x) q(\mathrm{d}s\,\mathrm{d}x)\Big\|\Big] \leq 2 \int_0^t \int_A E[\|f(s,x)\|] \nu(\mathrm{d}s\,\mathrm{d}x) \quad (3.9)$$

Remark 3.16. By definition of Bochner integral and a Theorem of S.Bochner (see e.g. [46], Chap. V, §5); $f \in M_\nu^{T,1}(E/F)$ if and only if $f \in M^T(E/F)$ and f is Bochner integrable w.r.t. $\nu \otimes P$. Moreover, from Definition 3.10 and Theorem 3.15 it also follows that f is strong 1-integrable, iff $f \in M^T(E/F)$ and f is Bochner integrable w.r.t. $\nu \otimes P$.

Theorem 3.17. *Let $f \in M_\nu^{T,1}(E/F) \cap M_\nu^{T,2}(E/F)$, then f is strong $1 \wedge 2$-integrable w.r.t. $q(\mathrm{d}t, \mathrm{d}x)$ on $(0,t] \times A$, for any $0 < t \leq T$, $A \in \mathcal{B}(E \setminus \{0\})$. Moreover (3.9) holds true and the following inequality holds*

$$E\Big[\Big\|\int_0^t \int_A f(s,x) q(\mathrm{d}s\,\mathrm{d}x)\Big\|^2\Big] \leq 16 \Big(\int_0^t \int_A E[\|f(s,x)\|] \nu(\mathrm{d}s\,\mathrm{d}x)\Big)^2 + 4 \int_0^t \int_A E[\|f(s,x)\|^2] \nu(\mathrm{d}s\,\mathrm{d}x) \quad (3.10)$$

The strong $1 \wedge 2$-integral of f coincides with the strong 1-integral of f on $(0,t] \times A$.

We recall here the definition of type 2 Banach spaces (see e.g. [1, 25])

Definition 3.18. A separable Banach space F is of type 2, if there is a constant K_2, such that if $\{X_i\}_{i=1}^n$ is any finite set of centered independent F-valued random variables, such that $E[\|X_i\|^2] < \infty$, then

$$E\Big[\Big\|\sum_{i=1}^n X_i\Big\|^2\Big] \leq K_2 \sum_{i=1}^n E[\|X_i\|^2] \quad (3.11)$$

We remark that any separable Hilbert space is a Banach space of type 2. Moreover, a Banach space is of type 2 as well as of cotype 2 if and only if it is isomorphic to a Hilbert space [21], where a Banach space of cotype 2 is defined by putting $\geq$ instead of $\leq$ in (3.11) (see [1], or [25]).

Theorem 3.19. *Suppose that F is a separable Banach space of type 2. Let $f \in M_\nu^{T,2}(E/F)$, then f is strong 2 -integrable w.r.t. $q(\mathrm{d}t\,\mathrm{d}x)$ on $(0,t] \times A$, for any $0 < t \leq T$, $A \in \mathcal{B}(E \setminus \{0\})$. Moreover*

$$E\Big[\Big\|\int_0^t \int_A f(s,x)q(\mathrm{d}s\,\mathrm{d}x)\Big\|^2\Big] \leq 4K_2 \int_0^t \int_A E[\|f(s,x)\|^2]\nu(\mathrm{d}s\,\mathrm{d}x) \quad (3.12)$$

where K_2 is the constant in the Definition 3.18 of type 2 Banach spaces.

Theorem 3.20. *Suppose $(F, \mathcal{B}(F)) := (H, \mathcal{B}(H))$ is a separable Hilbert space. Let $f \in M_\nu^{T,2}(E/H)$, then f is strong 2-integrable w.r.t. $q(\mathrm{d}t\,\mathrm{d}x)$ on $(0,t] \times A$, for any $0 < t \leq T$, $A \in \mathcal{B}(E \setminus \{0\})$. Moreover*

$$E\Big[\Big\|\int_0^t \int_A f(s,x)q(\mathrm{d}s\,\mathrm{d}x)\Big\|^2\Big] = \int_0^t \int_A E[\|f(s,x)\|^2]\nu(\mathrm{d}s\,\mathrm{d}x). \quad (3.13)$$

Proof of Theorem 3.15. First we prove that under the hypothesis of Theorem 3.15 f is strong 1-integrable. We start remarking that given $f \in \Sigma(E/F)$ the inequality (3.9) holds. In fact, given f as in (3.4), it follows from Theorem 2.3

$$\begin{aligned}
&E\Big[\Big\|\int_0^t \int_A f(s,x)q(\mathrm{d}s\,\mathrm{d}x)\Big\|\Big] \\
&\leq E\Big[\sum_{k=1}^{n-1}\sum_{l=1}^{m_k} \|a_{k,l}\| \mathbf{1}_{F_{k,l}} |N((t_k, t_{k+1}] \cap (0,t] \times A_{k,l} \cap A) \\
&\qquad\qquad - \nu((t_k, t_{k+1}] \cap (0,t] \times A_{k,l} \cap A)|\Big] \\
&\leq E\Big[\sum_{k=1}^{n-1}\sum_{l=1}^{m_k} \|a_{k,l}\| \mathbf{1}_{F_{k,l}} N((t_k, t_{k+1}] \cap (0,t] \times A_{k,l} \cap A) \\
&\qquad\qquad + \nu((t_k, t_{k+1}] \cap (0,t] \times A_{k,l} \cap A)]\Big] \\
&= 2\sum_{k=1}^{n-1}\sum_{l=1}^{m_k} \|a_{k,l}\| P(F_{k,l}) \nu((t_k, t_{k+1}] \cap (0,t] \times A_{k,l} \cap A)) \\
&= 2\int_0^t \int_A E[\|f(s,x)\|]\nu(\mathrm{d}s\,\mathrm{d}x). \qquad (3.14)
\end{aligned}$$

Let $f \in M_\nu^{T,1}(E/F)$, and let $\{f_n\}_{n\in\mathbb{N}} \in \Sigma(E/F)$ be a sequence L^1-approximating f in $(0,T] \times A \times \Omega$ w.r.t. $\nu \otimes P$. Then

$$\begin{aligned}
&E\Big[\Big\|\int_0^t \int_A (f_n(s,x) - f_m(s,x))q(\mathrm{d}s\,\mathrm{d}x)\Big\|\Big] \\
&\qquad \leq 2\int_0^t \int_A E[\|f_n(s,x) - f_m(s,x)\|]\nu(\mathrm{d}s\,\mathrm{d}x) \qquad (3.15)
\end{aligned}$$

so that $\int_0^t \int_A f_n(s,x)q(\mathrm{d}s\,\mathrm{d}x)$ is a Cauchy sequence in $L_1^F(\Omega,\mathcal{F},P)$ and the limit (3.8) exists for $p=1$. (Moreover the limit does not depend on the choice of the sequence $\{f_n\}_{n\in\mathbb{N}}$).

Using the triangle inequality it can be proven that inequality (3.9) holds for any $f \in M_\nu^{T,1}(E/F)$. □

Proof of Theorem 3.17. Let us prove that given $f \in \Sigma(E/F)$ the inequality (3.10) holds. In fact, given f as in (3.4) it follows

$$
\begin{aligned}
&E\Big[\Big\|\int_0^t\int_A f(t,x,\omega)q(\mathrm{d}t\,\mathrm{d}x)\Big\|^2\Big]\\
&\le E\Big[\Big(\sum_{k=1}^{n-1}\sum_{l=1}^{m}\|a_{k,l}\|\mathbf{1}_{F_{k,l}}|q((t_k,t_{k+1}]\cap(0,t]\times A_{k,l}\cap A)|\Big)^2\Big]\\
&= E\Big[\sum_{k=1}^{n-1}\sum_{l=1}^{m}\sum_{h=1}^{n-1}\sum_{j=1}^{m}\|a_{k,l}\|\mathbf{1}_{F_{k,l}}|q((t_k,t_{k+1}]\cap(0,t]\times A_{k,l}\cap A)\\
&\qquad\qquad\times\|a_{h,j}\|\mathbf{1}_{F_{h,j}}|q((t_h,t_{h+1}]\cap(0,t]\times A_{h,j}\cap A)\Big]\\
&= \sum_{k=1}^{n-1}\sum_{l=1}^{m}\sum_{\substack{h=1\\ h\neq k}}^{n-1}\sum_{\substack{j=1\\ j\neq l}}^{m}\|a_{k,l}\|P(F_{k,l})E[|q((t_k,t_{k+1}]\cap(0,t]\times A_{k,l}\cap A)]\\
&\qquad\qquad\times\|a_{h,j}\|P(F_{h,j})E[|q((t_h,t_{h+1}]\cap(0,t]\times A_{h,j}\cap A)]\\
&\quad+\sum_{k=1}^{n-1}\sum_{l=1}^{m}\|a_{k,l}\|^2P(F_{k,l}))E[|q((t_k,t_{k+1}]\cap(0,t]\times A_{k,l}\cap A)|^2]\\
&\le\Big(\sum_{k=1}^{n-1}\sum_{l=1}^{m}\|a_{k,l}\|P(F_{k,l})2\nu((t_k,t_{k+1}]\cap(0,t]\times A_{k,l}\cap A)\Big)^2\\
&\quad+\sum_{k=1}^{n-1}\sum_{l=1}^{m}\|a_{k,l}\|^2P(F_{k,l}))\nu((t_k,t_{k+1}]\cap(0,t]\times A_{k,l}\cap A)]\\
&=4\Big(\int_0^t\int_A E[\|f(t,x,\omega)\|]\nu(\mathrm{d}t\,\mathrm{d}x)\Big)^2\\
&\qquad+\int_0^t\int_A E[\|f(t,x,\omega)\|^2]\nu(\mathrm{d}t\,\mathrm{d}x).
\end{aligned}
\tag{3.16}
$$

Let $f \in M_\nu^1(E/F)\cap M_\nu^2(E/F)$, and let $\{f_n\}_{n\in\mathbb{N}} \in \Sigma(E/F)$ be a sequence L^1- and L^2-approximating f on $(0,t]\times A\times\Omega$ w.r.t. $\nu\otimes P$,

$A \in \mathcal{B}(E \setminus \{0\})$. We obtain

$$E\Big[\Big\|\int_0^t \int_A (f_n(s,x) - f_m(s,x))q(\mathrm{d}s\,\mathrm{d}x)\Big\|^2\Big]$$
$$\leq 4\Big(\int_0^t \int_A E[\|f_n(s,x) - f_m(s,x)\|]\nu(\mathrm{d}s\,\mathrm{d}x)\Big)^2$$
$$+ \int_0^t \int_A E[\|f_n(s,x) - f_m(s,x)\|^2]\nu(\mathrm{d}s\,\mathrm{d}x) \tag{3.17}$$

so that $\int_0^t \int_A f_n(s,x)q(\mathrm{d}s\,\mathrm{d}x)$ is a Cauchy sequence in $L_2^E(\Omega, \mathcal{F}, P)$ and the limit (3.8), for $p = 2$, exists. That inequality (3.10) holds for all $f \in M_\nu^1(E/F) \cap M_\nu^2(E/F)$ is proven in a similar way than inequality (3.9) in theorem 3.15 using the triangle inequality.

That the strong $1 \wedge 2$-integral coincides with the strong 1-integral follows from Remark 3.13. □

Proof of Theorem 3.19. The proof is similar to the proof of Theorem 3.17, but uses however the property of type 2 Banach spaces. □

Proof of Theorem 3.20. The proof is similar to the proof of Theorem 3.17, but uses however the scalar product of the Hilbert space H to prove equality (3.13). □

In [38] we introduce the "simple p-integrals" w.r.t. the compensated Poisson random measures $q(\mathrm{d}t\,\mathrm{d}x)$.

Definition 3.21. Let $p \geq 1$, $T > 0$, $A \in \mathcal{B}(E \setminus \{0\})$. Let $f \in M^T(E/F)$ and f bounded $\nu \otimes P$-a.s. on $(0,T] \times A \times \Omega$. f is simple-p-integrable on $(0,T] \times A \times \Omega$, if for all sequences $\{\delta_n\}_{n \in \mathbb{N}}$, $\delta_n \in \mathbb{R}_+$, such that $\lim_{n\to\infty} \delta_n = 0$, the limit

$$\lim_{n\to\infty}^p \sum_{0<s\leq T} \mathbf{1}_{\Delta_s \in \Lambda_{\delta_n} \cap A}\, f(s, \Delta X_s, \omega) - \int_0^T \int_{\Lambda_{\delta_n} \cap A} f(t,x,\omega)\nu(\mathrm{d}t\,\mathrm{d}x) \tag{3.18}$$

with

$$\Lambda_{\delta_n} := \{x \in E \setminus \{0\} : \delta_n < \|x\|\} \tag{3.19}$$

exists and does not depend on the choice of the sequences $\{\delta_n\}_{n \in \mathbb{N}}$, satisfying the above properties.

We call the limit (3.18) the simple-p integral of f on $(0,T] \times A$ w.r.t. the compensated Poisson random measure $N(\mathrm{d}t\,\mathrm{d}x) - \nu(\mathrm{d}t\,\mathrm{d}x)$, and denote it with

$$\int_0^T \int_A f(t,x,\omega)[N(\mathrm{d}t\,\mathrm{d}x)(\omega) - \nu(\mathrm{d}t\,\mathrm{d}x)]. \tag{3.20}$$

Remark 3.22. If f is both r- and q-simple integrable on $(0,T] \times A$, $p, q \geq 1$, then the limit in (3.18) with $p = r$ coincides P-a.s. with the limit in (3.18) with $p = q$, as there exist in both cases a subsequence of a sequence $\{\delta_n\}_{n\in\mathbb{N}}$, such that $\lim_{n\to\infty} \delta_n = 0$, for which the convergence (3.18) holds also P- a.s.

Remark 3.23. Let f, g be simple integrable on $(0,T] \times A$. For any $\alpha, \beta \in \mathbb{R}$, $\alpha f + \beta g$ is simple integrable on $(0,T] \times A$ and

$$\begin{aligned}\alpha \int_0^T \int_A & f(t,x)[N(\mathrm{d}t\,\mathrm{d}x) - \nu(\mathrm{d}t\,\mathrm{d}x)] \\ &+ \beta \int_0^T \int_A g(t,x)[N(\mathrm{d}t\,\mathrm{d}x) - \nu(\mathrm{d}t\,\mathrm{d}x)] \\ &= \int_0^T \int_A (\alpha f(t,x) + \beta g(t,x))[N(\mathrm{d}t\,\mathrm{d}x) - \nu(\mathrm{d}t\,\mathrm{d}x)].\end{aligned} \tag{3.21}$$

In [38]we prove the following results, Theorems 3.24-3.25.

Theorem 3.24. *Let $A \in \mathcal{B}(E \setminus \{0\})$, $f \in M_\nu^{T,1}(E/F)$, and assume f is left continuous in the time interval $(0,T]$ for every $x \in A$ and P-a.e. $\omega \in \Omega$ fixed. Then f is simple 1-integrable on $(0,T] \times A$ and the simple 1-integral coincides with the strong 1-integral.*

Moreover, if $f \in M_\nu^{T,1}(E/F) \cap M_\nu^{T,2}(E/F)$, then f is simple 2-integrable on $(0,T] \times A$ and the simple 2-integral coincides with the strong 1 and strong $1 \wedge 2$-integral.

Theorem 3.25. *Let F be a separable Banach space of type 2. Let $A \in \mathcal{B}(E \setminus \{0\})$, $f \in M_\nu^{T,2}E/F)$, and let f be left continuous in the time interval $(0,T]$ for every $x \in A$ and P-a.e. $\omega \in \Omega$ fixed.*

Then f is simple 2-integrable on $(0,T] \times A$. The simple 2-integral coincides with the strong 2-integral.

Remark 3.26. In the proof of the Lévy-Itô decomposition on Banach spaces in [8] it is proven, that the function $f(x) = x$ is (in our terminology) always p-simple integrable, for $p \geq 1$. It is however known that there are examples (see e.g. [1, 25, 27]) of separable Banach spaces F, and Lévy measures ν, s.t. $x \notin M_\nu^{T,1}(E/F)$, so that from Remark 3.16 it follows that on such Banach spaces $f(x) = x$ is not strong-1 integrable w.r.t. the compensated Poisson random measure associated to a Lévy process with Lévy measure ν.

4. Stochastic integration of Banach valued deterministic functions

Prior to the results in [38], listed in §3, we defined and studied in [2] stochastic integrals of deterministic $\mathcal{B}(E \setminus \{0\})/\mathcal{B}(F)$-measurable functions $f\colon E \to F$ w.r.t. the random compensated measures $N_t(\mathrm{d}x) - \nu_t(\mathrm{d}x)$ at time t fixed (defined in Definition 2.10), however associated to Banach valued Lévy-processes $(X_t)_{t\geq 0}$. (In this case $\nu_t(\mathrm{d}x) = t\nu(\mathrm{d}x)$, with $\nu(\mathrm{d}x)$ a Lévy-measure on $(E\setminus\{0\}, \mathcal{B}(E\setminus\{0\}))$). In this Section we state the results obtained in [2], concerning the stochastic integrals of deterministic functions $f\colon E \to F$, for the more general case where the random measures $N_t(\mathrm{d}x) - \nu_t(\mathrm{d}x)$ are associated to Banach valued additive processes $(X_t)_{t\geq 0}$ (Definition 2.1). We shall see that these part of the results in [2] can be seen (a posteriori!) as a straight consequence of the results of [38] and Remarks 4.2, 4.4 and 4.5 below. We present part of these results in a slightly different way than in [2], with the scope of unifying the definitions of [2] and [38], and studying stochastic integrals of deterministic functions w.r.t. $N_t(\mathrm{d}x) - \nu_t(\mathrm{d}x)$, as a particular case of the analysis of the stochastic integrals of random functions w.r.t. $q(\mathrm{d}t\,\mathrm{d}x)$, analyzed in [38] and discussed in Section 3. E.g., instead of the strong $p\wedge q$-integral of a deterministic function f w.r.t. $N_t(\mathrm{d}x) - \nu_t(\mathrm{d}x)$, defined in Definition 4.3 below, we introduced in [2] the concept of boundedly strong p-stochastic integral (we refer to [2] for the definition of "boundedly strong p-integral"). The boundedly strong p-integral is obtained by approximation of a class of natural integrals of finite valued functions. This class is a smaller class than the class of natural integrals approximating the strong $p \wedge q$-integral, $1 < q \leq p$. The definition of boundedly strong p-integral can however not be applied to random functions.

Definition 4.1. Suppose $t > 0$, $A \in \mathcal{B}(E\setminus\{0\})$, and $f\colon E\setminus\{0\} \to F$ are given. A sequence $\{f_n\}_{n\in\mathbb{N}}$ of $\mathcal{B}(E \setminus \{0\})\ /\mathcal{B}(F)$ measurable functions is L^p-approximating f on A w.r.t. ν_t, if f_n is ν_t-a.s. converging to f, when $n \to \infty$, and

$$\lim_{n\to\infty} \int_A \|f_n(x) - f(x)\|^p \, d\nu_t = 0, \tag{4.1}$$

i.e. $\|f_n - f\|$ converges to zero in $L^p(A, \nu_t)$, when $n \to \infty$.

Remark 4.2. From the definition in Section 2 of the measures $\nu_t(\mathrm{d}x)$ on $\mathcal{B}(E\setminus\{0\})$ and $\nu(\mathrm{d}t\,\mathrm{d}x)$ on $\mathcal{B}(\mathbb{R}_+ \times E\setminus\{0\})$ and the definition of Bochner integral (see e.g. [46], Chap. V, §5) it follows that, if $f\colon E \setminus \{0\} \to F$ is $\mathcal{B}(E \setminus \{0\})\ /\mathcal{B}(F)$ measurable, then

$$\int_A f(x)\nu_t(\mathrm{d}x) = \int_0^t \int_A f(x)\nu(\mathrm{d}s\,\mathrm{d}x) \tag{4.2}$$

for all $A \in \mathcal{B}(E\setminus\{0\})$, $t > 0$. In fact, if f is finite valued, takes values on the sets $A_1, \ldots, A_N \in \mathcal{F}(E \setminus \{0\})$ (defined in (2.6)) and has the form

$$f(x) = \sum_{n=1}^{N} a_k \mathbf{1}_{A_k} \tag{4.3}$$

$$a_k \in F\,,\ A_k \in \mathcal{F}(E \setminus \{0\}) \text{ and } A_k \cap A_j = \emptyset \text{ if } j \neq k$$

then

$$\int_A f(x)\nu_t(\mathrm{d}x) = \sum_{n=1}^{N} a_k \nu_t(A_k \cap A) \tag{4.4}$$

$$= \int_0^t \int_A f(x)\nu(\mathrm{d}s\,\mathrm{d}x). \tag{4.5}$$

Definition 4.3. Let $t > 0$, $p \geq 1$ (resp. $1 \leq q < p$). We say that $f \colon E \to F$, which is $\mathcal{B}(E \setminus \{0\})/\mathcal{B}(F)$ measurable, is strong p (resp. strong $p{\wedge}q$)-integrable on A w.r.t. $N_t(\mathrm{d}x)(\omega) - \nu_t(\mathrm{d}x)$, $A \in \mathcal{B}(E\setminus\{0\})$, if there is a sequence $\{f_n\}_{n\in\mathbb{N}}$, of finite valued $\mathcal{F}(E\setminus\{0\})/\mathcal{B}(F)$-measurable functions (i.e. of the form (4.3)), such that f_n is L^p (resp. L^p and L^q)-approximating f on A w.r.t. ν_t, and for any such sequence the limit of the natural integrals of f_n exists in $\in L_p^F(\Omega, \mathcal{F}, P)$, for $n \to \infty$, i.e.

$$\int_A f(x)(N_t(\mathrm{d}x)(\omega) - \nu_t(\mathrm{d}x)) := \mathrm{lim}^p_{n\to\infty} \int_A f_n(x)(N_t(\mathrm{d}x)(\omega) - \nu_t(\mathrm{d}x)) \tag{4.6}$$

exists. Moreover, the limit (4.6) does not depend on the sequence $\{f_n\}_{n\in\mathbb{N}}$, of finite valued $\mathcal{F}(E \setminus \{0\})\ /\mathcal{B}(F)$-measurable functions which is L^p (resp. L^p and L^q)-approximating f on A w.r.t. ν_t.

We call the limit in (4.6) the strong p (resp. strong $p \wedge q$)-integral of f w.r.t. $N_t(\mathrm{d}x)(\omega) - \nu_t(\mathrm{d}x)$ on A.

Remark 4.4. From Remark 4.2 it follows that if a function $f \colon E \to F$ is strong p- (resp. strong $p \wedge q$-) integrable w.r.t. $q(\mathrm{d}s\,\mathrm{d}x)(\omega)$ on $(0,t] \times A$, then f is strong p- (resp. strong $p \wedge q$-) integrable w.r.t. $N_t(\mathrm{d}x)(\omega) - \nu_t(\mathrm{d}x)$ and the integrals coincide.

Remark 4.5. Let $p \geq 1$ and $f \colon E \to F$. $f \in M_\nu^{t,p}(E/F)$ if and only if $f \in M^p_{\nu_t}(E/F)$ with

$$M^p_{\nu_t}(E/F) := \{f \colon E \to F,\ f \text{ is } \mathcal{B}(E \setminus \{0\})/\mathcal{B}(F)\text{-measurable and } \int \|f(x)\|^p\, \nu_t(\mathrm{d}x) < \infty\}. \tag{4.7}$$

The following Theorems 4.6-4.8 are consequences of Theorems 3.15-3.20, Theorems 3.24-3.25 and Remarks 4.2, 4.4, 4.5. Theorems 4.6 and 4.7 have been previously proven in [2], however for the case were $N_t(\mathrm{d}x)(\omega) - \nu_t(\mathrm{d}x)$ is associated to a Lévy process $(X_t)_{t\geq 0}$. Also for this case, Theorem 4.7 has been proven in [2], however in a slightly different manner: using the concept of (the less general) "boundedly strong p-integral" instead of strong $1\wedge 2$-integral (we refer to [2] for the definition of "boundedly strong p-integral").

Theorem 4.6. *Let $f \in M^1_{\nu_t}(E/F)$, then f is simply 1-integrable and strong 1-integrable w.r.t. $N_t(\mathrm{d}x)(\omega) - \nu_t(\mathrm{d}x)$. The simple 1-integral coincides with the strong 1-integral. Moreover*

$$E\left[\left\|\int_A f(x)(N_t(\mathrm{d}x)(\omega) - \nu_t(\mathrm{d}x))\right\|\right] \leq 2\int_A \|f(x)\|\nu_t(\mathrm{d}x). \tag{4.8}$$

Theorem 4.7. *Assume $f \in M^1_{\nu_t}(E/F) \cap M^2_{\nu_t}(E/F)$. Then f is simply 2-integrable and strong $2\wedge 1$-integrable w.r.t. $N_t(\mathrm{d}x)(\omega) - \nu_t(\mathrm{d}x)$. The simple 2-integral coincides with the strong $2\wedge 1$-integral and strong 1 integral. Moreover*

$$\begin{aligned} &E\left[\left\|\int_A f(x)(N_t(\mathrm{d}x)(\omega) - \nu_t(\mathrm{d}x))\right\|^2\right] \\ &\quad\leq 16\left(\int_A \|f(x)\|(N_t(\mathrm{d}x)(\omega) - \nu_t(\mathrm{d}x))\right)^2 \\ &\qquad + 4\int_A \|f(x)\|^2(N_t(\mathrm{d}x)(\omega) - \nu_t(\mathrm{d}x)). \end{aligned} \tag{4.9}$$

Theorem 4.8. *Assume $f \in M^2_{\nu_t}(E/F)$ and F is a separable Banach space of type 2. Then f is simply 2-integrable and strong 2-integrable w.r.t. $N_t(\mathrm{d}x)(\omega)-\nu_t(\mathrm{d}x)$. The simple 2-integral coincides with the strong 2-integral. Moreover*

$$\begin{aligned} &E\left[\left\|\int_A f(x)(N_t(\mathrm{d}x)(\omega) - \nu_t(\mathrm{d}x))\right\|^2\right] \\ &\quad\leq 4K_2\int_A \|f(x)\|^2(N_t(\mathrm{d}x)(\omega) - \nu_t(\mathrm{d}x)). \end{aligned} \tag{4.10}$$

where K_2 is the constant in the Definition 3.18 *of type* 2 *Banach spaces. If F is a separable Hilbert space then*

$$E\left[\left\|\int_A f(x)(N_t(\mathrm{d}x)(\omega) - \nu_t(\mathrm{d}x))\right\|^2\right] = \int_A E[\|f(x)\|^2]\nu(\mathrm{d}x) \tag{4.11}$$

5. The Lévy-Itô decomposition theorem

The Lévy-Itô decomposition theorem on $\mathbb{R}^d$ was coinceved By Lévy [23, 24] and proved by Itô [17]. It states that any additive process on $\mathbb{R}^d$ is decomposed into a jump semimartingale and a continuous semimartingale driven by a Brownian motion. Its generalization to infinite dimensional state spaces are given in [18, 42, 43], where the decomposition is proven for the case where the state space is a (co-) nuclear space, and in [8], for the case of Banach spaces. In all these mentioned works the pure jump martingale part is given in terms of (what we call) the simple p stochastic-integral of the function $f(x) = x$ (see also the final Remark in [2]). In [2] we prove the Lévy-Itô decomposition theorem for Lévy processes on separable Banach spaces, however with the pure jump martingale part being a strong p-integral, $p = 1, 2$. Here we state the Lévy-Itô decomposition in this stronger form for general additive processes. We remark that for this stronger form of the decomposition we need the Lévy measure to satisfy condition i) or ii) in Theorem 5.1 below. On any separable Hilbert space E condition ii) is always satisfied. It follows in particular that on such spaces (and hence also on $\mathbb{R}^d$) the pure jump martingale part in the Lévy-Itô decomposition is a strong p-integral.

Theorem 5.1 (Lévy-Itô decomposition theorem on separable Banach spaces). *Let $(X_s)_{s\geq 0}$ be an additive process on a separable Banach space $(E, \mathcal{B}(E))$, and ν_t the corresponding Lévy measure at time $t > 0$. Suppose $N_t(\omega, \mathrm{d}x)$ is the associated Poisson random measure at time $t > 0$. Suppose one of the two following conditions hold*

i)

$$\int_{\{E\setminus 0\}} \min(1, \|x\|)\, \nu_t(\mathrm{d}x) < \infty\,, \tag{5.1}$$

ii) *F is a separable Banach space of type 2 and*

$$\int_{\{E\setminus 0\}} \min(1, \|x\|^2)\, \nu_t(\mathrm{d}x) < \infty\,. \tag{5.2}$$

Then for all K with $0 < K < \infty$, there is $\alpha_{K,t} \in E$, such that

$$X_t = B_t + \int_{\|x\|\leq K} x(N_t(\mathrm{d}x) - \nu_t(\mathrm{d}x)) + \alpha_{K,t} + \int_{\|x\|>K} xN_t(\mathrm{d}x) \quad P\text{-a.s.} \tag{5.3}$$

(we omit for simplicity to write the dependence on $\omega \in \Omega$), where $(B_t)_{t\geq 0}$ is an E-valued Brownian motion with 0-mean and $\int_{\|x\|>K} xN_t(\mathrm{d}x) =$

$\sum_{0<s\leq t} \Delta X_s \mathbf{1}_{\|\Delta X_s\|>K}$. *For all* $\Lambda \in \mathcal{F}(E \setminus \{0\})$, $(B_t)_{t\geq 0}$ *is independent of* $(N_t^\Lambda)_{t\geq 0}$, *(with the notation* $N_t^\Lambda(\omega) := N_t(\omega, \Lambda)$*).*

If condition ii) *is satisfied, the integral* $\int_{\|x\|<K} x(N_t(\mathrm{d}x) - \nu_t(\mathrm{d}x))$ *is the strong* 2*-integral of the function* $f(x) = x$ *w.r.t.* $N_t(\mathrm{d}x) - \nu_t(\mathrm{d}x)$.

If condition i) *is satisfied, then it is the strong* 1*- (and strong* $2 \wedge 1$*-) integral of the function* $f(x) = x$ *w.r.t.* $N_t(\mathrm{d}x) - \nu_t(\mathrm{d}x)$.

Proof. It is proven in [8] that X_t is decomposed like in (5.3), with

$$\int_{\|x\|<K} x(N_t(\mathrm{d}x) - \nu_t(\mathrm{d}x))$$

being the p-simple integral of x w.r.t. $(N_t(\mathrm{d}x) - \nu_t(\mathrm{d}x))$, with $p \geq 1$. (The proof in [8] is rather sketchy, but in the final Remark of [2] we give the precise references related to it.) It follows from Theorem 4.7 and 4.8 that $\int_{\|x\|<K} x(N_t(\mathrm{d}x) - \nu_t(\mathrm{d}x))$ is a strong 1- (and strong $2\wedge 1$-) integral, if condition i) is satisfied, and resp. a strong 2-integral, if condition ii) is satisfied. □

Remark 5.2. We remark that a more direct proof is proposed in [2], for the case where $(X_s)_{s\geq 0}$ is a Lévy process. In this case we can prove directly the decomposition in its strong formulation in Theorem 5.1, without using first the proof of [8].

Acknowledgments

I thank S. Albeverio for the joy of collaboration in [2] and S. Albeverio, V. Mandrekar and L. Tubaro for very useful discussions and comments related to this article and [38].

References

[1] Araujo A., Giné E.: *The central limit theorem for real and Banach valued random variables*, Wiley series in probability and mathematical statistics, New York, Chichester, Brisbane, Toronto, 1980.

[2] Albeverio S., Rüdiger B.: The Lévy-Itô decomposition theorem and stochastic integrals on separable Banach spaces, submitted, BiBoS preprint 2002.

[3] Albeverio S., Rüdiger B.: Infinite dimensional Stochastic Differential Equations obtained by subordination and related Dirichlet forms, *J. Funct. Anal.* **204** (2003) 122–156.

[4] Albeverio S., Rüdiger B., Wu J.-L.: Analytic and Probabilistic Aspects of Lévy Processes and Fields in Quantum Theory, in *Lévy Processes*, Barndorff-Nielen O. E., Mikosch T., Resnick S. I., eds., 187–224, Birkhäuser Boston, Inc., Boston, 2001.

[5] Bauer, H.: *Wahrscheinlichkeitstheorie und Grundzüge der Masstheorie*, 2nd edition, de Gruyter Lehrbuch, Walter de Gruyter, Berlin, New York, 1974.

[6] Bensoussan, A., Lions J.-L.: *Contrôle impulsionnel et inéquations quasi-variationnelles* (French) [*Impulse control and quasivariational inequalities*], Méthodes Mathématiques de l'Informatique, 11, Gauthier-Villars, Paris, 1982.

[7] Björk T., Di Masi G., Kabanov Y., Runggaldier W.: Towards a general theory of bond markets, *Finance Stoch.* **1** (1997) 141–174.

[8] Dettweiler E.: Banach space valued processes with independent increments and stochastic integrals, in *Probability in Banach spaces IV (Oberwolfach 1982)*, 54–83, Springer, Berlin, 1982.

[9] Dinculeanu N.: Stochastic Integral of Process Measures in Banach Spaces I. Process Measures with Integrable Variation, *Rend. Accad. Naz. Sci. XL Mem. Mat. Appl. (5)* **22** (1998) 85–128.

[10] Dinculeanu N.: Stochastic Integral of Process Measures in Banach Spaces II. Process Measures with Integrable Semivariation, *Rend. Accad. Naz. Sci. XL Mem. Mat. Appl. (5)* **22** (1998) 129–164.

[11] Dinculeanu N., Muthia M.: Stochastic Integral of Process Measures in Banach Spaces III. Square Integrable Martingale Measures, *Rend. Accad. Naz. Sci. XL Mem. Mat. Appl. (5)* **23** (1999) 1–56.

[12] Dynkin E. B.: *Die Grundlagen der Theorie der Markoffschen Prozesse*, Springer Verlag, Berlin-Göttingen-Heidelberg-New York, 1982.

[13] Frisch U., Shlesinger F., Zaslavski F. (eds.): Lévy flights and Related Topics in Physics, *Proceedings of the International Workshop (Nice, France, 27-30 June 1994)*.

[14] Gihman I. I., Skorohod A. V.: *The theory of stochastic processes II*, Springer, Berlin-Heidelberg-New York, 1975.

[15] Gilbarg D., Trudinger N. S.: *Elliptic Partial Differential Equations of Second Order*, Reprint of the 1998 Edition, Springer-Verlag, Berlin, Heidelberg, 2001.

[16] Ikeda N., Watanabe S.: *Stochastic Differential Equations and Diffusion Processes*, 2nd edition, North-Holland Mathematical Library, Vol. 24, North Holland Publishing Company, Amsterdam, Oxford, New York, 1989.

[17] Itô K.: On stochastic processes I (Infinitely divisible laws of probability), *Jap. J. Math.* **18** (1942) 261–301.

[18] Itô K.: Continuous additive $\mathcal{S}'$-processes. *Lecture Notes in Control and Inform. Sci.* **25**, 36–46, Springer, New York, 1980.

[19] Klafter J., Shlesinger F., Zumofen G.: Beyond Brownian motion, *Physics Today* **49** (1996) 33–39.

[20] Kolmogorov A. N., Fomin S. V.: *Elementi di teoria delle funzioni e di analisi funzionale*, Edizione MIR, Moscau (1980) [Translation from the russian: *Elementy teorii funktsij i funktisianal'nogo*, NAUKA, Moscow].

[21] Kwapién S.: Isomorphic characterisations of inner product spaces by orthogonal series with vector valued coefficients, *Studia Math.* **44** (1972) 583–595.

[22] Kwapień S., Woycziński W.: *Random series and stochastic integrals: single and multiple*, Probability and its Applications, Birkhäuser, Boston, Inc., Boston, MA, 1992.

[23] Lévy P.: Sur les intégrales dont les éléments sont des variables aléatoires indépendantes, *Ann. Scuola Norm. Sup. Pisa (2)* **3** (1934) 337–366; **4** (1935)

217–218 (Reprinted in "Oeuvres de Paul Lévy, Vol. 4," Gauthier-Villars, Paris, 1980.)

[24] Lévy P.: *Théorie de l'addition de variables aléatoires*, Gauthier-Villars, Paris, 1937.

[25] Linde W.: *Infinitely Divisible and Stable Measures on Banach spaces*, Teubner-Texte zur Mathematik, Band 58, Berlin, 1983.

[26] Jurek Z. J., Mason J. D.: *Operator-Limit Distributions in Probability Theory*, Wiley-Interscience Publication, John Wiley and Sons, Inc. New York, Chichester, Brisbane, Toronto, Singapore (1993).

[27] Mandrekar V., Hamadani G. G.: Lévy-Khinchine representation and Banach spaces of type and cotype, *Studia Math.* **66** (1979/80) 299–306.

[28] Mandrekar V., Rüdiger B.: Existence and uniqueness of path wise solutions for stochastic integral equations driven by non Gaussian noise on separable Banach spaces, (submitted) preprint SFB 611, Bonn, february 2003.

[29] Mandrekar V., Rüdiger B.: in preparation.

[30] Marcus M. B., Rosinski J.: L^1-norms of infinitely divisible random vectors and certain stochastic integrals, *Electron. Comm. Probab.* **6** (2001) 15–29.

[31] Mikulevicius R., Rozovskii B. L.: Normalized stochastic integrals in topological vector spaces,*Séminaire de Probabilités XXXII*, eds Azéma J., Émery M., Ledoux M., Yor M., Lecture Notes in Math., **1686**, Springer Verlag, Berlin, 1998.

[32] Pratelli M.: Integration stochastique et géométrie des espaces de Banach, *Séminaire de Probabilités, XXII*, 129–137, Lecture Notes in Math., **1321**, Springer, Berlin, 1988.

[33] Protter P.: *Stochastic Integration and Differential Equations, A New Approach*, Applications of Mathematics **21**, third printing, Springer Verlag, Berlin, Heidelberg, New York, 1995.

[34] Privault N., Wu J. L.: Poisson stochastic integration on Hilbert spaces, *Ann. Math. Blaise Pascal* **6** (1999) 41–60.

[35] Rosinski J.: Bilinear random integrals, *Dissertationes Math. (Rozpawy Mat.)* **259** (1987) 71 pp.

[36] Rosinski J.: Random integrals of space valued functions, *Studia Math.* **LXXVIII** (1984) 15–38.

[37] Rosinski J., Rowecka E.: Random integrals and stable measures in Banach spaces, *Bull. Polish Acad. Sci. Math.* **32** (1984) 363–373.

[38] Rüdiger B.: Stochastic Integration with respect to compensated Poisson random measures on separable Banach spaces, submitted, preprint SFB 611, Bonn, February 2003.

[39] Rüdiger B., Ziglio G.: The Itô formula for Banach valued stochastic integrals obtained by integration with respect to compensated Poisson random measures, in preparation.

[40] Sato K.: *Lévy Processes and Infinitely Divisible Distributions*, Cambridge University Press, Cambridge, U.K., 1999.

[41] Skorohod A. V.: *Studies in the theory of random processes*, Addison-Wesley Publishing Company, Inc., Reading, MA, 1965 [Translated from the Russian by Scripta Technica, Inc.]

[42] Üstünel A. S.: Additive processes on nuclear spaces. *The Annals of Probability* **12** (1984) 858–868.

[43] Üstünel A. S.: Stochastic integration on nuclear spaces and its applications, *Ann. Inst. Henri Poincaré, Série B* **XVIII** (1982) 165–200.

[44] Walsh J. B.: An introduction to stochastic partial differential equations, *École d'Été de Probabilités de Saint-Flour XIV (1984)*, ed. Hennequin P.-L., 266–439, Lecture Notes in Math., **1180**, Springer Verlag, Berlin, Heidelberg, New York, Tokyo, 1986.

[45] West B. J.: *An Essay of the Importance of Being Non Linear*, Lecture Notes in Biomathematics **62**, Springer-Verlag, Berlin, 1985.

[46] Yosida K.: *Functional analysis*, sixth edition, Classics in Mathematics, Springer Verlag, Berlin, 1980.

EXPONENTIAL ERGODICITY OF CLASSICAL AND QUANTUM MARKOV BIRTH AND DEATH SEMIGROUPS

Raffaella Carbone
Università di Pavia, Dipartimento di Matematica, via Ferrata 1, 27100 Pavia, Italy
carbone@dimat.unipv.it

Franco Fagnola
Università di Genova, Dipartimento di Matematica, via Dodecaneso 35, 16146 Genova, Italy
fagnola@dima.unige.it

Abstract With a quantum Markov semigroup $(\mathcal{T}_t)_{t\geq 0}$ on $\mathcal{B}(h)$ with a faithful normal invariant state ρ we associate the semigroup $(T_t)_{t\geq 0}$ on Hilbert-Schmidt operators on h (the $L^2(\rho)$ space) defined by $T_t(\rho^{s/2}x\rho^{(1-s)/2}) = \rho^{s/2}\mathcal{T}_t(x)\rho^{(1-s)/2}$. This allows us to study the spectrum of the infinitesimal generator of $(T_t)_{t\geq 0}$ and deduce informations on the speed of convergence to equilibrium of the given semigroup. We apply this idea to show that some quantum Markov semigroups related to birth-and-death processes converge to equilibrium exponentially rapidly in $L^2(\rho)$. Moreover, through unitary transformations, we extend these results to other semigroups as, for instance, the quantum Ornstein-Uhlenbeck semigroup.

1. Introduction

Let h be a separable Hilbert space with an orthonormal basis $(e_n)_{n\geq 0}$ and denote $\mathcal{B}(h)$ the algebra of all bounded operators on h. A *quantum dynamical semigroup* (QDS) $\mathcal{T} = (\mathcal{T}_t)_{t\geq 0}$ on $\mathcal{B}(h)$ is a weak*-continuous semigroup of completely positive normal maps on $\mathcal{B}(h)$. It is markovian if it is also identity preserving (i.e. $\mathcal{T}_t(\mathbb{1}) = \mathbb{1}$, where we denote by $\mathbb{1}$ the identity operator on h.) We say that a state (i.e. a positive operator on h with trace 1) ρ is invariant for $\mathcal{T}$ if $\mathrm{tr}(\rho\mathcal{T}_t(x)) = \mathrm{tr}(\rho x)$ for any time t and any x in $\mathcal{B}(h)$; it is faithful if the condition $\mathrm{tr}(\rho x) = 0$, for a non-negative x, implies that x is zero.

S. Albeverio et al. (eds.),
Proceedings of the International Conference on Stochastic Analysis and Applications, 169–183.

QMS are the natural generalization of classical Markov semigroups and were introduced in physics to model the decay to equilibrium of quantum open systems, and in mathematics to construct the theory of quantum Markov processes. The equilibrium state is represented by an invariant state ρ of the semigroup and the decay to the equilibrium state ρ is expressed by $\lim_{t\to\infty} \mathrm{tr}(\sigma \mathcal{T}_t(x)) = \mathrm{tr}(\rho x)$ for every $x \in \mathcal{B}(h)$ and every positive operator σ with trace 1.

In the applications it is often interesting to investigate the speed of convergence and in particular to verify if the above convergence occurs at an exponential speed.

This kind of problems have been extensively studied for classical Markov semigroups (see, for example, [6] and [13] and references therein). One of the most powerful techniques consists in associating with a given Markov semigroup $(S_t)_{t\geq 0}$, on a space of bounded measurable functions over $\mathbb{R}^d$, say, with invariant density π, another semigroup $(\widetilde{S}_t)_{t\geq 0}$ on $L^2(\mathbb{R}^d)$ defined by

$$\widetilde{S}_t(\pi^{1/2} f) = \pi^{1/2} S_t(f). \tag{1.1}$$

Since $\|\widetilde{S}_t(\pi^{1/2} f)\|_{L^2} = \int_{\mathbb{R}^d} (S_t f)^2(x)\pi(x)\,\mathrm{d}x$, studying the convergence of $(\widetilde{S}_t)_{t\geq 0}$ in $L^2(\mathbb{R}^d)$, is equivalent to studying the convergence of the semigroup $(S_t)_{t\geq 0}$ in the space $L^2(\mathbb{R}^d, \pi)$. This allows to use spectral analysis to study the infinitesimal generator of $(\widetilde{S}_t)_{t\geq 0}$ and give estimates of the speed of convergence to equilibrium of the given semigroup $(S_t)_{t\geq 0}$.

In [3] and [5], we introduced an analogous non commutative technique for studying the convergence problem for some QMS with a faithful normal invariant state ρ. In the first Sections of this paper, we give a brief exposition of the main related results.

In (1.1) we essentially have an embedding $f \mapsto \pi^{1/2} f$ of bounded measurable functions over $\mathbb{R}^d$ in $L^2(\mathbb{R}^d)$. In the non-commutative context, there are several analogues, which are all the embeddings of the form $x \mapsto \rho^{s/2} x \rho^{(1-s)/2}$, with $s \in [0, 1]$.

In Section 2 we describe "quantum" versions of the embedding and the properties of the embedded semigroup. In Section 3 we study a special class of semigroups, which are the quantum generalization of classical birth and death semigroups, and we describe some results about their exponential convergence. In Sections 4 and 5 we concentrate on some applications and, in particular, on the quantum Ornstein-Uhlenbeck semigroup and on its unitary transforms. These semigroups arise in several models for quantum open systems (see [1, 2, 9, 16]).

2. Embedding in $L^2(h)$

We consider a Markov semigroup $\mathcal{T} = (\mathcal{T}_t)_{t\geq 0}$ on $\mathcal{B}(h)$ with a faithful invariant state ρ and we call $\mathcal{L}$ its infinitesimal generator. In the algebra $\mathcal{B}(h)$ it is difficult to study the convergence speed of a semigroup towards the invariant state. Like in the classical case, it is better to move to a Hilbert space; here we take the space $L^2(h)$ of the Hilbert-Schmidt operators on h endowed with the scalar product $\langle x, y\rangle = \mathrm{tr}(x^*y)$.

We introduce the embeddings

$$i_s \colon \mathcal{B}(h) \to L^2(h) \qquad i_s(x) = \rho^{s/2} x \rho^{(1-s)/2},$$

for $s \in [0, 1]$. The embedding corresponding to $s = 1/2$ will be called *symmetric.*

These maps are well defined on their domains and one can prove that *if ρ is a faithful state, i_s is an injective contraction with dense range. For $s = 1/2$ it is also positivity preserving.*

We now construct a semigroup T on $L^2(h)$ associated with $\mathcal{T}$ and the state ρ. For any $t \geq 0$, we consider the operator $\mathcal{T}_t \colon \mathcal{B}(h) \to \mathcal{B}(h)$ and we define a corresponding operator T_t by the relation

$$T_t \circ i_s = i_s \circ \mathcal{T}_t.$$

The operator T_t is well defined on $i_s(\mathcal{B}(h)) \subset L^2(h)$ since i_s is injective and it will obviously depend on s, even if we do not write it in order to make the notations simpler.

We have the following commutative diagram

$$\begin{array}{ccc} \mathcal{B}(h) & \xrightarrow{\mathcal{T}_t} & \mathcal{B}(h) \\ \big\downarrow i_s & & \big\downarrow i_s \\ L^2(h) & \xrightarrow{T_t} & L^2(h) \end{array}$$

where the operator T_t is initially only densely defined, but it can be seen that it has a bounded extension to the space $L^2(h)$.

We can easily prove that the family $(T_t)_{t\geq 0}$ is still a semigroup. A little more difficult is proving that this semigroup is contractive in L^2. In the classical case the L^2-contractivity is an easy consequence of Jensen inequality. In the quantum context it has been proved only with the help of some additional hypotheses, for example in [14] with KMS symmetry of the semigroup and in [15] with a detailed balance condition. We have proved it in the case the semigroup on $\mathcal{B}(h)$ is contractive, identity preserving, and enjoys the so-called Schwarz property, i.e.

$$\mathcal{T}_t(x^*)\mathcal{T}_t(x) \leq \mathcal{T}_t(x^*x), \qquad \text{for} \quad x \in \mathcal{B}(h).$$

This property, which is sometimes also called $(1 + 1/2)$-positivity, is a much weaker condition than complete positivity, but a little stronger than positivity (i.e. $\mathcal{T}_t(x)$ positive when x positive).

What we need is the following consequence

Theorem 2.1. *Fix s in $[0,1]$. If $\mathcal{T}$ is a QMS on $\mathcal{B}(h)$ with an invariant faithful state ρ, then there exists a unique strongly continuous contraction semigroup $T = (T_t)_{t\geq 0}$ on $L^2(h)$ such that, for x in $\mathcal{B}(h)$,*

$$T_t(i_s(x)) = i_s(\mathcal{T}_t(x)).$$

In particular, since $\mathcal{T}$ is identity-preserving, $\rho^{1/2}$ is an invariant vector for the semigroup T. Moreover, if L is the infinitesimal generator of the semigroup T, we can prove that, for x in $D(\mathcal{L})$, $i_s(x) \in D(L)$ and $L(i_s(x)) = i_s((\mathcal{L}(x))$, and, for $x = \mathbb{1}$, we obtain that $\rho^{1/2} \in D(L)$ and $L(\rho^{1/2}) = 0$.

So, by this embedding technique, we have constructed a strongly continuous contractive semigroup T defined on a Hilbert space and with an invariant vector $v = \rho^{1/2}$. For a semigroup with these properties we have a result which puts in strict relation the exponential convergence of the semigroup to the invariant vector and the spectral gap of the semigroup's infinitesimal generator.

Definition 2.2. If L is the infinitesimal generator of a strongly continuous contraction semigroup on a Hilbert space $\mathcal{K}$ with invariant vector v, we can define the spectral gap of L by

$$\operatorname{gap}(L) = \inf\left\{-\operatorname{Re}\langle x, Lx\rangle \mid x \in D(L),\ \langle v, x\rangle = 0,\ \|x\| = 1\right\}.$$

It is worth mentioning here that we will deal only with semigroups T with a unique invariant vector v. If this is not the case, it would be more useful to define $\operatorname{gap}(L)$ as the infimum of the same quantities over x orthogonal to the subspace generated by all the invariant vectors. We remark that, since L is the generator of a contractive semigroup, $\operatorname{gap}(L)$ is always non-negative. When L is self-adjoint, L is a negative operator and $\operatorname{gap}(L)$ is the infimum of non-zero eigenvalues of $(-L)$.

The following proposition can be proved as in [13] by a simple argument.

Proposition 2.3. *Let $(T_t)_{t\geq 0}$ be a strongly continuous contraction semigroup on a Hilbert space $\mathcal{K}$ with infinitesimal generator L and invariant vector v, then the spectral gap of the operator L is the maximum positive value ε satisfying, for any x in $\mathcal{K}$ and $t \geq 0$,*

$$\|T_t x - \langle v, x\rangle v\| \leq \mathrm{e}^{-\varepsilon t}\|x - \langle v, x\rangle v\|.$$

The importance of this result consists in highlighting the link between exponential convergence and spectral gap. Always remembering that, in our case, $\mathcal{K} = L^2(h)$ and $v = \rho^{1/2}$, it tells that $T_t(x)$ converges exponentially fast to its projection on $\rho^{1/2}$ for any x if and only if the spectral gap of L is strictly positive and, if this is positive, it coincides with the exact rate of (exponential) convergence. For this reason we can concentrate our study on the estimation of the spectral gap.

3. The birth and death class

Here we want to study the speed of convergence to equilibrium of a remarkable class of semigroups by applying the ideas introduced in the previous sections, so by looking for a method which allows us to estimate the spectral gap of its generators.

We now choose h to be the Hilbert space $l^2(\mathbb{N})$ of square-summable sequences of complex numbers and we denote by $(e_n)_{n\geq 0}$ its canonical basis. We define

the *right shift* operator S, with domain $D(S) = h$, by $Se_n = e_{n+1}$,
the *number* operator N, by

$$Ne_n = ne_n, \text{ with domain } D(N) = \{u \in h \mid \sum_{n\geq 0} n^2 |u_n|^2 < \infty\}.$$

The triplet $(l^2(\mathbb{N}), N, S)$ is sometimes called the "harmonic oscillator."

We take λ and μ two positive functions defined on $\mathbb{N}$, with $\mu(0) = 0$, and $(\lambda(n))_{n\geq 0}$ and $(\mu(n))_{n\geq 1}$ sequences of strictly positive real numbers. We consider the QMS on $\mathcal{B}(h)$, with infinitesimal generator of the form

$$\begin{aligned} \pounds(x) = &-\frac{1}{2}\left\{\mu^2(N)x - 2\mu(N)SxS^*\mu(N) + x\mu^2(N)\right\} \\ &-\frac{1}{2}\left\{\lambda^2(N)x - 2\lambda(N)S^*xS\lambda(N) + x\lambda^2(N)\right\} \end{aligned} \tag{3.1}$$

where S^* is the adjoint of S. This infinitesimal generator must be understood in the form sense, i.e., $\pounds(x)$ is a densely defined sesquilinear form on h for each $x \in \mathcal{B}(h)$.

The existence of QDS $(\mathcal{T}_t)_{t\geq 0}$ with infinitesimal generator which is a restriction of $\pounds$ has been proved (see for example [10]). Among these semigroups, we consider the minimal one, whose restriction to the commutative algebra generated by N is the infinitesimal generator of a birth-and-death process with infinitesimal rates $\lambda^2(n)$ and $\mu^2(n)$ (as shown in [8]).

Indeed, if we denote by M_f the multiplication operator by f, with $f = (f(n))_{n\geq 0}$ in $l^\infty(\mathbb{N})$ (i.e. $M_f = \sum_{n\geq 0} f(n)|e_n\rangle\langle e_n|$), it can be easily seen that $\mathcal{L}(M_f) = M_g$ where

$$g(n) \doteq \mu^2(n)(f(n-1) - f(n)) + \lambda^2(n)(f(n+1) - f(n)), \quad n \geq 0$$

and $f(-1)$ can be defined arbitrarily since $\mu(0) = 0$.
We remember that, according to Dirac notation, for u and v in h, $|u\rangle\langle v|$ is the map

$$|u\rangle\langle v| \colon h \longrightarrow h, \qquad |u\rangle\langle v|(w) = \langle v, w\rangle u.$$

For this diagonal semigroup it is known that it is identity preserving (i.e. Markov) if and only if

$$\sum_{n\geq 0} \frac{\lambda^2(0)\cdots\lambda^2(n-1)}{\mu^2(1)\cdots\mu^2(n)} \sum_{k=0}^{n} \frac{\mu^2(1)\cdots\mu^2(k)}{\lambda^2(0)\cdots\lambda^2(k)} = \infty \tag{3.2}$$

(see [12]). Therefore, since the identity operator belongs to the diagonal algebra, also the minimal quantum semigroup is Markov under the same condition.

Moreover it is well-known that the classical restriction has an invariant measure if and only if

$$\sum_{n\geq 1} \frac{\lambda^2(0)\cdots\lambda^2(n-1)}{\mu^2(1)\cdots\mu^2(n)} < \infty. \tag{3.3}$$

Thus, when (3.3) holds, an invariant (diagonal) state for our QMS is

$$\rho = c\left(|e_0\rangle\langle e_0| + \sum_{n\geq 1} \frac{\lambda^2(0)\cdots\lambda^2(n-1)}{\mu^2(1)\cdots\mu^2(n)} |e_n\rangle\langle e_n|\right),$$

with c a suitable normalization constant.

Therefore, when conditions (3.2) and (3.3) hold, we have the necessary hypotheses for embedding the semigroup in $L^2(h)$. Among all possible embeddings i_s, here we choose to use only the symmetric one ($s = 1/2$) since it seems to give better results for the class we are studying. We will find a semigroup $T = (T_t)_{t\geq 0}$ on $L^2(h)$ with infinitesimal generator L.

If we now introduce the subspace $\mathcal{M}$ of $L^2(h)$ generated by the elements $|e_j\rangle\langle e_k|$, starting from (3.1), for x in $\mathcal{M}$, we can explicitely write the action of the operator L

$$
\begin{aligned}
L(x) = &-\frac{1}{2}\sum_{j,k\geq 0}\Big(\lambda^2(j)+\lambda^2(k)+\mu^2(j)+\mu^2(k)\Big)\,x_{jk}\,|e_j\rangle\langle e_k| \\
&+\sum_{j,k\geq 0}\sqrt{\lambda(j)\lambda(k)\mu(j+1)\mu(k+1)}\,x_{j+1,k+1}\,|e_j\rangle\langle e_k| \\
&+\sum_{j,k\geq 1}\sqrt{\lambda(j-1)\lambda(k-1)\mu(j)\mu(k)}\,x_{j-1,k-1}|e_j\rangle\langle e_k|. \qquad (3.4)
\end{aligned}
$$

It is easily verified that such a generator L is always symmetric on $\mathcal{M}$. This is one of the reasons why we use symmetric embedding: with all other embeddings, $L_{|\mathcal{M}}$ is symmetric only under further conditions on the infinitesimal rates λ and μ.
Since, for the computation of the spectral gap, we have to minimize terms of the form $(-\operatorname{Re}\langle x, Lx\rangle)$, that can be written explicitely only for x in $\mathcal{M}$, it would be useful to know if $\mathcal{M}$ is a core for L and if L is self-adjoint. Unfortunately, this is not always true (see [3] for a counterexample), but, if we introduce a simple growth condition on birth rates and death rates, we have the following result.

Theorem 3.1. *Let us assume that the conditions (3.2) for the identity preservation and (3.3) for the existence of the invariant state hold and that there exists a real positive number c such that, for any k in $\mathbb{N}$,*

$$\lambda^2(k)+\mu^2(k)\leq c(1+k); \qquad (3.5)$$

then L is self-adjoint and $\mathcal{M}$ is a core for it.

So, when the rates λ and μ verify the hypotheses of Theorem 3.1, one can easily see that $\rho^{1/2}$ is the unique fix point in $L^2(h)$ of the semigroup T and

$$\operatorname{gap}(L)=\inf\Big\{-\langle x, Lx\rangle \,|\, x\in\mathcal{M},\ \|x\|^2=1,\ \langle\rho^{1/2},x\rangle=0\Big\}.$$

We now need some ideas in order to face this minimum problem and make it simpler. A suitable decomposition of the space $L^2(h)$ will turn out to be useful.
We introduce, for k in $\mathbb{N}$, the linear spaces

$$\mathcal{F}_k=\Big\{\textstyle\sum_{n\geq 0}\left(y_n|e_n\rangle\langle e_{n+k}|+z_n|e_{n+k}\rangle\langle e_n|\right)\,\Big|\,y,z\in l^2(\mathbb{N})\Big\}, \qquad (3.6)$$

i.e. $\mathcal{F}_0$ is the subset of the diagonal elements of $L^2(h)$ and $\mathcal{F}_k$, with $k>0$, is the subset of the elements of $L^2(h)$ which have non-zero entries only

on the k-th diagonals. It can be easily seen that $L^2(h) = \oplus_{k\geq 0}\mathcal{F}_k$. This actually is an orthogonal decomposition since $\mathrm{tr}(x^*y) = 0$ for $x \in \mathcal{F}_k$ and $y \in \mathcal{F}_m$ with $m \neq k$.
Indeed we can write any x in $L^2(h)$ in the form $x = \sum_{k\geq 0}\xi_k$, where ξ_k is the projection of x on $\mathcal{F}_k$, i.e.

$$\xi_0 = \sum_{n\geq 0} x_{nn}|e_n\rangle\langle e_n|,$$

$$\xi_k = \sum_{n\geq 0}(x_{n,n+k}|e_n\rangle\langle e_{n+k}| + x_{n+k,n}|e_{n+k}\rangle\langle e_n|), \quad k \geq 1.$$

The operators L associated with the class (3.1) fit with this decomposition in the sense that, for any ξ_k in $\mathcal{F}_k$, $L(\xi_k)$ still is in $\mathcal{F}_k$. Moreover, if ξ_k is a selfadjoint element of $\mathcal{F}_k$ with $k > 0$, i.e. $\xi_k = \sum_{n\geq 0}(u_n|e_n\rangle\langle e_{n+k}| + \bar{u}_n|e_{n+k}\rangle\langle e_n|)$, for some $u = (u_n)_{n\geq 0}$ in $l^2(\mathbb{N})$, then

$$L(\xi_k) = \sum_{n\geq 0}(w_n|e_n\rangle\langle e_{n+k}| + \bar{w}_n|e_{n+k}\rangle\langle e_n|)$$

for some $w = (w_n)_{n\geq 0}$ in $l^2(\mathbb{N})$. This correspondence defines new operators $L_k\colon l^2(\mathbb{N}) \to l^2(\mathbb{N})$, $L_k(u) = w$, and, by using (3.4), one can see that L_k is a tridiagonal linear operator. For $k = 0$ we obtain the diagonal restriction L_0 of L, which is the embedding in $l^2(\mathbb{N})$ of the generator of a classical birth and death semigroup.
After a little work, we can prove the following result.

Theorem 3.2. *Under the hypotheses of Theorem* 3.1,

$$\begin{aligned}\mathrm{gap}(L) &= \inf_{x\perp\rho^{1/2}, x\in D(L)\setminus\{0\}} -\frac{\langle x, Lx\rangle}{\|x\|^2}\\ &= \left(\inf_{k>0}\inf_{\xi_k\in\mathcal{F}_k\setminus\{0\}} -\frac{\langle \xi_k, L\xi_k\rangle}{\|\xi_k\|^2}\right) \wedge \left(\inf_{\xi_0\in\mathcal{F}_0\setminus\{0\},\xi_0\perp\rho^{1/2}} -\frac{\langle \xi_0, L\xi_0\rangle}{\|\xi_0\|^2}\right)\\ &= \left(\inf_{k>0}\inf_{u\in l^2(\mathbb{N})\setminus\{0\}} -\frac{\langle u, L_k u\rangle}{\|u\|^2}\right) \wedge \mathrm{gap}(L_0).\end{aligned}$$

So we have reduced the original minimum problem on an operator space to a countable sequence of easier problems: the computation of the spectral gap of a classical birth-and-death process ($\mathrm{gap}(L_0)$) and the minimum problems for a sequence of quadratic forms on $l^2(N)$. These quadratic forms are represented by tridiagonal operators that can be regarded as perturbations of L_0. Therefore their lower bounds can be estimated by using classical techniques for the study of the spectral gap and perturbation theory.

We refer to [4] and [5] for a detailed account of results obtained by this method.

4. Some examples

In this section we want to show the results we have obtained in computing or estimating the spectral gap of some generators in the birth and death class. Obviously any choice of the rates λ and μ verifying (3.2), (3.3) and (3.5) is allowed, but we have started considering some cases which seem to be interesting for the applications.

M/M/1 generator (corresponding to $\lambda(n) = \lambda$ for $n \geq 0$, $\mu(0) = 0$ and $\mu(n) = \mu$ for $n \geq 1$), whose restriction to the diagonal algebra gives the generator of an M/M/1 queue,

M/M/∞ generator, (corresponding to $\lambda(n) = \lambda$ and $\mu(n) = \mu\sqrt{n}$ for $n \geq 0$), whose restriction to the diagonal algebra gives the infinitesimal generator of an M/M/∞ queue.

Ornstein-Uhlenbeck generator (corresponding to $\lambda(n) = \lambda\sqrt{n+1}$, $\mu(n) = \mu\sqrt{n}$ for $n \geq 0$)

The spectral gap of the related classical birth-and-death processes was often known. For the estimates of the spectral gap in this case, we mainly refer to [6] by M.F. Chen and [13] by T. Liggett.

We have computed explicitely or estimate only the (quantum) spectral gap of these generators. We remark that the quantum gap of the Ornstein-Uhlenbeck generator had already been studied in [7] with a different method (see also [2], Sect.VII, and [5]).

The results, for the three generators mentioned above, are written in the following table, together with the estimate of the gap for another generator (the last in the table), which has a birth rate a little more general than the Ornstein-Uhlenbeck's one.

$\lambda(n)$	$\mu(n)$	classical gap	quantum gap
λ	μ	$(\mu-\lambda)^2$	$\text{gap}(L)=\begin{cases}\frac{1}{2}\mu^2-\lambda^2 & \text{if } 2\lambda<\mu\\ (\mu-\lambda)^2 & \text{if } \lambda<\mu<2\lambda\end{cases}$
λ	$\mu\sqrt{n}$	μ^2	$\frac{\mu^2}{2}(\sqrt{1+\theta^2}-\theta)^2 \leq \text{gap}(L) < \frac{\mu^2}{2}\frac{1-e^{-\theta^2}}{\theta^2}$
$\lambda\sqrt{n+1}$	$\mu\sqrt{n}$	$\mu^2-\lambda^2$	$\text{gap}(L)=\frac{1}{2}(\mu^2-\lambda^2)$
$\lambda\sqrt{n+\beta}$	$\mu\sqrt{n}$	$\mu^2-\lambda^2$	$\text{gap}(L) \geq \frac{1}{2}\mu^2\left(1-\frac{\lambda^2}{\mu^2}\right)^\beta$
β, λ, μ positive constants with $\lambda<\mu$, $\theta=\frac{\lambda}{\mu}$			

Starting from the results in the table, with some simple comparison techniques that we have studied, we have proved the exponential ergodicity (and estimate the spectral gap) also for other quite interesting

semigroups in the class.
The first is a semigroup arising in Quantum Optics, in the Jaynes-Cummings model (see [9]). It is characterized by the choice of rates

$$\lambda(n) = \sqrt{\lambda^2(n+1) + R^2 \sin^2(\phi\sqrt{n+1})}, \qquad \mu(n) = \mu\sqrt{n},$$

for $n \geq 0$, with λ, μ, R, ϕ positive constants.

The second is the Ornstein-Uhlenbeck semigroup on the q-deformed oscillator algebra. The corresponding rates are

$$\lambda(n) = \lambda\sqrt{\frac{1-q^{n+1}}{1-q}} \qquad \mu(n) = \mu\sqrt{\frac{1-q^n}{1-q}}$$

for $n \geq 0$, q in $(-1, 1)$, λ and μ positive constants. In [11] one can find a classification of the q-deformed algebras that highlights the importance of this q-deformed oscillator algebra. Here we only recall that this algebra can be represented by the triplet $(l^2(\mathbb{N}), N_q, S)$ where S is the usual right shift operator and N_q is the q-number operator, with domain $l^2(\mathbb{N})$, defined on the canonical basis by $N_q e_n = (1-q^n)/(1-q)e_n$.

5. Unitary transforms of the Ornstein-Uhlenbeck generator

The Ornstein-Uhlenbeck semigroup is the most important example for physical applications. It appears in several models in quantum open systems where it gives the evolution of observables. Many quantum Markov semigroups with an apparently more complicated generator (see, for instance, [1, 9, 16]) can be obtained via a unitary conjugation of the Ornstein-Uhlenbeck generator.

Let U be a unitary operator on h. For an operator $\mathcal{R}$ on $\mathcal{B}(h)$, with domain $D(\mathcal{R})$, we can define the U-transform of $\mathcal{R}$ as the operator $\mathcal{R}_U$ that satisfies, for all x in its domain $D(\mathcal{R}_U) = U^*D(\mathcal{R})U$, $\mathcal{R}_U(x) = U^*\mathcal{R}(UxU^*)U$. If now $\mathcal{L}$ (*Lindbladian*) is the infinitesimal generator of a QDS $\mathcal{T}$, then the U-transform of $\mathcal{T}$, that is the semigroup $\mathcal{T}_U$ such that $\mathcal{T}_{U,t}(x) = U^*\mathcal{T}_t(UxU^*)U$ for all x in $\mathcal{B}(h)$ and $t \geq 0$, is a QDS and its infinitesimal generator is the U-transform $\mathcal{L}_U$ of $\mathcal{L}$. This unitary equivalence allows to study the main properties like markovianity, existence of invariant states, ergodicity, ... on the simpler among $\mathcal{T}$ (resp. $\mathcal{L}$) and $\mathcal{T}_U$ (resp. $\mathcal{L}_U$). In particular, if ρ is an invariant faithful state for $\mathcal{T}$, then $U^*\rho U$ is an invariant faithful state for $\mathcal{T}_U$; so, by using the technique described in the previous sections, we can embed both the semigroups in $L^2(h)$ and we obtain that the embedded generators have the same spectral gap.

We will show that some apparently more complicated Lindbladians (see, e.g., [16], §6.1) are U-transforms of the quantum Ornstein-Uhlenbeck generator where U is a Weyl or squeeze operator. Therefore the results on the quantum Ornstein-Uhlenbeck generator hold also for these Lindbladians.

In this section, $\mathcal{L}_0$ will always denote the Ornstein-Uhlenbeck operator on the harmonic oscillator, that can be written in a slightly more general way than above

$$\begin{aligned}\mathcal{L}_0(x) = &-(\mu^2/2)\{a^*ax - 2a^*xa + xa^*a\}\\ &- (\lambda^2/2)\{aa^*x - 2axa^* + xaa^*\} + i\omega[N,x]\end{aligned}$$

with λ and μ positive constants, $a = SN^{1/2}$ (annihilation operator) and $a^* = N^{1/2}S^*$ (creation operator). Here we have added the term with the anti-commutation with N, but it is easy to see that this does not change the invariant state and the spectral gap, so, if $0 < \lambda < \mu$, $\mathcal{T}$ has a unique invariant state $\rho = (1-\lambda^2/\mu^2)\sum_{n\geq 0}(\lambda^2/\mu^2)^n\,|e_n\rangle\langle e_n|$, which is faithful, and the $L^2(\rho)$-spectral gap of $\mathcal{L}_0$ is equal to $(\mu^2-\lambda^2)/2$. For a more detailed description of these arguments and of the properties of the quantum Ornstein-Uhlenbeck operator, always see [5] and [7]. (Several results presented here were exposed also in the Ph.D. thesis of the first author).

We now introduce some remarkable unitary transforms (by using the so-called Weyl and Squeeze operators) and make explicit computations for the U-transform of the Ornstein-Uhlenbeck operator.

Weyl-transform. Let us consider, for z in $\mathbb{C}$, the Weyl operators

$$W_z = \exp(za^* - \bar{z}a);$$

these operators are unitary for all z and $W_z^* = W_{-z}$. It can be easily verified that

$$\begin{aligned}&W_z^*aW_z = a + z, \qquad W_z^*a^*W_z = a^* + \bar{z},\\ &W_z^*NW_z = N + \bar{z}a + za^* + |z|^2.\end{aligned} \tag{5.1}$$

The Weyl-transform of an operator $\mathcal{L}$ is the operator $\mathcal{L}_z$ defined by

$$\mathcal{L}_z(x) = W_z^*\mathcal{L}(W_zxW_z^*)W_z \text{ with domain } D(\mathcal{L}_z) = W_z^*D(\mathcal{L})W_z.$$

Proposition 5.1. *All the operators of the form*

$$\mathcal{L}(x) = \mathcal{L}_0(x) + i[\bar{\gamma}a + \gamma a^*, x] \tag{5.2}$$

with γ in $\mathbb{C}$, can be obtained as unitary transforms of the operator $\mathcal{L}_0$.

Proof. By using relations (5.1), with an easy computation, we can get

$$\mathcal{L}_z(x) = \mathcal{L}_0(x)+i\left[\left(\omega - i(\mu^2-\lambda^2)/2\right)za^* + \left(\omega + i(\mu^2-\lambda^2)/2\right)\bar{z}a, x\right].$$

So we just need to choose z in $\mathbb{C}$ such that $(\omega - i(\mu^2-\lambda^2)/2)\,z = \gamma$, that is $z = \frac{2\gamma[2\omega+i(\mu^2-\lambda^2)]}{4\omega^2+(\mu^2-\lambda^2)^2}$. Then $\mathcal{L}$ is the Weyl-transform of $\mathcal{L}_0$. □

Squeeze-transform. For $s \in \mathbb{C}$, we introduce the unitary squeeze operators

$$S(s) = \exp\frac{sa^{*2} - \bar{s}a^2}{2}.$$

These are unitary operators such that $S(s)^* = S(-s)$ and, if $s = re^{i\phi}$ ($r \geq 0$, $\phi \in \mathbb{R}$),

$$\begin{aligned} S(s)^*aS(s) &= \cosh(r)a + e^{i\phi}\sinh(r)a^*, \\ S(s)^*a^*S(s) &= \cosh(r)a^* + e^{-i\phi}\sinh(r)a, \\ S(s)^*NS(s) &= \cosh^2(r)N + \sinh^2(r)(N+1) \\ &\quad + \sinh(r)\cosh(r)(e^{i\phi}a^{*2} + e^{-i\phi}a^2). \end{aligned} \tag{5.3}$$

It is not hard to check the above identities since the function $t \to \langle e_\ell, S(ts)^*aS(ts)e_m\rangle$ ($\ell, m \geq 0$) is analytic and, it is easy to see by induction that, for $n \geq 0$ we have

$$\begin{aligned} &\frac{d^{2n}}{dt^{2n}}\langle e_\ell, S(ts)^*aS(ts)e_m\rangle = r^{2n}\langle e_\ell, S(ts)^*aS(ts)e_m\rangle, \\ &\frac{d^{2n+1}}{dt^{2n+1}}\langle e_\ell, S(ts)^*aS(ts)e_m\rangle = e^{i\phi}r^{2n+1}\langle e_\ell, S(ts)^*aS(ts)e_m\rangle. \end{aligned}$$

Summing the Taylor series we find the first identity. The second one follows by taking the adjoint and the third by multiplication.

The Squeeze-transform of an operator $\mathcal{L}$ is the operator $\mathcal{L}^{(s)}$ defined by

$$\mathcal{L}^{(s)}(x) = S(s)^*\mathcal{L}(S(s)xS(s)^*)S(s) \text{ with domain } S(s)^*D(\mathcal{L})S(s).$$

Proposition 5.2. *All the operators of the form*

$$\begin{aligned} &\mathcal{L}(x) \\ &= -\frac{\alpha}{2}(Nx - 2a^*xa + xN) - \frac{\beta}{2}((N+1)x - 2axa^* + x(N+1)) \\ &\quad - \frac{\bar{\gamma}}{2}(a^2x - 2axa + xa^2) - \frac{\gamma}{2}(a^{*2}x - 2a^*xa^* + xa^{*2}) \\ &\quad + i\delta\left[(\alpha+\beta)N + \gamma a^{*2} + \bar{\gamma}a^2, x\right] \end{aligned} \tag{5.4}$$

can be obtained as unitary transforms of the operator $\mathcal{L}_0$ if and only if the constants α, β, δ, γ ($\alpha,\beta,\delta\in\mathbb{R}$, $\gamma\in\mathbb{C}$) verify $\alpha\beta>|\gamma|^2$.

Proof. By using relations (5.3), we obtain that the Squeeze transform $\mathcal{L}^{(s)}$ of $\mathcal{L}_0$ verifies

$$\begin{aligned}
2\mathcal{L}^{(s)}(x) &= -(\mu^2\cosh^2(r)+\lambda^2\sinh^2(r))(Nx-2a^*xa+xN)\\
&-(\mu^2\sinh^2(r)+\lambda^2\cosh^2(r))\{(N+1)x-2axa^*+x(N+1)\}\\
&-(\mu^2+\lambda^2)\sinh(r)\cosh(r)\\
&\quad\times\{\mathrm{e}^{\mathrm{i}\phi}(a^{*2}x-2a^*xa^*+xa^{*2})+\mathrm{e}^{-\mathrm{i}\phi}(a^2x-2axa+xa^2)\}\\
&+2i\omega\cosh(2r)[N,x]+i\omega\sinh(2r)[\mathrm{e}^{\mathrm{i}\phi}a^{*2}+\mathrm{e}^{-i\phi}a^2,x].
\end{aligned}$$

So we have to search for some λ, μ, s and ω which verify

$$\begin{aligned}
&\alpha=\mu^2\cosh^2(r)+\lambda^2\sinh^2(r), && \beta=\mu^2\sinh^2(r)+\lambda^2\cosh^2(r),\\
&2\gamma=(\lambda^2+\mu^2)\sinh(2r)\mathrm{e}^{\mathrm{i}\phi}, && \delta(\alpha+\beta)=\omega\cosh(2r),\\
&2\gamma\delta=\omega\sinh(2r)\mathrm{e}^{\mathrm{i}\phi},
\end{aligned}$$

that is

$$\begin{aligned}
\lambda^2 &= \frac{1}{2}\left(\sqrt{(\alpha+\beta)^2-4|\gamma|^2}-\alpha+\beta\right),\\
\mu^2 &= \frac{1}{2}\left(\sqrt{(\alpha+\beta)^2-4|\gamma|^2}+\alpha-\beta\right),\\
s &= \frac{\gamma}{4|\gamma|}\lg\frac{\alpha+\beta+2|\gamma|}{\alpha+\beta-2|\gamma|},\quad \omega=\delta\sqrt{(\alpha+\beta)^2-4|\gamma|^2}. \qquad (5.5)
\end{aligned}$$

With this choice of the parameters, $\mathcal{L}$ is the squeeze transform of $\mathcal{L}_0$. □

Remarks. 1. The generators obtained by unitarily transforming the Ornstein-Uhlenbeck operator all have the same spectral gap, that is $(\mu^2-\lambda^2)/2$ for the operator given by (5.2) and $(\alpha-\beta)/2$ for the one given by (5.4).

2. By using both transforms together, we can obtain all the operators $\bar{\mathcal{L}}$ of the form

$$\bar{\mathcal{L}}(x)=\mathcal{L}(x)+\mathrm{i}\,[\nu a^*+\bar{\nu}a,x]$$

where $\mathcal{L}$ is the operator described in (5.4), with $\alpha>\beta>0$, $\delta\in\mathbb{R}$, γ and ν in $\mathbb{C}$ such that $|\gamma|^2<\alpha\beta$. Indeed

$$\begin{aligned}
(\mathcal{L}_z)^{(s)}(x)=\mathcal{L}^{(s)}(x)+\mathrm{i}\Bigg[\left(\omega-\mathrm{i}\frac{\mu^2-\lambda^2}{2}\right)zS(s)^*a^*S(s)\\
+\left(\omega+\mathrm{i}\frac{\mu^2-\lambda^2}{2}\right)\bar{z}S(s)^*aS(s),x\Bigg]
\end{aligned}$$

By relations (5.3) and the previous proposition, we have that the constants λ, μ, s, ω can be chosen as in (5.5), while z has to satisfy

$$\begin{cases} \nu_1 = [\omega\cosh r + \sinh r\,(\omega\cos\phi - g\sin\phi)]\, z_1 \\ \qquad + [g\cosh r + \sinh r\,(\omega\sin\phi + g\cos\phi)]\, z_2 \\ \nu_2 = [\sinh r\,(\omega\sin\phi + g\cos\phi) - g\cosh r]\, z_1 \\ \qquad + [\omega\cosh r - \sinh r\,(\omega\cos\phi - g\sin\phi)]\, z_2 \end{cases}$$

where ν_1 and ν_2 [resp. z_1 and z_2] are the real and complex parts of ν [resp. z], and $g = (\alpha - \beta)/2 = (\mu^2 - \lambda^2)/2$ is the spectral gap. Such a z always exists since the coefficient matrix has rank 2.

Acknowledgments

We wish to thank the Organizing Committee for a very nice week in Tunisia.

References

[1] Accardi L., Lu Y. G., Volovich I. V.: *Quantum Theory and its Stochastic Limit*, Springer Verlag, Berlin, 2002.

[2] Alicki R., Lendi K.: *Quantum Dynamical Semigroups and Applications*, Lecture Notes in Physics, 286, Springer, 1987.

[3] Carbone R., Exponential L^2-Convergence of Some Quantum Markov Semigroups Related to Birth-and-Death Processes, In: "Stochastic Analysis and Mathematical Physics (Santiago 1998)", 1–22, Trends Math., Birkhäuser, Boston, MA, 2000.

[4] Carbone R.: Exponential Ergodicity of Some Quantum Markov Semigroups, Ph. D. Thesis, Dipartimento di Matematica dell'Università di Milano, 2000.

[5] Carbone R., Fagnola F.: Exponential L^2-convergence of quantum Markov semigroups on $\mathcal{B}(h)$, *Math. Notes* **68**(3-4) (2000) 452–463.

[6] Chen M. F.: *From Markov chains to non equilibrium particle systems*, World Scientific, 1992.

[7] Cipriani F., Fagnola F., Lindsay J. M.: Spectral analysis and Feller property for quantum Ornstein-Uhlenbeck semigroups, *Comm. Math. Phys.* **210**(1) (2000) 85–105.

[8] Fagnola F., Rebolledo R.: An Ergodic Theorem in Quantum Optics, *Contributions in probability, Atti del convegno "in memoria di A. Frigerio", Udine, 13-14 settembre 1994*, 73–86, Udine, 1995.

[9] Fagnola F., Rebolledo R., Saavedra C.: Quantum flows associated to master equations in quantum optics, *J. Math. Phys.* **35** (1994) 1–12.

[10] Fagnola F.: Quantum Markov Semigroups and Quantum Markov Flows, *Proyecciones* **18**(3) (1999) 1–144.

[11] Guichardet A.: On the Representations of a Q-oscillator Algebra, *J. Math. Phys.* **39**(9) (1998) 4965–4969.

[12] Karlin S., Taylor H. M.: *A First Course on Stochastic Processes*, Academic Press, 1975.

[13] Liggett T.: Exponential L_2 convergence of attractive reversible nearest particle systems, *Ann. Probab.* **17** (1989) 403–432.

[14] Goldstein S., Lindsay J. M.: KMS-symmetric semigroups, *Math. Z.* **219** (1995) 591–608.

[15] Majewsky W. A., Streater R. F.: Detailed balance and quantum dynamical maps, *J. Phys. A* **31** (1998) 7981–7995.

[16] Walls D. F., Milburn G. J.: *Quantum Optics*, Springer Verlag, 1995.

ASYMPTOTIC FLUX ACROSS HYPERSURFACES FOR DIFFUSION PROCESSES

Andrea Posilicano
Dipartimento di Scienze, Università dell'Insubria, via Valleggio 11, I-22100 Como, Italy
andreap@uninsubria.it

Stefania Ugolini
Dipartimento di Scienze, Università dell'Insubria, via Valleggio 11, I-22100 Como, Italy
ugolini@mat.unimi.it

Abstract We suggest a rigorous definition of the pathwise flux across the boundary of a bounded open set for transient finite energy diffusion processes. The expectation of such a flux has the property of depending only on the current velocity v, the nonsymmetric (with respect to time reversibility) part of the drift. In the case where the diffusion has a limiting velocity we define the asymptotic flux across subsets of the sphere of radius R, when R tends to infinity, and compute its expectation in terms of v.

Keywords: Diffusion processes, (random) flux across surfaces, current velocity

1. Introduction

In a previous paper [13] the authors gave a pathwise probabilistic versions of the Scattering-into-Cones and Flux–across–Surfaces theorems in Quantum Mechanics and then recovered the known analytical results by taking suitable expectations.

Here we extend the main probabilistic results contained in [13] to a large class of Markovian diffusions with no a priori connection with Quantum Mechanics.

We consider diffusions on $\mathbb{R}^d$ which are weak solutions of s.d.e.'s of the form

$$\mathrm{d}X_t = b(t, X_t)\,\mathrm{d}t + \mathrm{d}W_t\,, \tag{1.1}$$

S. Albeverio et al. (eds.),
Proceedings of the International Conference on Stochastic Analysis and Applications, 185–197.

where W_t is a standard Wiener process and the drift vector field $b_t(x) \equiv b(t,x)$ satisfies

$$\forall T > 0, \qquad \mathbb{E} \int_0^T \|b(t,X_t)\|^2 \, dt < +\infty . \tag{1.2}$$

The condition (1.2) is a *finite energy* condition in the sense of Föllmer [6]. By Girsanov theory one proves that (1.2) is equivalent to a *finite entropy* condition: the probability measure describing the weak solution of (1.1) has finite relative entropy with respect to the Wiener measure. By definition of relative entropy, this fact implies absolute continuity and therefore the distribution of X_t, $t > 0$ is absolutely continuous (with respect to Lebesgue measure) with some density function ρ_t.

A very peculiar consequence of (1.2) is that not only the Markovian property is preserved under time reversal, but also the diffusion one. We will call this property *time reversibility of the diffusive character.* The class of diffusion processes individuated by this invariance property was firstly proposed by Nelson in 1966 [10] in the framework of Stochastic Mechanics. Such diffusions were then studied by Zheng and Meyer who called them *semimartingales dans les deux directions du temps* [14]. Within this class, Carlen in 1984 [2] (see [4] for an alternative proof) solved the existence problem of weak solutions of (1.1) in the case of (unbounded) drift fields satisfying (1.2). Successively, Föllmer [6] gave a very elegant characterization based on the relative entropy approach. Under time reversal the process solution of (1.1) is again a solution of a s.d.e. of type (1.1) with some dual drift field b_t^* (and, of course, another standard Wiener process) which satisfies the relation:

$$b_t(x) - b_t^*(x) = \nabla \log \rho_t(x) \tag{1.3}$$

The duality relation (1.3) allows to introduce a relevant decomposition of the drift field as the sum of two vector fields:

$$v_t = \frac{b_t + b_t^*}{2}, \qquad u_t = \frac{b_t - b_t^*}{2}$$

called respectively current and osmotic velocity [11]. In the symmetric case, $v_t = 0$ and thus $b_t = u_t$ is of gradient type according to (1.3). Therefore the current velocity v_t represents the non symmetric part of the drift field. We will see that only the current velocity is involved in the expression of the flux across surfaces by the diffusion paths.

A useful consequence of the time reversibility of the diffusion property is then the validity (in the weak sense) of the continuity equation for the couple (ρ, v):

$$\frac{\partial}{\partial t} \rho_t = -\nabla \cdot (\rho_t v_t) . \tag{1.4}$$

In this paper we give a rigorous definition of the pathwise flux across the boundary of a bounded set by a transient Markov diffusion process solution of (1.1) and satisfying the energy condition (1.2).

Given a bounded open set D let us consider the function

$$N_{\partial D} := N^+_{\partial D} - N^-_{\partial D}\,, \tag{1.5}$$

where $N^+_{\partial D}(\gamma)$ (resp. $N^-_{\partial D}(\gamma)$) denotes the number of inward (resp. outward) crossing by the path $t \mapsto X_t(\gamma)$ of the boundary ∂D, γ being the point in the probability space. It is not a local time because we need to distinguish outward from inward crossing in order to have the pathwise analogue of a net flux. The problem is that almost surely the diffusion X_t intersects ∂D on a set of times that has no isolated points and is uncountable. Therefore the definition of $N_{\partial D}$ given above makes no sense in general. However, by a suitable redefinition of $N_{\partial D}$ as the total mass of the almost surely compactly supported random distribution $-\frac{d}{dt}\chi_D(X_t)$, where χ_D is the characteristic function of the set D (see section 3 for details), we can give a rigorous definition of the pathwise flux across ∂D. Then, by using the continuity equation (1.4), in the case where ∂D is a regular hypersurface we can compute (see theorem 3.2) the expectation $\Phi_{\partial D}$ of the pathwise flux $N_{\partial D}$ in terms of the current velocity v, obtaining

$$\Phi_{\partial D} = \int_0^{+\infty}\int_{\partial D} \rho(t,x)\, v(t,x)\cdot n(x)\, d\sigma(x)\, \mathrm{d}t\,, \tag{1.6}$$

where n denotes the outward unit normal vector along ∂D and σ is the surface measure. We interpret $\Phi_{\partial D}$ as the flux of X_t across ∂D. Note that, by our choice of signs in (1.5), D is a source if $\Phi_{\partial D} > 0$ and is a sink if $\Phi_{\partial D} < 0$.

Of course the definition of flux given above does not extend to the case of a hypersurface which is not a boundary. This restriction can be avoided, at least asymptotically, in the case where the diffusion X_t has a limiting velocity, i.e.

$$\lim_{t\uparrow\infty} \frac{1}{t} X_t = v_\infty \tag{1.7}$$

exists almost surely for some non zero random vector v_∞. A simple but general condition giving (1.7), which is again a finite entropy condition, was obtained by Carlen [3] (see Theorem 4.3 below).

Suppose that Σ is an open subset of the unit sphere with $\partial\Sigma$ a finite union of C^1 manifolds. In order to define the asymptotic flux across Σ we consider the cone $C_\Sigma := \left\{\lambda x \in \mathbb{R}^d \mid x \in \Sigma,\ \lambda > 0\right\}$. Using (1.7) (see section 5 for the details) we can then define N^a_Σ, the asymptotic pathwise flux across Σ, by the limit $R \uparrow \infty$ of the mass N_{Σ_R} of the almost surely compactly supported random distribution $\frac{d}{dt}\chi_{C\cap B^c_R}(X_t)$, B_R being the closed ball of radius R. Note the change of sign in the definition of N_{Σ_R} with respect to $N_{\partial D}$. This is consistent with the fact that the exterior normal to $\Sigma_R := C_\Sigma \cap S_R$, S_R being the sphere of radius R, coincides with the interior normal to the boundary of $C \cap B^c_R$. No confusion can arise between the two different definitions since Σ_R is never the boundary of an open subset of $\mathbb{R}^d$.

We show (see theorem 5.2) that N^a_Σ is well defined since almost surely one has

$$N^a_\Sigma := \lim_{R\uparrow\infty} N_{\Sigma_R} = \chi_{C_\Sigma}(v_\infty)\,. \tag{1.8}$$

Moreover, if as before we define the flux by taking the expectation of the corresponding pathwise object, we have (see theorem 5.3)

$$\Phi^a_\Sigma = \lim_{R\uparrow\infty} \int_0^{+\infty}\int_{\Sigma_R} \rho(t,x)\,v(t,x)\cdot n(x)\,\mathrm{d}\sigma(x)\,\mathrm{d}t\,. \tag{1.9}$$

The conditions required in order to obtain the stated results are, besides (1.1) and (1.2), that ρ_t and v_t belong to the Sobolev space $H^1(\mathbb{R}^3)$ in order to obtain (1.6), (1.7) to get (1.8), and moreover we assume

$$\mathbb{E}\int_0^T \|\nabla v(t,X_t)\|^2\,\mathrm{d}t < +\infty$$

for some $T > 0$, to get (1.9).

Finally let us remark that our definition of a pathwise flux across ∂D is heuristically equivalent to the ill-defined Stratonovich stochastic integral

$$\int_0^{+\infty}\left(\int_{\partial D}\delta(X_t - x)\,n(x)\,\mathrm{d}\sigma(x)\right)\circ\mathrm{d}X_t\,.$$

Indeed, proceeding heuristically, one obtains

$$\begin{aligned} N_{\partial D} &= \int_0^{+\infty} -\frac{\mathrm{d}}{\mathrm{d}t}\chi_D(X_t)\,\mathrm{d}t = -\int_0^{+\infty}\frac{\mathrm{d}X_t}{\mathrm{d}t}\cdot\nabla\chi_D(X_t)\,\mathrm{d}t \\ &= \int_0^{+\infty}\frac{\mathrm{d}X_t}{\mathrm{d}t}\cdot n(X_t)\,\delta_{\partial D}(X_t)\,\mathrm{d}t \\ &= \int_0^{+\infty}\left(\int_{\partial D}\delta(X_t - x)\,n(x)\,\mathrm{d}\sigma(x)\right)\circ\mathrm{d}X_t\,. \end{aligned}$$

2. The class of finite energy diffusion processes

Consider the measurable space $(\Omega, \mathcal{F})$, with $\Omega = C(\mathbb{R}_+;\mathbb{R}^d)$, $\mathcal{F}$ the Borel σ–algebra, and let $(\Omega, \mathcal{F}, \mathcal{F}_t, X_t)$ be the evaluation stochastic process $X_t(\gamma) := \gamma(t)$, $\gamma \in \Omega$, with $\mathcal{F}_t = \sigma(X_s, 0 \leq s \leq t)$ the natural filtration.

Let us suppose that:

H1) there exists a Borel probability measure $\mathbb{P}$ on $(\Omega, \mathcal{F})$ such that:

— $(\Omega, \mathcal{F}, \mathcal{F}_t, X_t, \mathbb{P})$ is a Markov process;

— $W_t := X_t - X_0 - \int_0^t b(s, X_s)\,\mathrm{d}s$ is a $(\mathbb{P}, \mathcal{F}_t)$–Wiener process, i.e. $\mathbb{P}$ is a weak solution of the stochastic differential equation

$$\mathrm{d}X_t = b(t, X_t)\,\mathrm{d}t + \mathrm{d}W_t \tag{2.1}$$

with initial distribution $\mu = \mathbb{P} \circ X_0^{-1}$, $\mu \ll \lambda$, λ the Lebesgue measure on $\mathbb{R}^d$;

H2) the adapted process b_t satisfies

$$\forall T > 0\,, \qquad \mathbb{E}\int_0^T \|b(t, X_t)\|^2\,\mathrm{d}t < +\infty \tag{2.2}$$

where $\mathbb{E}$ denotes the expectation with respect to the measure $\mathbb{P}$.

Remark 2.1. Since

$$H_{\mathcal{F}_T}(\mathbb{P}, \mathbb{P}^W) := \mathbb{E}\left(\log \left.\frac{\mathrm{d}\mathbb{P}}{\mathrm{d}\mathbb{P}^W}\right|_{\mathcal{F}_T}\right) = \frac{1}{2}\,\mathbb{E}\int_0^T \|b(t, X_t)\|^2\,\mathrm{d}t$$

where $\mathbb{P}^W := \int \mu(x)\mathbb{P}^W_x$, with $\mathbb{P}^W_x$ denoting the Wiener measure starting from x, see [6], the finite energy condition (2.2) is equivalent to a finite relative entropy condition. Thus (2.2) implies that $\mathbb{P}$ is absolutely continuous with respect to $\mathbb{P}^W$, and so X_t admits a density function ρ_t for any $t \geq 0$.

As a consequence of H1 and H2, the Markovian diffusion X_t preserves the diffusion property under time reversal. Indeed by Föllmer [6] one has:

Lemma 2.2. *Under the hypothesis H1, H2, defining* $\hat{\mathbb{P}} := \mathbb{P} \circ R$, *where* R *is the pathwise time reversal on* $C([0,T];\mathbb{R}^d)$, $R(\gamma)(t) := \gamma(T - t)$, *there exists an adapted process* $\hat{b}_t$ *such that:*

$$\hat{W}_t := X_t - X_0 - \int_0^t \hat{b}(s, X_s)\,\mathrm{d}s \tag{2.3}$$

is a $(\hat{\mathbb{P}}, \mathcal{F}_t)$*–Wiener process.*

Remark 2.3. Lemma 2.2 states that the finite energy (entropy) condition (2.2) is a sufficient condition for the time reversibility of the diffusion property. The proof is based on the fact that the finite entropy condition is invariant under time reversal. The extension to the infinite dimensional case is in [7]. Sufficient conditions are also given in [8]. Sufficient and necessary conditions for reversibility of diffusion property, in the case of Lipschitz drift fields, are investigated in [9].

It is well known that the drift field can be seen as a stochastic forward derivative in the sense of Nelson [10], [11]. In particular from (2.1) and (2.2) it follows that (see [6]):

$$b_t = \lim_{h\downarrow 0} \frac{1}{h}\mathbb{E}[X_{t+h} - X_t \mid \mathcal{F}_t] \quad \text{in } L^2(\mathbb{P})\,.$$

Analogously, from (2.3) one also has:

$$\hat{b}_t = \lim_{h\downarrow 0} \frac{1}{h}\hat{\mathbb{E}}[X_{t+h} - X_t \mid \mathcal{F}_t] \quad \text{in } L^2(\hat{\mathbb{P}})\,,$$

where $\hat{\mathbb{E}}$ denotes the expectation with respect to the measure $\hat{\mathbb{P}}$. For our approach it is convenient to work with the same probability measure $\mathbb{P}$ as proposed by Nelson [10]. To this end we write:

$$\hat{b}_t = \lim_{h\downarrow 0} -\frac{1}{h}\mathbb{E}[X_{T-t} - X_{T-t-h} \mid \hat{\mathcal{F}}_{T-t}] \circ R \quad \text{in } L^2(\mathbb{P})\,,$$

where $\hat{\mathcal{F}}_t = \sigma(X_s,\, s \geq t)$ is the natural future filtration.

Since the Markov property is preserved under time reversal also the dual drift is given by some measurable function $\hat{b}_t(\gamma) = \hat{b}(t, (X_t(\gamma))$. Let us define $b^*_t(x) = -\hat{b}_{T-t}(x)$ so that, as already obtained in [11]:

$$b^*_t = \lim_{h\downarrow 0} -\frac{1}{h}\mathbb{E}[X_t - X_{t-h} \mid \hat{\mathcal{F}}_t] \quad \text{in } L^2(\mathbb{P})\,,$$

and the following relation holds:

$$b_t(x) - b^*_t(x) = \nabla \log \rho_t(x) \tag{2.4}$$

between the drift field and its dual (see [2] and for the non Markovian case [6]).

The duality relation (2.4) allows to introduce the decomposition

$$b_t = u_t + v_t, \qquad b^*_t = -u_t + v_t$$

where

$$v_t = \frac{b_t + b^*_t}{2}, \qquad u_t = \frac{b_t - b^*_t}{2}$$

are called current and osmotic velocity respectively [10].

In the symmetric case, $v_t = 0, b_t = u_t, b_t^* = -u_t$, thus the drift field coincides up to the sign with its dual and, according to (2.4), it is of gradient type. Therefore the current velocity v_t represents the not symmetric (with respect to time reversal) part of the drift field. We will see that only the current velocity is involved in the expression of the flux across surfaces.

An important consequence of the time reversibility of the diffusion property is the validity of the continuity equation (in the weak sense) for the couple (ρ, v).

Indeed, recalling the Fokker-Planck equation associated with (2.1):

$$\frac{\partial}{\partial t}\rho_t = -\nabla \cdot (\rho_t b_t) + \Delta \rho_t$$

and the Fokker-Planck equation associated with (2.3):

$$\frac{\partial}{\partial t}\rho_t = -\nabla \cdot (\rho_t b_t^*) - \Delta \rho_t$$

and putting together the two equations one has:

$$\frac{\partial}{\partial t}\rho_t = -\nabla \cdot (\rho_t v_t) \tag{2.5}$$

where the definition of the current velocity has been used.

3. The pathwise flux across a boundary.

Given an open set D we want now to define the flux across ∂D by the path of a diffusion.

In order to do this we would like to introduce a pathwise analogous of the flux as the function

$$N_{\partial D}(\gamma) := N_{\partial D}^{+}(\gamma) - N_{\partial D}^{-}(\gamma) \ ,$$

where $N_{\partial D}^{+}(\gamma)$ (resp. $N_{\partial D}^{-}(\gamma)$) denotes the number of inward (resp. outward) crossing by $[0, +\infty) \ni t \mapsto \gamma(t)$ of ∂D. The problem is that the above definition makes no sense since $\mathbb{P}$–a.s. the set $\{t : X_t \in \partial D\}$ has no isolated points and is uncountable. Therefore we are forced to proceed in an alternative way.

Let us observe that if $\#\{t \mid \gamma(t) \in \partial D\} < +\infty$ then $N_{\partial D}(\gamma)$ is the total mass of the random signed measure

$$\sum_{t \in \{s \mid \gamma(s) \in \partial D\}} c(t)\, \delta_t \ ,$$

where $c(t) = +1$ if t corresponds to an outward crossing and $c(t) = -1$ if t corresponds to an inward crossing. Therefore

$$\sum_{t \in \{s | \gamma(s) \in \partial D\}} c(t)\, \delta_t = -\frac{\mathrm{d}}{\mathrm{d}t} \chi_D(\gamma(t)),$$

where the derivative has to be intended in distributional sense, and thus we give the following

Definition 3.1. Given any open domain D, we define the random distribution

$$\mu_D \colon \Omega \to \mathcal{D}'(\mathbb{R})$$

by

$$\langle \mu_D(\gamma), \phi \rangle := \chi_D(\gamma(0))\, \phi(0) + \int_0^{+\infty} \chi_D(\gamma(t))\, \dot{\phi}(t)\, \mathrm{d}t,$$

where $\phi \in \mathcal{D}(\mathbb{R}) \equiv C_c^\infty(\mathbb{R})$.

Supposing now that D is bounded and that X_t is transient, we have that $\mathbb{P}$-almost surely the random distribution μ_D has a compact support and so its mass, which we denote by $N_{\partial D}$, is well defined. We define then the flux across ∂D by $\Phi_{\partial D} := \mathbb{E}(N_{\partial D})$.

By the continuity equation (2.5) such an expectation can be explicitly calculated in terms of the current velocity v (use [13], theorem 7):

Theorem 3.2. *Let $(\Omega, \mathcal{F}, \mathbb{P}_t, X_t, \mathbb{P})$ satisfy H1 and H2, with $\rho_t \in H^1(\mathbb{R}^3)$ and $v_t \in H^1(\mathbb{R}^3)$ for any $t \geq 0$. For any open bounded domain D, with ∂D a finite union of C^1 manifolds, one has*

$$\Phi_{\partial D} = \int_0^{+\infty} \int_{\partial D} \rho(t,x)\, v(t,x) \cdot n(x)\, \mathrm{d}\sigma(x)\, \mathrm{d}t\ ,$$

where n denotes the outward unit normal vector along ∂D and σ is the surface measure.

4. Diffusion with an asymptotic velocity.

Since our goal is to define an asymptotic flux across hypersurfaces, we need to impose a condition on the time evolution of the process $\frac{1}{t} X_t$.

Definition 4.1. We say that the diffusion paths admit an asymptotic velocity when

H3)

$$\lim_{t \uparrow \infty} \frac{1}{t} X_t = v_\infty \neq 0 \qquad \mathbb{P}\text{–a.s.}$$

and moreover $\mu_\infty \ll \lambda$, where λ denotes the Lebesgue measure on $\mathbb{R}^d$ and μ_∞ is the distribution of v_∞.

From now on by an open cone C_Σ we will mean a set of the form

$$\left\{\lambda x \in \mathbb{R}^3 \mid x \in \Sigma,\ \lambda > 0\right\},$$

where Σ is an open subset of the unit sphere with $\partial\Sigma$ a finite union of C^1 manifolds.

Remark 4.2. For any open cone C_Σ, any ball B_R of radius R, and for any diffusion $(\Omega, \mathcal{F}, \mathbb{P}_t, X_t, \mathbb{P})$ satisfying H1, H2 and H3, one has

$$\lim_{t\uparrow\infty} \chi_{C_\Sigma \cap B_R^c}(X_t) = \lim_{t\uparrow\infty} \chi_{C_\Sigma}(X_t) = \chi_{C_\Sigma}(v_\infty) \qquad \mathbb{P}\text{–a.s.}\ .$$

See [13] for a two-line proof. Thus hypotheses H3 requires that the limiting velocity is non negligible and such that asymptotically the paths have the same direction as their limiting velocity. For example the Brownian motion in $\mathbb{R}^3$ is transient but it has no limiting velocity according to our definition because of the S.L.L.N. Only a Brownian motion with drift could satisfy the requirement of our definition.

A simple but general condition giving the existence of a limiting velocity, which is again a finite relative entropy condition (now on the full σ-algebra and with respect to $\widetilde{\mathbb{P}}$, see the proof below), is given in the following (see [3]):

Theorem 4.3. *Let $(\Omega, \mathcal{F}, \mathbb{P}_t, X_t, \mathbb{P})$ satisfy H1 and H2. If moreover one has:*

$$\mathbb{E}\int_{t_0}^{+\infty} \|b(t, X_t) - X_t/t\|^2\, \mathrm{d}t < +\infty \qquad t_0 > 0, \tag{4.1}$$

then:

$$\lim_{t\uparrow\infty} \frac{1}{t} X_t = v_\infty \qquad \mathbb{P}\text{–a.s.}$$

for some random variable v_∞.

Proof. The condition (4.1) implies, by [5], prop. 2.11, that $\mathbb{P} \ll \widetilde{\mathbb{P}}$ on $\sigma(X_s, t_0 \le s < +\infty)$, where $\widetilde{\mathbb{P}}$ is the weak solution of the simple stochastic differential equation

$$\mathrm{d}X_t = \frac{1}{t} X_t\, \mathrm{d}t + \, d\widetilde{W}_t\ .$$

with $\widetilde{W}_t$ a standard Wiener process and is such that $\mathbb{P} \circ X_{t_0}^{-1} = \widetilde{\mathbb{P}} \circ X_{t_0}^{-1}$. Therefore:

$$\mathrm{d}\left(\frac{1}{t}\,X_t\right) = -\frac{1}{t^2}\,X_t\,\mathrm{d}t + \frac{1}{t}\,\mathrm{d}X_t = \frac{1}{t}\,\widetilde{W}_t\,.$$

and so

$$\frac{1}{t}\,X_t = \frac{1}{t_0}\,X_{t_0} + \int_{t_0}^{t} \frac{1}{s}\,\mathrm{d}\widetilde{W}_s\,.$$

Since

$$\widetilde{\mathbb{E}}\left(\int_{t_0}^{+\infty} \frac{1}{s}\,\mathrm{d}\widetilde{W}_s\right)^2 = \int_{t_0}^{+\infty} \frac{1}{s^2}\,\mathrm{d}s < +\infty$$

by Doob's martingale convergence theorem one gets $\widetilde{\mathbb{P}}$–a.s. convergence of $\frac{1}{t}\,X_t$. The theorem then follows by absolutely continuity. □

Remark 4.4. Under the same hypotheses of Theorem 4.3 it is possible to prove (see [3]) that the random variable v_∞ generates the tail σ–algebra

$$\mathcal{T} := \bigcap_{t>0} \sigma(X_s,\ s \geq t)\,.$$

5. The asymptotic flux across hypersurfaces.

Let us consider hypersurfaces of the following type:

$$\Sigma_R = C_\Sigma \cap S_R\,,$$

where S_R is the sphere of radius R. We will define the pathwise flux across Σ_R in the limit when $R \uparrow \infty$.

Suppose at first that $\#\,\{t\,:\,\gamma(t) \in \Sigma_R\} < +\infty$. Since, by H3, $t \mapsto \gamma(t)$ is definitively either in C_Σ or in $\bar{C}_\Sigma^c$, if R is sufficiently large one has

$$\sum_{t\in\{s|\gamma(s)\in\Sigma_R\}} c(t)\,\delta_t = \sum_{t\in\left\{s|\gamma(s)\in\Sigma_R\cup(\partial C_\Sigma\cup B_R^c)\right\}} c(t)\,\delta_t = \frac{\mathrm{d}}{\mathrm{d}t}\,\chi_{C_\Sigma\cap B_R^c}(\gamma(t))\,.$$

We are therefore lead to give the following

Definition 5.1. The asymptotic pathwise flux across Σ is defined by

$$N_\Sigma^a := \lim_{R\uparrow\infty} N_{\Sigma_R}\,,$$

where N_{Σ_R} is the total mass of the random distribution $-\mu_{C_\Sigma\cap B_R^c}$.

The following result shows that the above definition makes sense.

Theorem 5.2. *Let $(\Omega, \mathcal{F}, \mathbb{P}_t, X_t, \mathbb{P})$ satisfy H1, H2 and H3. Then $\mathbb{P}$-almost surely the random distribution $\mu_{C_\Sigma \cap B_R^c}$ has compact support and so its mass N_{Σ_R} is well defined. Moreover one has*

$$\lim_{R\uparrow\infty} N_{\Sigma_R} = \chi_{C_\Sigma}(v_\infty) \qquad \mathbb{P}\text{–}a.s..$$

Proof. Let

$$\tau_R(\gamma) := \sup\{t \geq 0 \mid \|\gamma(t)\| < R\} .$$

By H3, $\tau_R < +\infty$ $\mathbb{P}$–a.s. Thus $\mu_{C\cap B_R^c}$ has compact support $\mathbb{P}$–a.s.

Let $\phi_\gamma \in \mathcal{D}(\mathbb{R})$ such that $\phi_\gamma = 1$ on a neighbourhood of $[0, \tau_R(\gamma)]$. By the definition of $\mu_{C_\Sigma\cap B_R^c}$ one has

$$\begin{aligned}\langle \mu_{C_\Sigma\cap B_R^c}(\gamma), \phi_\gamma\rangle &= -\,\chi_{C_\Sigma\cap B_R^c}(\gamma(0)) - \chi_{C_\Sigma}(v_\infty(\gamma)) \int_{\tau_R(\gamma)}^{+\infty} \dot{\phi}_\gamma(t)\, dt \\ &= -\,\chi_{C_\Sigma\cap B_R^c}(\gamma(0)) + \chi_{C_\Sigma}(v_\infty(\gamma))\,,\end{aligned}$$

and the thesis then immediately follows by taking the limit $R \uparrow \infty$. □

The next theorem shows that the definition of asymptotic flux across Σ by $\Phi_\Sigma^a := \mathbb{E}(N_\Sigma^a)$ is consistent with the result given in theorem 3.2 in the case of the flux across a boundary:

Theorem 5.3. *Let $(\Omega, \mathcal{F}, \mathbb{P}_t, X_t, \mathbb{P})$ satisfy H1, H2 and H3 and suppose $\rho_t \in H^1(\mathbb{R}^3)$, $v_t \in H^1(\mathbb{R}^3)$ for any $t \geq 0$ and*

$$\mathbb{E}\int_0^T \|\nabla v(t, X_t)\|^2\, dt < +\infty \tag{5.1}$$

for some $T > 0$. Then

$$\Phi_\Sigma^a \equiv \mathbb{E}(\chi_{C_\Sigma}(v_\infty)) = \lim_{R\uparrow\infty} \int_0^{+\infty}\int_{\Sigma_R} \rho(t,x)\, v(t,x)\cdot n(x)\, d\sigma(x)\, dt\,.$$

Proof. Proceeding as is [13] one has

$$\mathbb{E}(\chi_{C_\Sigma}(v_\infty)) = \lim_{R\uparrow\infty} \int_0^{+\infty}\int_{\Sigma_R\cup(\partial C_\Sigma\cap B_R^c)} \rho(t,x)\, v(t,x)\cdot n(x)\, d\sigma(x)\, dt\,.$$

The proof is then concluded by proving that:

$$\lim_{R\uparrow\infty} \int_0^{+\infty}\int_{\partial C_\Sigma\cap B_R^c} \rho(t,x)\, v(t,x)\cdot n(x)\, d\sigma(x)\, dt \;= 0\,. \tag{5.2}$$

By the monotone convergence theorem, (5.2) follows from

$$\int_0^T \int_{\partial(C_\Sigma \cap B_R^c)} |\rho(t,x)\, v(t,x) \cdot n(x)|\, \mathrm{d}\sigma(x)\, \mathrm{d}t \; < +\infty \tag{5.3}$$

for some $T > 0$. Since

$$\begin{aligned}
&\int_0^T \int_{\partial(C_\Sigma \cap B_R^c)} |\rho(t,x)^{1/2} \rho(t,x)^{1/2}\, v(t,x) \cdot n(x)|\, \mathrm{d}\sigma(x)\, \mathrm{d}t \\
&\le \int_0^T \Bigg(\bigg(\int_{\partial(C_\Sigma \cap B_R^c)} \rho(t,x)\, \mathrm{d}\sigma(x) \bigg)^{1/2} \times \\
&\quad \times \bigg(\int_{\partial(C_\Sigma \cap B_R^c)} \rho(t,x))\, \|v(t,x)\|^2\, \mathrm{d}\sigma(x) \bigg)^{1/2} \Bigg) \mathrm{d}t\,,
\end{aligned}$$

by trace estimates on functions in $H^1(\mathbb{R}^3)$ (see e.g. [1], chap. 5) one has

$$\begin{aligned}
&\int_0^T \int_{\partial(C_\Sigma \cap B_R^c)} |\rho(t,x)\, v(t,x) \cdot n(x)|\, \mathrm{d}\sigma(x)\, \mathrm{d}t \\
&\le \int_0^T \bigg(\int_{\mathbb{R}^3} \Big(\rho(t,x)\, \mathrm{d}x + \int_{\mathbb{R}^3} \|\nabla \rho^{1/2}(t,x)\|^2\, \mathrm{d}x \bigg)^{1/2} \times \\
&\quad \times \bigg(\int_{\mathbb{R}^3} \rho(t,x) \|v(t,x)\|^2\, \mathrm{d}x + \int_{\mathbb{R}^3} \|\nabla(\rho^{1/2} v)(t,x)\|^2\, \mathrm{d}x \bigg)^{1/2} \Bigg)\, \mathrm{d}t\,.
\end{aligned}$$

From (2.4) we have $u = \frac{\nabla \rho}{2\rho}$, hence one has $\nabla \rho^{1/2} = u\rho^{1/2}$ and $\nabla(\rho^{1/2} v) = uv\rho^{1/2} + \rho^{1/2}\nabla v$. Therefore in order to obtain (5.3) it is sufficient to have (5.1) and

$$\mathbb{E} \int_0^T \Big(\|u(t,X_t)\|^2 + \|v(t,X_t)\|^2 \Big)\, \mathrm{d}t < +\infty$$

which is equivalent to (2.2). □

6. Conclusion

The present paper introduces the notion of flux across the boundary of a bounded open set D in Euclidean space. For regular compact domains it is the expectation of the total mass of the (almost surely compactly supported) random distribution given by the distributional time-derivative of the functional

$$-\chi_D(X_t)\,,$$

where the diffusion X_t, which satisfies the stochastic differential equation $\mathrm{d}X_t = b_t(X_t)\,\mathrm{d}t + \mathrm{d}W_t$, is transient. If, moreover, the limit

$$\lim_{t\uparrow\infty} \frac{1}{t}\, X_t$$

exists almost surely, then to the notion of asymptotic flux is given a sense as well. The flux can be expressed in terms of ρ_t, the density of the distribution of X_t, and the current velocity $v_t = b_t - \frac{1}{2}\nabla \log \rho_t$.

References

[1] Burenkov V.I.: *Sobolev Spaces on Domains*, Teubner, Stuttgart, Leipzig, 1998.

[2] Carlen E.: Conservative Diffusions, *Commun. Math. Phys.* **94** (1984) 293–315.

[3] Carlen E.: The Pathwise Description of Quantum Scattering in Stochastic Mechanics, *Lecture Notes in Physics* **262**, 139–147, Springer-Verlag, Berlin, Heidelberg, New York, 1986.

[4] Carmona R.: Probabilistic Construction of Nelson Processes, *Probabilistic Methods in Mathematical Physics (Katata/Kyoto 1985)*, 55–81, Academic Press, Boston,1987.

[5] Ershov M.: On the Absolute Continuity of Measures Corresponding to Diffusion Processes, *Theory of Prob. and Appl.* **17** (1972) 169–174.

[6] Föllmer H.: Time Reversal on Wiener Space, *Lecture Notes in Mathematics* **1158** 119–129, Springer-Verlag, Berlin, Heidelberg, New York, 1986.

[7] Föllmer H., Wakolbinger A.: Time Reversal of Infinite-dimensional Diffusions, *Stochastic Process. Appl.* **22** (1986) 59–77.

[8] Haussmann U., Pardoux É.: Time Reversal of Diffusions, *Ann. Probab.* **14** (1986) 1188–1205.

[9] Millet A., Nualart D., Sanz M.: Integration by Parts and Time Reversal of Diffusion Processes, *Ann. Probab.* **17** (1989) 208–238.

[10] Nelson E.: Derivation of the Schrödinger Equation from Newtonian Mechanics. *Phys. Rev.* **150** (1966) 1079–1085.

[11] Nelson E.: *Dynamical Theories of Brownian Motion*, Princeton Univ. Press, Princeton, 1967.

[12] Nelson E.: *Quantum Fluctuations*, Princeton Univ. Press, Princeton, 1985.

[13] Posilicano A., Ugolini S.: Scattering into Cones and Flux across Surfaces in Quantum Mechanics: a Pathwise Probabilistic Approach, *J. Math. Phys.* **43** (2002) 5386–5399.

[14] Zheng W., Meyer P.-A.: Quelques résultats de Mécanique Stochastique, *Lecture Notes in Mathematics* **1059**, 223–244. Springer-Verlag, Berlin, Heidelberg, New York, 1985.

REFLECTED BACKWARD STOCHASTIC DIFFERENTIAL EQUATION WITH SUPER-LINEAR GROWTH

Khaled Bahlali

UFR Sciences, UVT, B.P. 132, 83957 La Garde Cedex, France. & CPT, CNRS Luminy, Case 907, 13288 Marseille Cedex 9, France

bahlali@univ-tln.fr

El Hassan Essaky

Université Cadi Ayyad, Faculté des Sciences Semlalia, Département de Mathématiques, B.P. 2390, 40000 Marrakech, Morocco

essaky@ucam.ac.ma

Boubakeur Labed

Université Mohamed Khider, Département de Mathématiques, B.P. 145 Biskra, Algérie

Abstract We deal with reflected backward stochastic differential equations (RBSDE) in a d-dimensional convex region with super-linear growth coefficient. We prove, in this setting, various existence and uniqueness results. This is done with an unbounded terminal data.

Keywords: Backward Stochastic Differential Equation

1. Introduction

Backward stochastic differential equations (BSDEs) have been intensively studied in the last years due to their connections with mathematical finance, see El Karoui *et al.* [5], stochastic control and stochastic games, see Hamadène and Lepeltier [10]. Since the first existence and uniqueness result established by Pardoux and Peng [17] several works have attempted to relax the Lipschitz condition and the growth of the generator, see for instance [4, 8, 9, 12, 13, 19]. In one dimensional case, the existence of solutions for BSDEs with continuous coefficient can be proved by comparison techniques, see Hamadène [8, 12], N'zi [14] and

S. Albeverio et al. (eds.),
Proceedings of the International Conference on Stochastic Analysis and Applications, 199–216.

N'zi-Ouknine [15] and the references therein. However the question of uniqueness still remains open. In [3] a topological approach, for multidimensional BSDE with continuous coefficient, is given. It turns out, roughly speaking, that "most" BSDE with continuous coefficient and a square integrable terminal data have a unique solution.

The existence and uniqueness for reflected backward stochastic differential equation (RBSDE) in a convex domain can be proved via a penalization or by a Picard's iteration method, see for instance [7, 16, 6, 11, 18, 19, 21].

The first result which gives an existence and uniqueness of the solutions for multidimensional BSDE with local assumptions (in the two spatial variables y, z) on the coefficient and an only square integrable terminal data has been established in [1]. However in [1], the growth of the coefficient is strictly sublinear. See also [2] for more developments.

In the present paper, we build a result of [1] and extend it in two senses. First, we assume that the coefficient is of super-linear growth. For example, it can behave as $|y|\sqrt{|\log|y||}$ or $|y||\log||y|\ldots$ Second, we consider a reflected BSDE in a convex domain. We prove in this setting an existence and uniqueness of the solution. We don't impose any boundedness condition on the terminal data. It will be assumed square integrable only, while the Lipschitz constant L_N behaves as $\sqrt{\log(N)}$. In the case where the coefficient is uniformly Lipschitz in its gradient variable z, the Lipschitz constant L_N behaves then as $\log(N)$. The techniques of comparison of the solutions used in one dimensional case do not work in higher dimension. Noticing also that the Picard approximation method, as well as the usual localization by stopping time fail in our situation. Our proofs are based on an approximation of the coefficient by a sequence (f_n) of Lipschitz functions and by using a suitable alternative localization which seems more adapted to the BSDEs than the usual one.

The following example is, in our knowledge, not covered by the previous papers. For $i = 1, \ldots, d$ let $h_i\colon \mathbb{R}^d \to \mathbb{R}$ be the function defined by $h_i(y) = -\frac{1}{e}\mathbf{1}_{\{|y|\le\frac{1}{e}\}} + |y|\log|y|\mathbf{1}_{\{|y|\ge\frac{1}{e}\}}$ and define the function g by $g(t, y, z) := (h_1(y) + |z|, \ldots, h_d(y) + |z|)$. It is not difficult to check that the function g satisfies the assumptions of Corollary 3.8 and hence if $\mathbb{E}(|\xi|^4) < \infty$ then the BSDE (1)-(5) has a unique solution.

The paper is organized as follows. The definitions, notations, and assumptions are collected in Section 2. The main results together with the proofs are given in Section 3.

2. Preliminaries

Let $(W_t)_{0\leq t\leq T}$ be a r-dimensional Wiener process defined on a complete probability space $(\Omega, \mathcal{F}, \mathbb{P})$. Let $(\mathcal{F}_t)_{0 \leq t\leq T}$ denote the natural filtration of (W_t) such that $\mathcal{F}_0$ contains all $\mathbb{P}$-null sets of $\mathcal{F}$, and ξ be an $\mathcal{F}_T$-measurable d-dimensional random variable. Let $f\colon [0,T]\times\Omega\times\mathbb{R}^d\times\mathbb{R}^{d\times r}$ be an $\mathcal{F}_t$-progressively measurable process. Consider the following assumptions:

(A.1)

(i) f is continuous in (y,z) for almost all (t,ω) and measurable in $(t\omega)$ for all (y,z).

(ii) There exists a constant $M>0$ such that,

$$\langle y, f(t,\omega,y,z)\rangle \leq M(1+|y|^2+|y|\cdot|z|) \quad \mathbb{P}\text{-a.s., a.e. } t\in[0,1].$$

(iii) There exist $M>0$ and $\alpha\in[0,1[$ such that,

$$|f(t,\omega,y,z)| \leq M(1+|y|\sqrt{|\log|y||}+|z|^\alpha) \quad \mathbb{P}\text{-a.s., a.e. } t\in[0,1].$$

(iv) There exist $M>0$ such that,

$$|f(t,\omega,y,z)| \leq M(1+|y|\sqrt{|\log|y||}+|z|) \quad \mathbb{P}\text{-a.s., a.e. } t\in[0,1].$$

(v) For each $N>0$, there exist L_N such that:

$$|f(t,\omega,y,z)-f(t,\omega,y,z')| \leq L_N\left(|y-y'|+|z-z'|\right)$$
$$|y|,|y'|,|z|,|z'|\leq N \quad \mathbb{P}\text{-a.s., a.e. } t\in[0,1].$$

(vi) There exist $M>0$ and $\alpha\in[0,1[$ such that,

$$|f(t,\omega,y,z)| \leq M(1+|y|\cdot|\log|y||+|z|^\alpha) \quad \mathbb{P}\text{-a.s., a.e. } t\in[0,1].$$

(vii) There exists $M>0$ such that

$$|f(t,\omega,y,z)| \leq M(1+|y|\cdot|\log|y||+|z|) \quad \mathbb{P}\text{-a.s., a.e. } t\in[0,1].$$

(A.2) An open and convex subset Θ of $\mathbb{R}^d$.

We denote by $\mathbb{L}$ the set of $\mathbb{R}^d\times\mathbb{R}^{d\times r}$-valued processes (Y,Z) defined on $\mathbb{R}_+\times\Omega$ which are $\mathcal{F}_t$-adapted and such that:

$$\|(Y,Z)\|^2 = \mathbb{E}\left(\sup_{0\leq t\leq 1}|Y_t|^2+\int_0^1|Z_s|^2\mathrm{d}s\right) < +\infty.$$

The couple $(\mathbb{L}, \|\cdot\|)$ is then a Banach space.

Let us introduce our reflected BSDE: Given a data (f, ξ) we want to solve the following reflected backward stochastic differential equation:

$$Y_t = \xi + \int_t^1 f(s, Y_s, Z_s)\mathrm{d}s - \int_t^1 Z_s \mathrm{d}W_s + K_1 - K_t, \quad 0 \le t \le 1 \quad (2.1)$$

Definition 2.1. A solution of the reflected BSDE (2.1) is a triple

$$(Y_t, Z_t, K_t), \quad 0 \le t \le 1$$

of progressively measurable processes with values in $\mathbb{R}^d \times \mathbb{R}^{d\times r} \times \mathbb{R}^d$ and satisfying:

(1) $(Y, Z) \in \mathbb{L}$

(2) $Y_t = \xi + \int_t^1 f(s, Y_s, Z_s)\mathrm{d}s - \int_t^1 Z_s \mathrm{d}W_s + K_1 - K_t$, $0 \le t \le 1$

(3) the process Y is continuous

(4) K is absolutely continuous, $K_0 = 0$, and $\int_0^{\cdot} (Y_t - \alpha_t)\, \mathrm{d}K_t \le 0$ for every progressively measurable process α_t which is continuous and takes values into $\overline{\Theta}$.

(5) $Y_t \in \overline{\Theta}$, $0 \le t \le 1$ a.s.

3. Reflected BSDE with locally Lipschitz coefficient

The main results are the following.

Theorem 3.1. *Let* **(A.1)** (i)-(iii), (v), *and* **(A.2)** *be satisfied. Assume moreover that* $\mathbb{E}\left(|\xi|^5\right) < \infty$. *Then the reflected BSDE* (1)-(5) *has a unique solution* $\{(Y_t, Z_t, K_t); 0 \le t \le 1\}$ *if the following condition is satisfied:*

$$\lim_{N\to\infty} \frac{1}{L_N^2}\left(\frac{1}{N} + \frac{1}{N^{2(1-\alpha)}} + \frac{1}{N^2}\right) \exp(2L_N^2) = 0.$$

Theorem 3.2. *Let* **(A.1)** (i)-(ii), (iv)-(v), *and* **(A.2)** *be satisfied. Assume moreover that* $\mathbb{E}\left(|\xi|^4\right) < \infty$. *Then the reflected BSDE* (1)-(5) *has a unique solution* $\{(Y_t, Z_t, K_t); 0 \le t \le 1\}$ *if the following condition is satisfied:*

$$\lim_{N\to\infty} \frac{1}{\sqrt{N}} \exp(2L_N^2) = 0.$$

The following corollary is consequence of Theorem 3.2.

Corollary 3.3. *Let conditions of Theorem 3.2 hold. Assume moreover that L_N satisfies the following assumption*

$$\exists L \geq 0, \quad L_N^2 \leq L + \log N.$$

Then the reflected BSDE (1)-(5) has a unique solution.

Define the family of semi norms $(\rho_n(f))_n$

$$\rho_n(f) = (\mathbb{E}\int_0^1 \sup_{|y|,|z|\leq n} |f(s,y,z)|^2 \mathrm{d}s)^{\frac{1}{2}}.$$

In order to prove Theorems 3.1, 3.2 and Corollary 3.3 we need the following lemmas.

Lemma 3.4. *Let f be a process which satisfies assumptions of Theorem 3.1. Then there exists a sequence of processes (f_n) such that,*

(i) *For each n, f_n is globally Lipschitz in (y,z) a.e. t and $\mathbb{P}$-a.s. ω.*

(ii) *For every $N \in \mathbb{N}^*$,*

$$|f_n(t,\omega,y,z) - f_n(t,\omega,y',z')| \leq L_{(N+\frac{1}{n})}\left(|y-y'| + |z-z'|\right),$$

for each (y,y',z,z') such that $|y|\leq N$, $|y'|\leq N$, $|z|\leq N$, $|z'|\leq N$.

(iii) *There exists a constant $K(M) > 0$ such that for each (y,z),*

$$\langle y, f_n(t,\omega,y,z)\rangle \leq K(M)(1+|y|^2+|y|\cdot|z|)$$

$\mathbb{P}$-a.s. and a.e. $t\in[0,1]$.

(iv) *There exists a constant $K(M) > 0$ such that for each (y,z),*

$$\sup_n |f_n(t,\omega,y,z|) \leq K(M)(1+|y|\sqrt{|\log|y||}+|z|^\alpha)$$

$\mathbb{P}$-a.s., a.e. $t\in[0,1]$.

(v) *For every N, $\rho_N(f_n - f) \to 0$ as $n\to\infty$.*

Proof. Let $\varphi_n\colon \mathbb{R}^d \to \mathbb{R}_+$ be a sequence of smooth functions with compact support which approximate the Dirac measure at 0 and which satisfy $\int \varphi_n(u)du = 1$. Let $\psi_n\colon \mathbb{R}^d \to \mathbb{R}_+$ be a sequence of smooth functions such that $0\leq\psi_n\leq 1$, $\psi_n(u) = 1$ for $|u|\leq n$ and $\psi_n(u) = 0$ for $|u|\geq n+1$. Likewise we define the sequence ψ'_n from $\mathbb{R}^{d\times r}$ to $\mathbb{R}_+$. We put, $f_{q,n}(t,y,z) = \int f(t,y-u,z)\varphi_q(u)\mathrm{d}u\,\psi_n(y)\psi'_n(z)$. For $n\in\mathbb{N}^*$, let $q(n)$ be an integer such that $q(n)\geq M[4n^2+10n+12]$. It is not difficult to see that the sequence $f_n := f_{q(n),n}$ satisfies all assertions (i)-(v). □

Consider for fixed (t,ω) the sequence $f_n(t,\omega,y,z)$ associated to f by Lemma 3.4. We get from Ouknine [16] that for each n there exists a unique triplet $\{(Y_t^n, Z_t^n, K_t^n; 0 \leq t \leq 1)\}$ of progressively measurable processes taking values in $\mathbb{R}^d \times \mathbb{R}^{d\times r} \times \mathbb{R}^d$ and satisfying:

(1′) Z^n is a predictable process and $\mathbb{E} \int_0^1 |Z_t^n|^2 dt < +\infty$

(2′) $Y_t^n = \xi + \int_t^1 f_n(s, Y_s^n, Z_s^n)\mathrm{d}s - \int_t^1 Z_s^n \mathrm{d}W_s + K_1^n - K_t^n$

(3′) the process Y^n is continuous

(4′) K^n is absolutely continuous $K_0^n = 0$, and $\int_0^{\cdot} (Y_t^n - \alpha_t)\,\mathrm{d}K_t^n \leq 0$ for every α_t progressively measurable process which is continuous and takes values into $\overline{\Theta}$

(5′) $Y_t^n \in \overline{\Theta}$, $0 \leq t \leq 1$ a.s.

The following lemma gives estimates for the processes (Y^n, Z^n, K^n).

Lemma 3.5. (a) *Let assumptions of Theorem* 3.1 *hold. Then there exists a constant C depending only on M and ξ, such that*

$$\mathbb{E}\left(\sup_{0\leq t\leq 1} |Y_t^n|^2 + \int_0^1 |Z_s^n|^2\,\mathrm{d}s\right) \leq C, \quad \forall\, n \in \mathbb{N}^*.$$

(b) *Assume moreover that* $\mathbb{E}\,(|\xi|^{2p}) < \infty$ *for some integer* $p > 1$. *Then, there exists a constant C depending only on M, p and ξ, such that*

$$\mathbb{E}\left(\sup_{0\leq t\leq 1} |Y_t^n|^{2p} + |K_1^n|^2\right) \leq C, \quad \forall\, n \in \mathbb{N}^*.$$

Proof. Assertion (a) follows from Itô's formula, assumption **(A.1)** (ii), Gronwall's lemma and Burkholder-Davis-Gundy inequality.

Let us prove (b). Using Itô's formula we obtain,

$$\begin{aligned} |Y_t^n|^2 + \int_t^1 |Z_s^n|^2\,\mathrm{d}s &= |\xi|^2 + 2\int_t^1 f_n(s, Y_s^n, Z_s^n) Y_s^n \mathrm{d}s \\ &\quad - 2\int_t^1 Z_s^n Y_s^n \mathrm{d}W_s + 2\int_t^1 Y_s^n \mathrm{d}K_s^n. \end{aligned} \tag{3.1}$$

Without loss of generality we can assume that $0 \in \Theta$. Hence by the relation (4′) we have

$$\int_t^1 Y_s^n \mathrm{d}K_s^n \leq 0.$$

We use **(A.1)** (ii) and the inequality $ab \leq \frac{a^2}{2} + \frac{b^2}{2}$ to obtain

$$|Y_t^n|^2 = |\xi|^2 + 2C + (2C + 2C^2) \int_t^1 |Y_s^n|^2 \,\mathrm{d}s - 2 \int_t^1 Z_s^n Y_s^n \mathrm{d}W_s$$

Taking the conditional expectation with respect to $\mathcal{F}_t$ in both sides we deduce

$$|Y_t^n|^2 \leq \mathbb{E}\left(|\xi|^2 + 2C + (2C + 2C^2) \int_t^1 |Y_s^n|^2 \,\mathrm{d}s / \mathcal{F}_t \right).$$

Jensen's inequality shows that for every $p > 1$,

$$\begin{aligned} \mathbb{E}\,|Y_t^n|^{2p} &\leq C_p\Big(\mathbb{E}\left[|\xi|^{2p}\right] + (2C)^p + (2C + 2C^2)^p \mathbb{E}\left[\int_t^1 |Y_s^n|^{2p} \mathrm{d}s \right] \Big) \\ &\leq C_p\Big(1 + \mathbb{E} \int_t^1 |Y_t^n|^{2p} \,\mathrm{d}s \Big). \end{aligned}$$

Gronwall's lemma implies that

$$\sup_{0 \leq t \leq 1} \mathbb{E}|Y_t^n|^{2p} < C, \tag{2.7}$$

where C is a constant not depending on n. Since for every n

$$\begin{aligned} &\mathbb{E} \sup_{0 \leq t \leq 1} \left| \int_t^1 Y_s^n Z_s^n \mathrm{d}W_s \right| \\ &\leq C\Big(\mathbb{E} \sup_{0 \leq t \leq 1} |Y_t^n|^2 \Big)^{\frac{1}{2}} \left(\mathbb{E} \sup_{0 \leq t \leq 1} \left| \int_0^t Z_s^n \mathrm{d}W_s \right|^2 \right)^{\frac{1}{2}}, \\ &\leq C\Big(\mathbb{E} \sup_{0 \leq t \leq 1} |Y_t^n|^2 \Big)^{\frac{1}{2}} \left(\mathbb{E} \int_0^1 |Z_s^n|^2 \mathrm{d}s \right)^{\frac{1}{2}} < \infty, \end{aligned}$$

then the local martingale, $\int_0^t Y_s^n, Z_s^n \mathrm{d}W_s, 0 \leq t \leq 1$ is a uniformly integrable martingale.

Coming back to equation (3.1) and applying Doob's maximal inequality for the martingale $\int_0^t Z_s^n Y_s^n \mathrm{d}W_s$, we obtain that there exists a universal constant C such that for every $n \in \mathbb{N}$,

$$\mathbb{E} \sup_{0 \leq t \leq 1} |Y_t^n|^{2p} < C.$$

From (2') we have

$$K_1^n - K_t^n = Y_t^n - \xi - \int_t^1 f_n(s, Y_s^n, Z_s^n) \mathrm{d}s + \int_t^1 Z_s^n \mathrm{d}W_s,$$

By assumption **(A.1)** (iii) we obtain

$$\mathbb{E}\,|K_1^n - K_t^n|^2 \leq C(\mathbb{E}\,|\xi|^2 + \mathbb{E}\,|Y_t^n|^2 + 1 + \mathbb{E}\int_t^1 |Y_s^n|^4 \mathrm{d}s + \mathbb{E}\int_t^1 |Z_s^n|^2\,\mathrm{d}s)$$

and from assertion (a) and (2.2), we deduce that

$$\sup_n \mathbb{E}\,|K_1^n|^2 \leq C, \quad \text{for all } n \in \mathbb{N}^*.$$

from which assertion (b) follows. Lemma 3.5 is proved. □

We shall prove the convergence of the sequence $(Y^n, Z^n, K^n)_{n\in\mathbb{N}^*}$.

Lemma 3.6. *Under assumptions of Theorem* 3.1, *there exists* (Y, Z, K) *such that*

$$\lim_{n\to\infty} \mathbb{E}\left\{\sup_{0\leq t\leq 1} |Y_t^n - Y_t|^2 + \sup_{0\leq t\leq 1} |K_t^n - K_t|^2 + \int_0^1 |Z_s^n - Z_s|^2\,\mathrm{d}s\right\} = 0.$$

Proof. It follows from Itô's formula that

$$\begin{aligned}
&|Y_t^n - Y_t^m|^2 + \int_t^1 |Z_s^n - Z_s^m|^2\,\mathrm{d}s \\
&= 2\int_t^1 (Y_s^n - Y_s^m)^*\,(f_n(s, Y_s^n, Z_s^n) - f_m(s, Y_s^m, Z_s^m))\mathrm{d}s \\
&\quad - 2\int_t^1 (Y_s^n - Y_s^m)^*\,(Z_s^n - Z_s^m)\,\mathrm{d}W_s \\
&\quad + 2\int_t^1 (Y_s^n - Y_s^m)\,(\mathrm{d}K_s^n - \mathrm{d}K_s^m).
\end{aligned}$$

Since $Y^n, Y^m \in \overline{\Theta}$ are progressively measurable and continuous processes, then from $(4')$ it follows that

$$\int_t^1 (Y_s^n - Y_s^m)\,\mathrm{d}K_s^n \leq 0 \text{ and } \int_t^1 (Y_s^m - Y_s^n)\,\mathrm{d}K_s^m \leq 0.$$

Hence

$$\begin{aligned}
\int_t^1 (Y_s^n - Y_s^m)\,(\mathrm{d}K_s^n - \mathrm{d}K_s^m) &= \int_t^1 (Y_s^n - Y_s^m)\,\mathrm{d}K_s^n \\
&\quad + \int_t^1 (Y_s^m - Y_s^n)\,\mathrm{d}K_s^m \leq 0.
\end{aligned}$$

For a given $N > 1$, let L_N be the Lipschitz constant of f in the ball $B(0, N)$. We put $B_{n,m,N} := \{(s,\omega) \mid |Y_s^n|^2 + |Z_s^n|^2 + |Y_s^m|^2 + |Z_s^m|^2 \geq N^2\}$, $B_{n,m,N}^c := \Omega \setminus B_{n,m,N}$.

Taking the expectation in the above equation, we deduce that

$$\begin{aligned}
&\mathbb{E}\,|Y_t^n - Y_t^m|^2 + \mathbb{E}\int_t^1 |Z_s^n - Z_s^m|^2 \mathrm{d}s \\
&\le \beta^2 \mathbb{E}\int_t^1 |Y_s^n - Y_s^m|^2 \mathrm{d}s \\
&\quad + \frac{1}{\beta^2}\mathbb{E}\int_t^1 |f_n(s, Y_s^n, Z_s^n) - f_m(s, Y_s^m, Z_s^m)|^2 \mathrm{d}s.
\end{aligned}$$

Hence

$$\begin{aligned}
&\mathbb{E}\,|Y_t^n - Y_t^m|^2 + \mathbb{E}\int_t^1 |Z_s^n - Z_s^m|^2 \mathrm{d}s \\
&\le \beta^2 \mathbb{E}\int_t^1 |Y_s^n - Y_s^m|^2 \mathrm{d}s \\
&\quad + \frac{1}{\beta^2}\mathbb{E}\int_t^1 |f_n(s, Y_s^n, Z_s^n) - f_m(s, Y_s^m, Z_s^m)|^2 1_{B_{n,m,N}} \mathrm{d}s \\
&\quad + \frac{4}{\beta^2}\mathbb{E}\int_t^1 |f_n(s, Y_s^n, Z_s^n) - f(s, Y_s^n, Z_s^n)|^2 1_{B^c_{n,m,N}} \mathrm{d}s \\
&\quad + \frac{2}{\beta^2}\mathbb{E}\int_t^1 |f(s, Y_s^n, Z_s^n) - f(s, Y_s^m, Z_s^m)|^2 1_{B^c_{n,m,N}} \mathrm{d}s \\
&\quad + \frac{4}{\beta^2}\mathbb{E}\int_t^1 |f(s, Y_s^m, Z_s^m) - f_m(s, Y_s^m, Z_s^m)|^2 1_{B^c_{n,m,N}} \mathrm{d}s.
\end{aligned}$$

Observe that for every $\varepsilon > 0$

$$|y|^2|\log|y|| \le |y|^2|y|^\varepsilon = |y|^{2+\varepsilon}, \tag{3.2}$$

then use the fact that f_n satisfies **(A.1)** (iii), inequality (3.2), Hölder's inequality, Chebychev's inequality and Lemma 3.5, we obtain

$$\begin{aligned}
&\mathbb{E}\int_t^1 |f_n(s, Y_s^n, Z_s^n) - f_m(s, Y_s^m, Z_s^m)|^2 1_{B_{n,m,N}} \mathrm{d}s \\
&\le M^2\, \mathbb{E}\int_t^1 \Big(2 + |Y_s^n|\sqrt{|\log|Y_s^n||} + |Z_s^n|^\alpha \\
&\qquad\qquad + |Y_s^m|\sqrt{|\log|Y_s^m||} + |Z_s^m|^\alpha\Big)^2 1_{B_{n,m,N}} \mathrm{d}s \\
&\le C(M)\, \mathbb{E}\int_t^1 \Big(1 + |Y_s^n|^{2+\varepsilon} + |Z_s^n|^{2\alpha} \\
&\qquad\qquad + |Y_s^m|^{2+\varepsilon} + |Z_s^m|^{2\alpha}\Big) 1_{B_{n,m,N}} \mathrm{d}s \\
&\le C(\xi, M)\left(\frac{1}{N^{2(1-\alpha)}} + \frac{1}{N} + \frac{1}{N^2}\right).
\end{aligned}$$

Since f is L_N-locally Lipschitz in the ball $B(0,N)$ we get

$$\begin{aligned}
&\mathbb{E}\,|Y_t^n - Y_t^m|^2 + \mathbb{E}\int_t^1 |Z_s^n - Z_s^m|^2 \mathrm{d}s \\
&\leq \beta^2 \mathbb{E}\int_t^1 |Y_s^n - Y_s^m|^2 \mathrm{d}s + \frac{C(\xi, M)}{\beta^2}\Big(\frac{1}{N^{2(1-\alpha)}} + \frac{1}{N} + \frac{1}{N^2}\Big) \\
&\quad + \frac{4}{\beta^2}\rho_N^2(f_n - f) + \frac{4}{\beta^2}\rho_N^2(f_m - f) \\
&\quad + \frac{2L_N^2}{\beta^2}\mathbb{E}\int_t^1 |Y_s^n - Y_s^m|^2 \mathrm{d}s + \frac{2L_N^2}{\beta^2}\mathbb{E}\int_t^1 |Z_s^n - Z_s^m|^2 \mathrm{d}s.
\end{aligned}$$

If we choose β such that $\frac{2L_N^2}{\beta^2} = 1$, we obtain

$$\begin{aligned}
\mathbb{E}\,|Y_t^n - Y_t^m|^2 &\leq \frac{4}{\beta^2}(\rho_N^2(f_n - f) + \rho_N^2(f_m - f)) \\
&\quad + \frac{C(\xi, M)}{\beta^2}\Big(\frac{1}{N^{2(1-\alpha)}} + \frac{1}{N} + \frac{1}{N^2}\Big) \\
&\quad + (1+\beta^2)\mathbb{E}\int_t^1 |Y_s^n - Y_s^m|^2 \mathrm{d}s.
\end{aligned}$$

It follows from Gronwall's lemma that, for every $t \in [0,1]$,

$$\begin{aligned}
&\mathbb{E}\,|Y_t^n - Y_t^m|^2 \\
&\leq \frac{C(\xi, M)}{L_N^2}\Big[\rho_N^2(f_n - f) + \rho_N^2(f_m - f) + \Big(\frac{1}{N^{2(1-\alpha)}} + \frac{1}{N} + \frac{1}{N^2}\Big)\Big] \\
&\quad \times \exp(2L_N^2(1-t)).
\end{aligned} \tag{3.3}$$

Since the process $\int_0^t Y_s^n, Z_s^n \mathrm{d}W_s$ is a martingale, then we use the Burkholder-Davis-Gundy inequality, to obtain

$$\begin{aligned}
&\mathbb{E}\sup_{0\leq t\leq 1} |Y_t^n - Y_t^m|^2 \\
&\quad \leq \frac{C}{L_N^2}\Big[\rho_N^2(f_n - f) + \rho_N^2(f_m - f) + \Big(\frac{1}{N^{2(1-\alpha)}} + \frac{1}{N} + \frac{1}{N^2}\Big)\Big] \\
&\quad \times \exp(2L_N^2),
\end{aligned}$$

$$\begin{aligned}
&\Big[\mathbb{E}\int_0^1 |Z_t^n - Z_t^m|^2 \,\mathrm{d}s\Big]^2 \\
&\quad \leq \frac{C}{L_N^2}\Big[\rho_N^2(f_n - f) + \rho_N^2(f_m - f) + \Big(\frac{1}{N^{2(1-\alpha)}} + \frac{1}{N} + \frac{1}{N^2}\Big)\Big] \\
&\quad \times \exp(2L_N^2).
\end{aligned}$$

Passing to the limit on n, m and on N, we show that $(Y^n, Z^n)_{n\in\mathbb{N}^*}$ is a Cauchy sequence in the Banach space $(\mathbb{L}, \|\cdot\|)$. We set

$$Y = \lim_{n\to+\infty} Y^n \quad \text{and} \quad Z = \lim_{n\to+\infty} Z^n.$$

If we return to the equation satisfied by the triple $(Y^n, Z^n, K^n)_{n\in\mathbb{N}^*}$, we can see that

$$\begin{aligned}\mathbb{E}\sup_{0\leq t\leq 1}|K_t^n - K_t^m|^2 \leq C\Big[&\mathbb{E}\sup_{0\leq t\leq 1}|Y_t^n - Y_t^m|^2\\ &+\mathbb{E}\int_0^1 |f_n(s, Y_s^n, Z_s^n) - f_m(s, Y_s^m, Z_s^m)|^2 \mathrm{d}s\\ &+\mathbb{E}\int_0^1 |Z_s^n - Z_s^m|^2 \mathrm{d}s\Big].\end{aligned}$$

We shall prove that the sequence of processes $f_n(., Y^n, Z^n)_n$ converges to $f(., Y, Z)$ in $L^2([0,1]\times\Omega)$

$$\begin{aligned}&\mathbb{E}\int_0^1 |f_n(s, Y_s^n, Z_s^n) - f(s, Y_s, Z_s)|^2 \mathrm{d}s\\ &\leq 2\mathbb{E}\int_0^1 |f_n(s, Y_s^n, Z_s^n) - f(s, Y_s^n, Z_s^n)|^2 \mathrm{d}s\\ &\quad + 2\mathbb{E}\int_0^1 |f(s, Y_s^n, Z_s^n) - f(s, Y_s, Z_s)|^2 \mathrm{d}s\\ &\leq 2\rho_N^2(f_n - f) + C(\xi, M)\Big(\frac{1}{N^{2(1-\alpha)}} + \frac{1}{N} + \frac{1}{N^2}\Big)\\ &\quad + 2L_N^2\mathbb{E}\int_0^1 (|Z_s^n - Z_s|^2 + |Y_s^n - Y_s|^2)\mathrm{d}s.\end{aligned}$$

Passing to the limits successively on n and N, we obtain

$$\mathbb{E}\int_0^1 |f_n(s, Y_s^n, Z_s^n) - f(s, Y_s, Z_s)|^2 \mathrm{d}s \to 0 \quad \text{as } n\to\infty.$$

Now

$$\mathbb{E}\sup_{0\leq t\leq 1}|K_t^n - K_t^m|^2 \to 0 \quad \text{as } n, m\to\infty.$$

Consequently there exists a progressively measurable process K such that

$$\mathbb{E}\sup_{0\leq t\leq 1}|K_t^n - K_t|^2 \to 0 \quad \text{as } n\to\infty,$$

and clearly (K_t) is an increasing (with $K_0 = 0$) and continuous process. □

Proof of Theorem 3.1: Existence. Combining Lemma 3.5 and Lemma 3.6 and passing to the limit in the RBSDE (2'), we show that the triplet $\{(Y_t, Z_t, K_t); 0 \leq t \leq 1\}$ is a solution satisfies (2). In order to finish the proof of the existence part of Theorem 3.1, it remains to check (1), (4), (5).
By Lemma 3.5, we have

$$\mathbb{E}\int_0^1 (|Y_s^n|^2 + |Z_s^n|^2)\mathrm{d}s \leq C,$$

from which (1) follows by using Lemma 3.6 and Fatou's Lemma.
Let α be a continuous process with values in $\overline{\Theta}$, it holds that

$$\langle Y^n(t) - \alpha(t), \mathrm{d}K^n(t)\rangle \leq 0,$$

by Shaisho [20], we obtain

$$\langle Y(t) - \alpha(t), \mathrm{d}K(t)\rangle \leq 0.$$

To finish the proof of the existence of solutions, we shall prove that

$$\mathbb{P}\left\{Y_t \in \overline{\Theta} \mid 0 \leq t < +\infty\right\} = 1.$$

Since the process (Y_t) is continuous, it suffices to prove that

$$\mathbb{P}\left\{Y_t \in \overline{\Theta}\right\} = 1 \quad \forall\, t \geq 0.$$

Since $Y^n \in \overline{\Theta}$ and Y^n converges to Y in $L^2([0,T] \times \Omega)$, there exists a subsequence Y^{n_k} such that $Y^{n_k} \to Y$ a.s. Hence $Y \in \overline{\Theta}$. The proof of the existence part is then completed. □

Proof of Theorem 3.1: Uniqueness. We prove uniqueness under condition (a), the proof under assumption (b) can be performed as that of Lemma 3.6, Step 2.

Let $\{(Y_t, Z_t, K_t)\ 0 \leq t \leq 1\}$ and $\{(Y_t', Z_t', K_t')\ 0 \leq t \leq 1\}$ be two solutions of our BSDE (1)–(5). Define

$$\{(\Delta Y_t, \Delta Z_t, \Delta K_t)\ 0 \leq t \leq 1\} = \{(Y_t - Y_t', Z_t - Z_t', K_t - K_t')\ 0 \leq t \leq 1\}.$$

It follows from Itô's formula that

$$\begin{aligned}
&\mathbb{E}\left[|\Delta Y_t|^2 + \int_t^1 |\Delta Z_s|^2\,\mathrm{d}s\right] \\
&= 2\mathbb{E}\int_t^1 \langle \Delta Y_s, f(s, Y_s, Z_s) - f(s, Y_s', Z_s')\rangle\,\mathrm{d}s \\
&\quad + 2\mathbb{E}\int_t^1 \langle \Delta Y_{s-}, \mathrm{d}\Delta K_s\rangle.
\end{aligned}$$

Since ΔK_s is continuous, it holds:

$$\int_t^1 \langle \Delta Y_{s-}, \mathrm{d}\Delta K_s \rangle = \int_t^1 \langle \Delta Y_s, \mathrm{d}\Delta K_s \rangle \quad a.s.$$

By Shaisho [20], we get

$$\mathbb{E} \int_t^1 \langle \Delta Y_s, \mathrm{d}\Delta K_s \rangle \leq 0.$$

For $N > 1$, define

$$A_N := \{(s, w) \mid |Y_s|^2 + |Y_s'|^2 + |Z_s|^2 + |Z_s'|^2 \geq N^2\},$$

$A_N^{\mathrm{c}} := \Omega \setminus A_N$ and denote by L_N the Lipschitz constant of f in the ball $B(0, N)$. Since $2|a| \cdot |b| \leq \beta^2 a^2 + \frac{1}{\beta^2} b^2$ for each $\beta \neq 0$, we get

$$\begin{aligned}
\mathbb{E} \left[|\Delta Y_t|^2 + \int_t^1 |\Delta Z_s|^2 \,\mathrm{d}s \right] &\leq \beta^2 \mathbb{E} \int_t^1 |\, \Delta Y_s \,|^2 \,\mathrm{d}s \\
&\quad + \frac{1}{\beta^2} \mathbb{E} \int_t^1 |f(s, Y_s, Z_s) - f(s, Y_s', Z_s')|^2 1_{A_N} \mathrm{d}s \\
&\quad + \frac{1}{\beta^2} \mathbb{E} \int_t^1 |f(s, Y_s, Z_s) - f(s, Y_s', Z_s')|^2 1_{A_N^{\mathrm{c}}} \mathrm{d}s.
\end{aligned}$$

As in the proof of Lemma 3.6, we can show that

$$\begin{aligned}
\mathbb{E} \left[|\Delta Y_t|^2 + \int_t^1 |\Delta Z_s|^2 \,\mathrm{d}s \right] &\leq \Big(\beta^2 + \frac{L_N^2}{\beta^2} \Big) \mathbb{E} \int_t^1 |\Delta Y_s|^2 \,\mathrm{d}s \\
&\quad + \frac{L_N^2}{\beta^2} \mathbb{E} \int_t^1 |\Delta Z_s|^2 \,\mathrm{d}s + \frac{C(\xi, M)}{\beta^2} \Big(\frac{1}{N^{2(1-\alpha)}} + \frac{1}{N} + \frac{1}{N^2} \Big).
\end{aligned}$$

If we choose β such that $L_N^2 / \beta^2 = 1$, and using Gronwall's Lemma and the Burkholder-Davis-Gundy inequality, we get

$$\mathbb{E} \sup_{0 \leq t \leq 1} |\Delta Y_t|^2 \leq \frac{C(\xi, M)}{L_N^2} \Big(\frac{1}{N^{2(1-\alpha)}} + \frac{1}{N} + \frac{1}{N^2} \Big) e^{L_N^2 + 1},$$

$$\mathbb{E} \int_0^1 |\Delta Z_s|^2 \,\mathrm{d}s \leq \frac{C(\xi, M)}{L_N^2} \Big(\frac{1}{N^{2(1-\alpha)}} + \frac{1}{N} + \frac{1}{N^2} \Big) e^{L_N^2 + 1},$$

from which the uniqueness follows. Theorem 3.1 is proved. □

Proof of Theorem 3.2. Arguing as in the proof of Theorem 3.1 we show that

$$\begin{aligned}
&\mathbb{E}\,|Y_t^n - Y_t^m|^2 + \mathbb{E}\int_t^1 |Z_s^n - Z_s^m|^2 \mathrm{d}s \leq \\
&\leq 2\mathbb{E}\int_t^1 \langle Y_s^n - Y_s^m, f_n(s, Y_s^n, Z_s^n) - f_m(s, Y_s^m, Z_s^m)\rangle 1_{B_{n,m,N}} \mathrm{d}s \\
&\quad + \beta^2 \mathbb{E}\int_t^1 |Y_s^n - Y_s^m|^2 1_{B^c_{n,m,N}} \mathrm{d}s \\
&\quad + \frac{4}{\beta^2}\mathbb{E}\int_t^1 |f_n(s, Y_s^n, Z_s^n) - f(s, Y_s^n, Z_s^n)|^2 1_{B^c_{n,m,N}} \mathrm{d}s \\
&\quad + \frac{2}{\beta^2}\mathbb{E}\int_t^1 |f(s, Y_s^n, Z_s^n) - f(s, Y_s^m, Z_s^m)|^2 1_{B^c_{n,m,N}} \mathrm{d}s \\
&\quad + \frac{4}{\beta^2}\mathbb{E}\int_t^1 |f(s, Y_s^m, Z_s^m) - f_m(s, Y_s^m, Z_s^m)|^2 1_{B^c_{n,m,N}} \mathrm{d}s.
\end{aligned}$$

We use Hölder's inequality, assumption **(A.1)** (iv), inequality (3.2), Chebychev's inequality and Lemma 3.5 to show that

$$\begin{aligned}
&\mathbb{E}\int_t^1 \langle Y_s^n - Y_s^m, f_n(s, Y_s^n, Z_s^n) - f_m(s, Y_s^m, Z_s^m)\rangle 1_{B_{n,m,N}} \mathrm{d}s \\
&\leq 2\mathbb{E}\int_t^1 |Y_s^n - Y_s^m||f_n(s, Y_s^n, Z_s^n) - f_m(s, Y_s^m, Z_s^m)| 1_{B_{n,m,N}} \mathrm{d}s \\
&\leq 2\Big[\mathbb{E}\int_t^1 |Y_s^n - Y_s^m|^2 1_{B_{n,m,N}} \mathrm{d}s\Big]^{\frac{1}{2}} \\
&\qquad \times \Big[\mathbb{E}\int_t^1 |f_n(s, Y_s^n, Z_s^n) - f_m(s, Y_s^m, Z_s^m)|^2 \mathrm{d}s\Big]^{\frac{1}{2}} \\
&\leq 2M\Big[\mathbb{E}\int_t^1 |Y_s^n - Y_s^m|^2 1_{B_{n,m,N}} \mathrm{d}s\Big]^{\frac{1}{2}} \\
&\qquad \times \Big[\mathbb{E}\int_t^1 \Big(2 + |Y_s^n|\sqrt{|\log|Y_s^n||} \\
&\qquad\qquad + |Z_s^n| + |Y_s^m|\sqrt{|\log|Y_s^m||} + |Z_s^m|\Big)^2 \mathrm{d}s\Big]^{\frac{1}{2}} \\
&\leq C(M,\xi)\Big[\mathbb{E}\int_t^1 |Y_s^n - Y_s^m|^4 \mathrm{d}s\Big]^{\frac{1}{4}} \Big[\mathbb{E}\int_t^1 1_{B_{n,m,N}} \mathrm{d}s\Big]^{\frac{1}{4}} \\
&\leq \frac{K(M,\xi)}{\sqrt{N}}.
\end{aligned}$$

Therefore

$$
\begin{aligned}
&\mathbb{E}\left(|Y_t^n - Y_t^m|^2\right) \\
&\quad \leq \left[\frac{2}{L_N^2}\left(\rho_N^2(f_n - f) + \rho_N^2(f_m - f)\right) + \frac{K(M,\xi)}{\sqrt{N}}\right] \\
&\quad \times \exp(2L_N^2(1-t)). \qquad (3.4)
\end{aligned}
$$

Passing to the limit first on n, m and next on N then using the Burkholder-Davis-Gundy inequality, we show that (Y^n, Z^n) is a Cauchy sequence in the Banach space $(\mathbb{L}, \|\cdot\|)$. The end of the proof can be performed as that of Theorem 3.1. Theorem 3.2 is proved. □

Proof of Corollary 3.3. Without loss of generality we assume $L = 0$ and $t < 1$. Coming back to equation (3.4) and suppose first that

$$L_N^2 \leq \frac{1}{4(1-t)} \log N,$$

then as in Theorem 3.2, passing to the limit successively on n, m and N we get the result. Assume now that

$$L_N \leq \sqrt{\log N},$$

let δ be a strictly positive number such that $\delta < \frac{1}{4}$ and $([t_{i+1}, t_i])$ be a subdivision of $[0,1]$ such that $|t_i - t_{i+1}| \leq \delta$. Iterating the previous arguments in all the subintervals $[t_{i+1}, t_i]$, we complete the proof of Corollary 3.3. □

The following corollaries give a weaker condition on L_N in the case where f is uniformly Lipschitz in the variable z and locally Lipschitz with respect to the variable y.

Corollary 3.7. *Let* **(A.1)** (i)-(ii), (vi) *and* **(A.2)** *be satisfied. Assume moreover that the generator f is locally L_N–Lipschitz in the variable Y and uniformly L-Lipschitz in the variable Z and $\mathbb{E}\left(|\xi|^5\right) < \infty$. Assume moreover that L_N satisfies the following condition*

$$\lim_{N\to\infty} \left(\frac{1}{N^{2(1-\alpha)}} + \frac{1}{N^2} + \frac{1}{N}\right) \exp(2L_N) = 0.$$

Then the reflected BSDE (1)-(5) *has a unique solution.*

Proof. The arguments used in the proof of Lemma 3.6 and inequality (3.2) lead to

$$\begin{aligned}
&\mathbb{E}\,|Y_t^n - Y_t^m|^2 + \mathbb{E}\int_t^1 |Z_s^n - Z_s^m|^2 \mathrm{d}s \\
&\leq \mathbb{E}\int_t^1 |Y_s^n - Y_s^m|^2 1_{B_{n,m,N}} \mathrm{d}s \\
&\quad + 2\mathbb{E}\int_t^1 |f_n(s, Y_s^n, Z_s^n) - f_m(s, Y_s^m, Z_s^m)|^2 1_{B_{n,m,N}} \mathrm{d}s \\
&\quad + \mathbb{E}\int_t^1 |Y_s^n - Y_s^m||f_n(s, Y_s^n, Z_s^n) - f_m(s, Y_s^m, Z_s^m)| 1_{B_{n,m,N}^c} \mathrm{d}s \\
&\leq \mathbb{E}\int_t^1 |Y_s^n - Y_s^m|^2 \mathrm{d}s + C(\xi, M)\left(\frac{1}{N^{2(1-\alpha)}} + \frac{1}{N} + \frac{1}{N^2}\right) \\
&\quad + 2\rho_N^2(f_n - f) + 2\rho_N^2(f_m - f) + \beta^2\mathbb{E}\int_t^1 |Y_s^n - Y_s^m|^2 \mathrm{d}s \\
&\quad + 2L_N\mathbb{E}\int_t^1 |Y_s^n - Y_s^m|^2 \mathrm{d}s + \frac{L^2}{\beta^2}\mathbb{E}\int_t^1 |Z_s^n - Z_s^m|^2 \mathrm{d}s.
\end{aligned}$$

Arguing as in the proof of Theorem 3.1, we conclude the proof of Corollary 3.7. □

Corollary 3.8. *Let* **(A.1)** (i)-(ii), (vii) *and* **(A.2)** *be satisfied. Assume moreover that the generator f is locally L_N-Lipschitz in the variable Y and globally L-Lipschitz in the variable Z and $\mathbb{E}(|\xi|^4) < \infty$. Assume moreover that L_N satisfies the following condition*

$$\lim_{N\to\infty}\left(\frac{1}{\sqrt{N}}\right)\exp(2L_N) = 0.$$

Then the reflected BSDE (1)-(5) *has a unique solution.*

Proof. Arguing as in the proof of Theorem 3.2 we obtain

$$\begin{aligned}
&\mathbb{E}\sup_{0\leq t\leq 1}|Y_t^n - Y_t^m|^2 \leq \\
&\leq C\left(2\rho_N^2(f_n - f) + 2\rho_N^2(f_m - f) + \frac{1}{\sqrt{N}}\right)\exp(2L_N)\exp(L^2+1).
\end{aligned}$$

$$\begin{aligned}
&\left(\mathbb{E}\int_0^1 |Z_s^n - Z_s^m|^2 \mathrm{d}s\right)^2 \\
&\leq C\Big(2\rho_N^2(f_n - f) + 2\rho_N^2(f_m - f) + \frac{1}{\sqrt{N}}\Big)\exp(2L_N)\exp(L^2+1).
\end{aligned}$$

From which the result follows. □

Arguing as in the proof of Corollary 3.3, one can prove the following result.

Corollary 3.9. *Let conditions of Corollary* 3.8 *hold. Assume moreover that* L_N *satisfies the following assumption*

$$\exists L' \geq 0, \quad L_N \leq L' + \log N.$$

Then the reflected BSDE (1)-(5) *has a unique solution.*

Remark 3.10. All the previous results remain valid if we replace Θ with $Domain(\phi)$, where ϕ is a convex, lower semi-continuous and proper function. Only minor modifications should be required in the proofs.

Remark 3.11. To clarify the presence of assumptions (iii)-(iv) and (vi)-(vii), we note that the condition $L_N = O(\sqrt{\log N})$ (resp. $L_N = O(\log N)$) in Theorems 3.1, 3.2, and Corollary 3.3 (resp. Corollaries 3.7, 3.8 and 3.9) forces the coefficient f to grow at most as $|y|\sqrt{|\log|y||}$ (resp. $|y|\,|\log|y||$).

Acknowledgments

We are grateful to the referee for suggestions which allowed us to improve the first version of the paper. We also wish to thank our friends M. Hassani and Y. Ouknine for various discussions on the BSDEs.

References

[1] Bahlali K.: Backward stochastic differential equations with locally Lipschitz coefficient, *C. R. Acad. Sci. Paris Sér. I Math.* **333** (2001) 481–486.

[2] Bahlali K.: Existence and uniqueness of solutions for BSDEs with locally Lipschitz coefficient. *Electron. Comm. Probab.* **7** (2002) 169–179.

[3] Bahlali K., Mezerdi B., Ouknine Y.: Some generic properties in backward stochastic differential equations with continuous coefficient, "Monte Carlo and probabilistic methods for partial differential equations (Monte Carlo, 2000)".

[4] Dermoune A., Hamadène S., Ouknine Y.: Backward stochastic differential equation with local time. *Stochastics Stochastics Rep.* **66** (1999) 103–119.

[5] El Karoui N., Peng S., Quenez M.-C.: Backward stochastic differential equations in finance. *Math. Finance* **7** (1997) 1–71.

[6] El Karoui N., Kapoudjian C., Pardoux É., Peng S., Quenez M.-C.: Reflected solutions of backward SDE's and related obstacle problems for PDE's, *Ann. Probab.* **25** (1997) 702–737.

[7] Gégout-Petit A., Pardoux E.: Équations différentielles stochastiques rétrogrades réfléchies dans un convexe, *Stochastics Stochastics Rep.* **57** (1996) 111–128.

[8] Hamadène S.: Multidimensional Backward SDE's with uniformly continuous coefficients, Preprint Université du Maine 00-3, Le Mans 2000.

[9] Hamadène S.: Équations différentielles stochastiques rétrogrades: le cas localement lipschitzien. *Ann. Inst. H. Poincaré Probab. Statist.* **32** (1996) 645–659.

[10] Hamadène S., Lepeltier J.-P.: Zero-sum stochastic differential games and backward equations, *Systems Control Lett.* **24** (1995) 259–263.

[11] Hamadène S., Ouknine Y.: Backward stochastic differential equations with jumps and random obstacle. To appear in *Stochastics Stochastics Rep.*

[12] Lepeltier J.-P., San Martin J.: Backward stochastic differential equations with continuous coefficient, *Statist. Probab. Lett.* **32** (1997) 425–430.

[13] Mao X.: Adapted solutions of backward stochastic differential equations with non-Lipschitz coefficients. *Stochastic Process. Appl.* **58** (1995) 281–292.

[14] N'zi M.: Multivalued backward stochastic differential equations with local Lipschitz drift. *Stochastics Stochastics Rep.* **60** (1997) 205–218.

[15] N'zi M., Ouknine Y.: Multivalued backward stochastic differential equations with continuous coefficients. *Random Oper. Stochastic Equations* **5** (1997) 59–68.

[16] Ouknine Y.: Reflected backward stochastic differential equations with jumps, *Stochastics Stochastics Rep.* **65** (1998) 111–125.

[17] Pardoux É., Peng S.: Adapted solution of a backward stochastic differential equation. *Systems Control Lett.* **14** (1990) 55–61.

[18] Pardoux É., Răşcanu A.: Backward stochastic differential equations with subdifferential operator and related variational inequalities. *Stochastic Process. Appl.* **76** (1998) 191–215.

[19] Rong S.: On solutions of backward stochastic differential equations with jumps and applications. *Stochastic Process. Appl.* **66** (1997) 209–236.

[20] Saisho Y.: Stochastic differential equations for multidimensional domain with reflecting boundary. *Probab. Theory Related Fields* **74** (1987) 455–477.

[21] Tang S.J., Li X.J.: Necessary conditions for optimal control of stochastic systems with random jumps, *SIAM J. Control Optim.* **32** (1994) 1447–1475.

SEMIDYNAMICAL SYSTEMS WITH THE SAME ORDER RELATION

Nedra Belhaj Rhouma
Institut Préparatoire aux études d'ingénieurs de Tunis,
2 rue Jawaher lel Nehru, 1008,
Montfleury-Tunis, Tunisia
Nedra.BelhajRhouma@ipeit.rnu.tn

Mounir Bezzarga
Institut Préparatoire aux études d'ingénieurs de Tunis,
2 rue Jawaher lel Nehru, 1008,
Montfleury-Tunis, Tunisia
Mounir.Bezzarga@ipeim.rnu.tn

Abstract Given a transient semidynamical system $(X, \mathcal{B}, \Phi, \omega)$, we introduce an additive functional $\mathcal{A}$ and we define the time changed semidynamical system $(X, \mathcal{B}, \Phi_{\mathcal{A}}, \omega)$. We study the properties of the new semidynamical system and we try to give a comparison with $(X, \mathcal{B}, \Phi, \omega)$. Conversely, we give the conditions fulfilled by two semidynamical system Φ and Φ' to get the existence of an additive functional $\mathcal{A}$ satisfying $\Phi' = \Phi_{\mathcal{A}}$.

1. Introduction

In this paper we consider a transient semidynamical system $(X, \mathcal{B}, \Phi, \omega)$ (cf. [2, 3]), then we introduce the definition of an additive functional $\mathcal{A} = (A_t)_{t \geq 0}$ defined on X. As in [5, 15, 32], we investigate the fundamental properties of such additive functional and we will illustrate it with some examples.

After that, we define the time changing τ associated to $\mathcal{A}$ and we show that by setting $\Phi_{\mathcal{A}}(t, x) = \Phi(\tau_t(x), x)$ for every $t \geq 0, x \in X$, we get a transient semidynamical system $(X, \mathcal{B}, \Phi_{\mathcal{A}}, \omega)$ called the time changed semidynamical system.

S. Albeverio et al. (eds.),
Proceedings of the International Conference on Stochastic Analysis and Applications, 217–237.

For $(X, \mathcal{B}, \Phi_{\mathcal{A}}, \omega)$, we define the order relation $\underset{\Phi_{\mathcal{A}}}{\leq}$, the lifetime $\rho_{\mathcal{A}}$, the Lebesgue measure $\Lambda^{\mathcal{A}}$ and the associated resolvent. We show particularly that "$\underset{\Phi}{\leq}$" = "$\underset{\Phi_{\mathcal{A}}}{\leq}$" and therefore the fine topologies $\mathcal{T}_{\Phi}$ and $\mathcal{T}_{\Phi_{\mathcal{A}}}$ defined respectively on $(X, \mathcal{B}, \Phi, \omega)$ and $(X, \mathcal{B}, \Phi_{\mathcal{A}}, \omega)$ are equal. This leads us to get that $\mathcal{E} = \mathcal{E}_{\mathcal{A}}$ where $\mathcal{E}$ (resp. $\mathcal{E}_{\mathcal{A}}$) is the set of excessive functions of Φ (resp. $\Phi_{\mathcal{A}}$).

Also, we consider the particular case when $\mathcal{A}$ is a contraction of time i.e. $A_t \leq t$ and we prove that there exists a measurable function $0 \leq f \leq 1$ such that $\Lambda^{\mathcal{A}} = f\Lambda$.

The last section of this paper is devoted to give sufficient and necessary conditions for two semidynamical systems Φ and Φ' to get the existence of an additive functional $\mathcal{A}$ such that $\Phi' = \Phi_{\mathcal{A}}$.

2. Preliminary

In this section, we will introduce some definitions which will be useful in the remainder of this paper. For more details see [2], [4] and [31].

In ([2, 4]), it is given the notion of a semidynamical system as being a separable measurable space $(X, \mathcal{B})$ with a distinguished point ω and a measurable map $\Phi : \mathbb{R}_+ \times X \longrightarrow X$ having the following properties:

(S_1) For any x in X, there exists an element $\rho(x)$ in $[0, \infty]$ such that $\Phi(t, x) \neq \omega$ for all $t \in [0, \rho(x))$ and $\Phi(t, x) = \omega$ for all $t \geq \rho(x)$,

(S_2) For any $s, t \in \mathbb{R}_+$ and any $x \in X$ we have

$$\Phi(s, \Phi(t, x)) = \Phi(s + t, x),$$

(S_3) $\Phi(0, x) = x$ for all $x \in X$,

(S_4) If $\Phi(t, x) = \Phi(t, y)$ for all $t > 0$, then $x = y$.

Definition 2.1. The collection $(X, \mathcal{B}, \Phi, \omega)$ with the above properties is called a semidynamical system in the meaning of [4] having ω as coffin state.

Set $X_0 = X \setminus \{w\}$, $\mathcal{B}_0 = \{A \in \mathcal{B}; A \subset X_0\}$ and for any $x \in X_0$, we denote by Γ_x the trajectory of x, i.e.:

$$\Gamma_x = \{\Phi(t, x); t \in [0, \rho(x))\}.$$

So for any $x, y \in X_0$, we put $x \leq_{\Phi} y$ if $y \in \Gamma_x$.

We also define the function Φ_x on $[0, \rho(x))$ by $\Phi_x(t) = \Phi(t, x)$.

Definition 2.2. We say (cf. [2]) that $(X, \mathcal{B}, \Phi, \omega)$ is a transient semidynamical system if moreover there exists a sequence $(A_n)_n \in \mathcal{B}^{\mathbb{N}}$ such that $X_0 = \bigcup_n A_n$ and for any $x \in X_0$, $\lambda(\Phi_x^{-1}(A_n)) < \infty$, where λ denotes the Lebesgue measure on $\mathbb{R}$.

Hypothesis 2.3. In what follows, we shall suppose that $(X, \mathcal{B}, \Phi, \omega)$ is a transient semidynamical system.

Under this assumption (cf. [2]) Φ_x becomes a measurable isomorphism between the interval $[0, \rho(x))$ and the trajectory Γ_x endowed with the trace measurable structures. Therefore "$\leq_\Phi$" becomes an order on X_0. Also X_0 becomes a space of injective trajectories.

Definition 2.4. A maximal trajectory is a totally ordered subset Γ of X_0 with respect to the above order, such that there is no minorant x_0 of Γ, $x_0 \in X_0 \setminus \Gamma$ and for any $x \in \Gamma$, we have $\Gamma_x \subset \Gamma$.

In [2] we have associated a proper submarkovian resolvent $\mathcal{V} = (V_\alpha)_{\alpha \in \mathbb{R}_+}$ of kernels on the measurable space $(X_0, \mathcal{B}_0)$, defined by

$$V_\alpha f(x) = \int_0^{\rho(x)} e^{-\alpha t} f(\Phi(t, x))dt, \ \forall x \in X_0,$$

where $\mathcal{B}_0 = \{U \in \mathcal{B}; U \subset X_0\}$.

The family $\mathcal{V}$ is the resolvent associated to the deterministic semigroup $\mathbb{H} = (H_t)_{t \in \mathbb{R}_+}$ given on $(X_0, \mathcal{B}_0)$ by

$$\varepsilon_x H_t = \begin{cases} \varepsilon_{\Phi(t,x)} & \text{if } t \in [0, \rho(x)), \\ 0 & \text{otherwise.} \end{cases}$$

The above semigroup was introduced by B.O.Koopmann (cf. [29]) in order to transform a nonlinear problem in finite dimension to a linear one in infinite dimension. The potential analysis of the semigroup $\mathbb{H}$ in the case of dynamical systems with real parameters is introduced by M.Hmissi in [23]. See also [24] and [27].

In the next, let us denote by $\mathcal{B}_0^\sigma$ the set of all countable union of maximal trajectories of X_0. Let (cf. [1]) $\mathcal{B}_0^*$ be the $\sigma-$algebra constituted by all subsets A of X_0 such that $A \cap M \in \mathcal{B}_0$ for any countable union M of trajectories of X_0. One can show that the resolvent family $\mathcal{V}$ may be considered on the measurable space $(X_0, \mathcal{B}_0^*)$ and we denote by $\mathcal{F}^*$ the set of all positive $\mathcal{B}_0^*-$measurable functions on X_0.

Definition 2.5 ([2]). The fine topology $\mathcal{T}_\Phi$ associated to the semidynamical system $(X, \mathcal{B}, \Phi, \omega)$, is the set of all subset D of X_0 having the following property:

$$x \in D \Rightarrow [\exists \varepsilon > 0 : \Phi(t, x) \in D, \forall t \in [0, \varepsilon) \cap [0, \rho(x))].$$

In [2], we characterized the set $\mathcal{E}$, of all $\mathcal{V}$-excessive functions on X_0, as being the set of all positive $\mathcal{B}_0$-measurable functions on X_0 which are decreasing with respect to the associated order "$\leq_\Phi$ " and $\mathcal{T}_\Phi$-continuous.

The fine topology was introduced in classical potential theory by M.Brelot (cf. [13]).

We consider also (cf. [4] chap III) the arrival time function $\Psi : X_0 \times X_0 \longrightarrow \mathbb{R}_+$ given by

$$\Psi(x,y) = \begin{cases} t & \text{if } \Phi(t,x) = y,\ t \in [0,\rho(x)), \\ +\infty & \text{otherwise.} \end{cases}$$

It is shown(cf. [3, 1]) that the arrival time function Ψ is measurable if we endow $X_0 \times X_0$ with the product measurable structure of the $\sigma-$algebra $\mathcal{B}_0^*$.

Starting with a transient semidynamical system $(X, \mathcal{B}, \Phi, \omega)$, we have associated in [2], a suitable topology $\mathcal{T}_\Phi^0$, namely the inherent topology in the meaning of [22] in the dynamical case, as being the set of all subset D of X_0 having the following property:

$$\Phi(t_0,x) \in D \Rightarrow [\exists \varepsilon > 0 : \Phi(t,x) \in D, \forall t \in (t_0 - \varepsilon, t_0 + \varepsilon) \cap [0,\rho(x))].$$

Proposition 2.6. *For each $x \in X_0$ the map Φ_x is an homeomorphism between the real interval $[0,\rho(x))$ and the trajectory Γ_x endowed with the trace topology of $\mathcal{T}_\Phi^0$. In particular the inverse $\Psi(x,.)$ is continuous on Γ_x.*

3. Additive functional

In this section, let $(X, \mathcal{B}, \Phi, \omega)$ be a fixed data transient semidynamical system and denote by ρ the lifetime associated defined on X and taking values in $[0,\infty]$.

Definition 3.1. A family $\mathcal{A}$=$\{A_t, t \in [0,\rho)\}$ of functions defined from X to $[0,+\infty]$ is called additive functional of $(X, \mathcal{B}, \Phi, \omega)$ provided the following conditions are satisfied:

(A_1) For each $x \in X_0$, the mapping : $t \rightarrow A_t(x)$ is nondecreasing, right continuous and satisfies $A_0(x) = 0$ for all $x \in X$,

(A_2) For each $x \in X_0$, for each $t \in [0,\rho(x))$, the mapping $x \rightarrow A_t(x)$ is measurable with respect to $\mathcal{B}_0^*$,

(A_3) For each $x \in X_0$, $t, s \geq 0$ such that $t + s \in [0,\rho(x))$,

$$A_{t+s}(x) = A_t(x) + A_s(\Phi(t,x))$$

(A_4) $A_t(\omega) = 0$, $\forall t \geq 0$.

Remark 3.2. Since $t \mapsto A_t(x)$ is nondecreasing, we denote

$$A_{\rho(x)^-}(x) = \lim_{t \to \rho(x)^-} A_t(x).$$

Thus, we can set

$$A_\infty(x) = A_{\rho(x)^-}(x) = \lim_{t \to \infty} A_t(x).$$

Remark 3.3. Let $\mathcal{A}$ be an additive functional and let $\mathcal{M} = \{M_t, t \in [0, \rho)\}$ be the family of functions defined on X by

$$M_t(x) = \exp(-A_t(x)), \text{ for all } x \in X.$$

So $\mathcal{M}$ verifies the functional equation

$$M_{s+t}(x) = M_t(x) M_s(\Phi(t, x)). \tag{E}$$

The analogous of $\mathcal{M}$ for a Markov process is termed a multiplicative functional. Such equation (E) was studied by M.Hmissi in [25] and [26].

Example 3.4. For every $x \in X$, let $A_t(x) = t \wedge \rho(x)$. Since the map $x \mapsto \rho(x)$ is measurable, then $\mathcal{A} = (A_t)_{t \geq 0}$ is an additive functional.

Example 3.5. Let $f \in \mathcal{F}^*$. For every $x \in X_0$ set

$$A_t(x) = \int_0^{t \wedge \rho(x)} f(\Phi(u, x)) du.$$

By setting $f(w) = 0$ for $t \geq \rho(x)$, we get

$$\begin{aligned} A_{t+s}(x) &= A_t(x) + \int_t^{t+s} f(\Phi(u, x)) du \\ &= A_t(x) + \int_0^s f(\Phi(u + t, x)) du \\ &= A_t(x) + \int_0^s f(\Phi(u, \Phi(t, x))) du \\ &= A_t(x) + A_s(\Phi(t, x)) \end{aligned}$$

for every $x \in X_0$, $t, s > 0$. Hence, $(A_t)_{t \geq 0}$ is an additive functional. Moreover, we have

$$A_\infty(x) = \int_0^\infty f(\Phi(u, x)) du = \int_0^{\rho(x)} f(\Phi(u, x)) du = V_0 f(x)$$

which is the potential of f.

Definition 3.6. Let $f \in \mathcal{F}^*$. For every $x \in X_0$, we define

$$I(f,t)(x) = \int_0^t f(\Phi(u,x))dA_u(x)$$

the integral of $f(\Phi(.,x))$ against the measure induced by the nondecreasing function $u \to A_u(x)$.

Note that $dA_u(x)$ attributes no mass to the interval $[\rho(x), \infty)$ by Remark 3.2.

Proposition 3.7. *Let B be an additive functional of $(X, \mathcal{B}, \Phi, \omega)$ and $f \in \mathcal{F}^*$, then*

$$A_t = \int_{(0,t]} f(\Phi(u,.))dB_u$$

defines an additive functional of $(X, \mathcal{B}, \Phi, \omega)$.

Proof. The conditions (A_1) and (A_2) are easy. To check (A_3), we consider

$$\begin{aligned} A_{t+s}(x) &= A_t(x) + \int_{(t,t+s]} f(\Phi(u,x))dB_u(x) \\ &= A_t(x) + \int_{(0,s]} f(\Phi(u+s,x))dB_{u+t}(x) \\ &= A_t(x) + \int_{(0,s]} f(\Phi(u,\Phi(s,x)))dB_u o\Phi(t,x) \\ &= A_t(x) + A_s(\Phi(t,x)). \end{aligned}$$

□

4. Time changed semidynamical system

This section presents the perturbation of a semidynamical system by means of an additive functional. The results are analogous to those from the transformation of a Markov process by an additive functional, developed by R.M.Blumental and R.K.Getoor in [5], C.Dellacherie and P.-A.Meyer in [15] and M.Sharpe in [32]. For Markov processes, the theory of time changes by the inverse of an additive functional $\mathcal{A}$ was used by K.Itô and H.P.McKean (cf. [28]). Time changes have been studied by many other authors like P.J.Fitzimmons (cf. [17]) and J.Glover (cf. [20, 21]).

Definition 4.1. Let $\mathcal{A}$ be an additive functional. For each $x \in X_0$ and every $0 \leq t < \infty$, we define

$$\tau_t(x) = \inf\{s : A_s(x) > t\}$$

where as usual, we set $\tau_t(x) = \infty$ if the set in braces is empty.

Note that in [5], τ is called the inverse of $\mathcal{A}$.

Remark 4.2. Note that if $t < A_s(x) = \inf\{u : \tau_u(x) > s\}$ then $\tau_t(x) \le s$, while if $t > A_s(x)$ then $\tau_t(x) \ge s$. Moreover, we have

$$A_s(x) = \inf\{t : \tau_t(x) > s\},\ 0 \le s < \infty. \tag{4.1}$$

Remark 4.3. If $t \mapsto A_t(x)$ is continuous, then $u \mapsto \tau_u(x)$ is strictly increasing on $[0, A_\infty(x)[$ and $\tau_t(x) = \max\{u : A_u(x) = t\}$ provided the set in braces is not empty. In particular $A_{\tau_u(x)}(x) = u$ if $\tau_u(x) < \infty$.

By a monotone class argument, we get the following result:

Proposition 4.4. *If f is a nonnegative Borel measurable function on $[0, \infty]$ vanishing at ∞ one has*

$$\int_{(0,\infty)} f(t)dA_t(x) = \int_0^\infty f(\tau_t(x))dt \tag{4.2}$$

Corollary 4.5. *If $f \in \mathcal{F}^*$, vanishing at ω one has*

$$\begin{aligned}\int_{(0,\infty)} f(\Phi(t,x))dA_t(x) &= \int_0^\infty f(\Phi(\tau_t(x),x)dt \\ &= \int_0^\infty f(\Phi_{\mathcal{A}}(t,x))dt.\end{aligned} \tag{4.3}$$

Definition 4.6. We call that $\mathcal{A}$ is a strict continuous additive functional (strict CAF) provided the following condition holds: for all $x \in X_0$, there exists $\rho_{\mathcal{A}}(x) > 0$ such that the mapping $t \to A_t(x)$ is a continuous bijection from $[0, \rho(x))$ to $[0, \rho_{\mathcal{A}}(x))$.
In the next, for every $t \ge \rho_{\mathcal{A}}(x)$, we will set $\tau_t(x) = \rho(x)$. Hence, $\Phi(\tau_t(x), x) = \omega$ for every $t \ge \rho_{\mathcal{A}}(x)$.

For the remainder of this paper, we assume that $\mathcal{A}$ is a strict CAF.

Proposition 4.7. *For each $s, t \ge 0$ one has*

$$\tau_{t+s}(x) = \tau_t(x) + \tau_s(\Phi(\tau_t(x), x)).$$

Proof. Let $t, v \ge 0$. By using (A_3) and Remark 4.3, we get

$$\begin{aligned}A_{\tau_t(x)+v}(x) &= A_{\tau_t(x)}(x) + A_v(\Phi(\tau_t(x), x)) \\ &= t + A_v(\Phi(\tau_t(x), x))\end{aligned}$$

Hence, $A_{\tau_t(x)+v}(x) > t+s$ if and only if $A_v(\Phi(\tau_t(x),x)) > s$. It follows that

$$\begin{aligned}\tau_{t+s}(x) &= \inf\{v : A_v(x) > t+s\} \\ &= \tau_t(x) + \inf\{v : A_{\tau_t(x)+v}(x) > t+s\} \\ &= \tau_t(x) + \inf\{v : A_v(\Phi(\tau_t(x),x)) > s\} \\ &= \tau_t(x) + \tau_s(\Phi(\tau_t(x),x))\end{aligned}$$

□

From now on, we consider the mapping

$$\begin{aligned}\Phi_{\mathcal{A}} \colon \mathbb{R}_+ \times X &\longrightarrow X \\ (t,x) &\longmapsto \Phi_{\mathcal{A}}(t,x) = \Phi(\tau_t(x),x).\end{aligned}$$

Remark 4.8. It is obvious that for each $x \in X_0$ we have $\Phi_{\mathcal{A}}(t,x) \neq \omega$ for $t \in [0, \rho_{\mathcal{A}}(x))$ and $\Phi_{\mathcal{A}}(t,x) = \omega$ for $t \geq \rho_{\mathcal{A}}(x)$. When $x = \omega$, we set $\rho_{\mathcal{A}}(\omega) = 0$. Hence, $\rho_{\mathcal{A}}(x)$ is the lifetime of x.

We then get the following result:

Theorem 4.9. *$(X, \mathcal{B}, \Phi_{\mathcal{A}}, \omega)$ is a semidynamical system which will be called the time changed semidynamical system.*

Proof. It is clear that the property (S_3) is obvious and that the property (S_1) is fulfilled by using Remark 4.8.

Next, we shall prove that $\Phi_{\mathcal{A}}$ is measurable. Indeed, let $\alpha \geq 0$, then by using standard facts on the pseudo-inverse of an increasing right continuous function, we get

$$\{(t,x) : \tau_t(x) < \alpha\} = \bigcup_{r \in \mathbb{Q} \cap [0,\alpha)} \{(t,x) : \tau_t(x) \leq r\}$$

is measurable. Hence, since Φ is measurable, we get that $\Phi_{\mathcal{A}}$ is measurable.

We shall prove (S_2). Let $t, s \geq 0$ and $x \in X$, then

$$\begin{aligned}\Phi_{\mathcal{A}}(s, \Phi_{\mathcal{A}}(t,x)) &= \Phi(\tau_s(\Phi_{\mathcal{A}}(t,x)), \Phi_{\mathcal{A}}(t,x)) \\ &= \Phi(\tau_s(\Phi(\tau_t(x),x)), \Phi(\tau_t(x),x)) \\ &= \Phi(\tau_s(\Phi(\tau_t(x),x)) + \tau_t(x), x) \\ &= \Phi(\tau_{s+t}(x), x) = \Phi_{\mathcal{A}}(s+t, x).\end{aligned}$$

Finally, to prove (S_4) let us consider $x, y \in X$ such that $\Phi_{\mathcal{A}}(t,x) = \Phi_{\mathcal{A}}(t,y)$ for all $t > 0$. Hence $\Phi(\tau_t(x),x) = \Phi(\tau_t(x),y)$ for all $t > 0$. Thus, using the right continuity of $t \hookrightarrow \tau_t(x)$ and by letting $t \to 0^+$, we get that $\Phi(0,x) = \Phi(0,y)$ which yields $x = y$. □

Notation 4.10. For every $x \in X$, we will denote by

$$\Gamma_x^{\mathcal{A}} = \{\Phi_{\mathcal{A}}(t,x) : t \in [0, \rho_{\mathcal{A}}(x)[\}.$$

Let us denote by $\Psi_{\mathcal{A}}$ the mapping defined on $X_0 \times X_0$ by

$$\Psi_{\mathcal{A}}(x,y) = \begin{cases} t & \text{if } \Phi_{\mathcal{A}}(t,x) = y,\ t \in [0, \rho_{\mathcal{A}}(x)), \\ +\infty & \text{otherwise.} \end{cases}$$

Then, we have the following result:

Lemma 4.11. *For every $x \in X_0$, we have $\Gamma_x = \Gamma_x^{\mathcal{A}}$. Particularly, the set $\Gamma_x^{\mathcal{A}}$ is measurable.*

Proof. Suppose that $y \in \Gamma_x^{\mathcal{A}}$. Then there exists $s \in [0, \rho_{\mathcal{A}}(x))$ such that $y = \Phi_{\mathcal{A}}(s,x) = \Phi(\tau_s(x), x) = y$ which yields that $\tau_s(x) < \rho(x)$ and $y \in \Gamma_x$. Conversely, let $y \in \Gamma_x$. Then $y = \Phi(t,x)$ for some $t \in [0, \rho(x))$. Hence, there exists $s > 0$ such that $t = \tau_s(x)$ and therefore $y = \Phi(\tau_s(x), x)$ which yields $y \in \Gamma_x^{\mathcal{A}}$. □

Proposition 4.12. *For every $y \in \Gamma_x$, we have*

$$\Psi(x,y) = \tau_{\Psi_{\mathcal{A}}(x,y)}(x).$$

Proof. Let $y \in \Gamma_x = \Gamma_x^{\mathcal{A}}$ and put $s = \Psi_{\mathcal{A}}(x,y)$.

Since $\Phi_{\mathcal{A}}(s,x) = \Phi(\tau_s(x), x) = y$, we get that $\tau_s(x) < \rho(x)$ and $\Psi(x,y) = \tau_s(x)$, i.e. $\Psi(x,y) = \tau_{\Psi_{\mathcal{A}}(x,y)}(x)$. □

An immediate consequence of Lemma 4.11 is the following useful result:

Proposition 4.13. *For every $x, y \in X$ we have*

$$x \underset{\Phi}{\leq} y \Leftrightarrow x \underset{\Phi_{\mathcal{A}}}{\leq} y.$$

5. Resolvent

As in [2], for every measurable function $f\colon X_0 \to \mathbb{R}_+$, for every $\alpha \geq 0$, we define

$$V_\alpha f(x) = \int_0^{\rho(x)} e^{-\alpha t} f \circ \Phi(t,x)dt$$

and

$$V_\alpha^{\mathcal{A}} f(x) = \int_0^{\rho_{\mathcal{A}}(x)} e^{-\alpha t} f \circ \Phi_{\mathcal{A}}(t,x)dt = \int_0^{\rho_{\mathcal{A}}(x)} e^{-\alpha t} f \circ \Phi(\tau_t(x), x)dt$$

Note that if we set $f(\omega) = 0$, then

$$V_\alpha f(x) = \int_0^\infty e^{-\alpha t} f \circ \Phi(t,x)dt$$

and

$$V_\alpha^{\mathcal{A}} f(x) = \int_0^\infty e^{-\alpha t} f \circ \Phi_{\mathcal{A}}(t,x)dt = \int_0^{\rho_{\mathcal{A}}(x)} e^{-\alpha t} f \circ \Phi(\tau_t(x),x)dt$$

As cited in the Preliminary it is known (cf. [2]) that the families $(V_\alpha)_{\alpha\geq 0}$ and $(V_\alpha^{\mathcal{A}})_{\alpha\geq 0}$ are resolvents. i.e. if $U = (V_\alpha)_\alpha$ ($U = (V_\alpha^{\mathcal{A}})_{\alpha\geq 0}$ resp.), we have

$$U_\alpha = U_\beta + (\beta - \alpha)U_\alpha U_\beta, \quad \forall \alpha, \beta \in \mathbb{R}_+,\ 0 \leq \alpha \leq \beta,$$
$$U_\alpha U_\beta = U_\beta U_\alpha,\ \forall \alpha, \beta \geq 0.$$

Using Corollary 4.5, we get

$$V_0^{\mathcal{A}} f(x) = \int_{(0,\infty)} f(\Phi(t,x))dA_t(x).$$

Definition 5.1. We say that the additive functional $\mathcal{A}$ is a contraction of time if

$$A_t(x) \leq t$$

for all $t \in [0, \rho(x))$ and $x \in X_0$.

Example 5.2. Let f be a measurable function such that $0 \leq f \leq 1$, then

$$A_t(x) = \int_0^{t\wedge\rho(x)} f(\Phi(s,x))ds$$

is a contraction of time.

We recall by [2] and [19], that the semidynamical system $(X, \mathcal{B}, \Phi, \omega)$ is transient if and only if there exists a measurable function $\varphi > 0$ on X_0 such that for all $x \in X_0$, we have

$$V_0\varphi(x) = \int_0^{\rho(x)} \varphi(\Phi(t,x))dt < \infty.$$

i.e. V_0 is a proper kernel.

Theorem 5.3. *If* $\mathcal{A}$ *is such that*

$$A_t(x) \leq \delta t$$

for some $\delta > 0$, then $(X, \mathcal{B}, \Phi_{\mathcal{A}}, \omega)$ is a transient semidynamical system.

The proof is based on the fact that if $\mathcal{A}$ and $\mathcal{A}'$ are two additive functionals associated to the same semidynamical system such that "$\mathcal{A} \leq \mathcal{A}'$" (i.e. $A_t \leq A'_t$, for any $t \geq 0$), then the associated potential kernels are in the same inequality.

6. The topology $\mathcal{T}_{\Phi_{\mathcal{A}}}$

In this section, we assume that $(X, \mathcal{B}, \Phi_{\mathcal{A}}, \omega)$ is transient.

Next, we prove the following general Theorem:

Theorem 6.1. *$(X, \mathcal{B}, \Phi, \omega)$ and $(X, \mathcal{B}, \Phi', \omega)$ be fixed data transient semidynamical systems such that for every $x, y \in X_0$ we have*

$$x \underset{\Phi}{\leq} y \Leftrightarrow x \underset{\Phi'}{\leq} y.$$

Then, the topologies $\mathcal{T}_\Phi$ and $\mathcal{T}_{\Phi'}$ are equal.

Proof. For each $x \in X_0$, let us denote by

$$\mathcal{V}_x = \{V \subset X_0 : \exists \alpha \in (0, \rho(x)) \text{ such that } \Phi(t, x) \in V, \forall t \in [0, \alpha)\}$$

and

$$\mathcal{V}'_x = \{V \subset X_0 : \exists \alpha \in (0, \rho'(x)) \text{ such that } \Phi'(t, x) \in V, \forall t \in [0, \alpha[\}$$

then $\mathcal{T}_\Phi$ ($\mathcal{T}_{\Phi'}$, resp.) is the topology for which $\mathcal{V}_x$ ($\mathcal{V}'_x$, resp.) generates all the neighborhoods of x (see [2]).

Let V be a neighborhood of x with respect to $\mathcal{T}_\Phi$. Since Φ is transient, then there exists $\varepsilon > 0$ such that $\Phi(\alpha, x) \in V$ for all $\alpha \leq \varepsilon$. Let $y = \Phi(\varepsilon, x)$, then $x \underset{\Phi}{\leq} y \Leftrightarrow x \underset{\Phi'}{\leq} y$ and so there exists $\delta > 0$ such that $y = \Phi'(\delta, x)$ which gives us that $\Phi'(\alpha, x) \in V$ for all $\alpha \leq \delta$. Hence $\mathcal{T}_\Phi \subset \mathcal{T}_{\Phi'}$ and by the same argument $\mathcal{T}_{\Phi'} \subset \mathcal{T}_\Phi$. □

Also, we have the following result:

Theorem 6.2. *$(X, \mathcal{B}, \Phi, \omega)$ and $(X, \mathcal{B}, \Phi', \omega)$ be fixed data transient semidynamical systems such that for every $x, y \in X_0$ we have*

$$x \underset{\Phi}{\leq} y \Leftrightarrow x \underset{\Phi'}{\leq} y.$$

Then, the topologies $\mathcal{T}^0_\Phi$ and $\mathcal{T}^0_{\Phi'}$ are equal.

Using Proposition 4.13 and Theorem 6.1, we get the following result:

Corollary 6.3. *The topologies $\mathcal{T}_\Phi$ and $\mathcal{T}_{\Phi_\mathcal{A}}$ are equal.*

Also, using Proposition 4.13 and Theorem 6.2, we get the following corollary:

Corollary 6.4. *The topologies $\mathcal{T}_\Phi^0$ and $\mathcal{T}_{\Phi_\mathcal{A}}^0$ are equal.*

In the remaining, we will simply write $\mathcal{T}$ to denote $\mathcal{T}_\Phi$ or $\mathcal{T}_{\Phi_\mathcal{A}}$.

One checks without difficulty that the topology $\mathcal{T}$ is generated by the collection of all sets in the form

$$[x, y[:= \{\Phi(t, x) : t \in [0, \Psi(x, y))\}.$$

In the next, we denote by $\mathcal{E}$ ($\mathcal{E}^\mathcal{A}$ resp.) the cone of excessive functions with respect to $(V_\alpha)_{\alpha\geq 0}$ ($(V_\alpha^\mathcal{A})_{\alpha\geq 0}$ resp.).

Now, we are ready to prove the following result:

Theorem 6.5. *We have $\mathcal{E} = \mathcal{E}^\mathcal{A}$.*

Proof. By [2], we know that $\mathcal{E}$ is the set of all measurable functions $f\colon X_0 \to \mathbb{R}_+$ nonincreasing with respect to "$\underset{\Phi}{\leq}$" and continuous with respect to $\mathcal{T}$.

By Proposition 4.13, we deduce that f is nonincreasing with respect to "$\underset{\Phi}{\leq}$" if and only if f is nonincreasing with respect to "$\underset{\Phi_\mathcal{A}}{\leq}$", so that $\mathcal{E} = \mathcal{E}^\mathcal{A}$. □

Theorem 6.6. *For each $t \geq 0$, the map*

$$\begin{aligned}(X_0, \mathcal{T}) &\longrightarrow \mathbb{R}\\ x &\longmapsto \tau_t(x)\end{aligned}$$

is continuous.

Proof. Let $x, y \in X_0$ such that $y \underset{\mathcal{T}}{\to} x$. Then $y = \Phi_\mathcal{A}(s, x)$ for some $s \in [0, \rho_\mathcal{A}(x))$. By Proposition 4.7, we have

$$\tau_{t+s}(x) = \tau_s(x) + \tau_t(y).$$

Since $y \underset{\mathcal{T}}{\to} x$ as $s \to 0^+$ we get

$$\tau_t(x) = \lim_{y \underset{\mathcal{T}}{\to} x} \tau_t(y).$$ □

Proposition 6.7. *Let $x \in X$. If the map $t \mapsto \tau_t(x)$ is continuous, then we have*

$$\rho(x) = \tau_{\rho_\mathcal{A}(x)}(x).$$

Proof. Since

$$\begin{aligned}\rho_{\mathcal{A}}(x) &= \sup\{t \geq 0 : \Phi_{\mathcal{A}}(t,x) \neq \omega\} \\ &= \sup\left\{\Psi_{\mathcal{A}}(x,y) : y \in \Gamma_x^{\mathcal{A}}\right\}\end{aligned}$$

then

$$\begin{aligned}\rho(x) &= \sup\{\Psi(x,y) : y \in \Gamma_x\} \\ &= \sup\left\{\tau_{\Psi_{\mathcal{A}}(x,y)}(x) : y \in \Gamma_x^{\mathcal{A}}\right\} \\ &= \tau_{\rho_{\mathcal{A}}(x)}(x).\end{aligned}$$

□

Proposition 6.8. *The mapping $x \to \rho_{\mathcal{A}}(x)$ is $\mathcal{B}$- measurable.*

Proof. Let $\alpha \geq 0$. Since the map $\widetilde{\Phi} : t \to \Phi(\tau_t(x), x)$ is right continuous, then we get

$$\begin{aligned}\{\rho_{\mathcal{A}} \geq \alpha\} &= \bigcap_{0 \leq t < \alpha} \{x : \Phi_{\mathcal{A}}(t,x) \neq \omega\} \\ &= \bigcap_{t \in \mathbb{Q} \cap [0,\alpha)} \{x : \Phi_{\mathcal{A}}(t,x) \neq \omega\} \\ &= \bigcup_{t \in \mathbb{Q} \cap [0,\alpha)} \widetilde{\Phi}^{-1}(\{w\}^C)\end{aligned}$$

which is measurable. □

7. The measure $\Lambda^{\mathcal{A}}$

In the next, we recall the following definitions (see [3]):

Definition 7.1. For each $x \in X_0$, we define a measure λ_x on Γ_x as follows

$$\lambda_x(B) = \lambda(\Phi_x^{-1}(B)) = \lambda\{t \geq 0 : \Phi(t,x) \in B\}$$

$\forall B \in \mathcal{B}_0, B \subset \Gamma_x$, where λ is the Lebesgue measure on $\mathbb{R}$.

We have particularly (see [3]) the following:

Proposition 7.2. *For every $x, y \in X_0$ and for every $B \in \mathcal{B}$ such that $B \subset \Gamma_x \cap \Gamma_y$, we have $\lambda_x(B) = \lambda_y(B)$.*

Proposition 7.3. *For every sequence $(\Gamma_{x_n})_n$ of trajectories in X_0 , there exists a unique measure λ_M on $M = \bigcup_n \Gamma_{x_n}$ such that : For every $n \in \mathbb{N}$,*

$$\lambda_M(B) = \lambda_{x_n}(B),$$

for every $B \in \mathcal{B}_0$, $B \subset \Gamma_{x_n}$.

In general, for every measurable subset B of X_0, we set

$$\Lambda(B) = \sup_M \lambda_M(B \cap M),$$

where M is any countable union of trajectories of X_0. This leads us to define a measure Λ on $\mathcal{B}_0$ called the Lebesgue measure associated to $(X, \mathcal{B}, \Phi, w)$. Similarly, we define $\lambda_x^{\mathcal{A}}, \lambda_M^{\mathcal{A}}$ and $\Lambda^{\mathcal{A}}$ for the semidynamical system $(X, \mathcal{B}, \Phi_{\mathcal{A}}, w)$.

We recall (cf. [1]) that in the same way Λ can be defined on the σ-algebra $\mathcal{B}_0^*$ which is the set of all subsets A of X_0 such that $A \cap M \in \mathcal{B}_0$ for any countable union M of trajectories of X_0.

Using the definition of the Lebesgue measure Λ and of the resolvent $\mathcal{V} = (V_\alpha)_{\alpha \in +}$ associated to $(X, \mathcal{B}, \Phi, \omega)$, we obtain, on the measurable space $(X_0, \mathcal{B}_0^*)$, that

$$\begin{aligned} V_\alpha f(x) &= \int_{\Gamma_x} e^{-\alpha\Psi(x,y)} f(y) d\Lambda(y) \\ &= \int e^{-\alpha\Psi(x,y)} G(x,y) f(y) d\Lambda(y), \quad \forall x \in X_0, \end{aligned}$$

where

$$G(x,y) = \begin{cases} 1 & \text{if } x \leq_\Phi y \\ 0 & \text{ortherwise,} \end{cases}$$

is the Green function associated to $(X, \mathcal{B}, \Phi, \omega)$.

We recall the following Lemma:

Lemma 7.4. *For each $[x, y[\subset X_0$, we have $\Lambda([x, y[) = \Psi(x, y)$. Moreover, if $[x, y] \subset \Gamma_{x_0}$ for some $x_0 \in X_0$, then*

$$\Lambda([x, y[) = \Lambda([x_0, y[) - \Lambda([x_0, x[). \tag{7.1}$$

Obviously, we get

Proposition 7.5. *Let $x_0, y \in X_0$ such that $[x_0, y] \subset \Gamma_{x_0}$. Then, we have*

$$\Lambda([x_0, y[) = \tau_{\Lambda^{\mathcal{A}}([x_0,y[)}(x_0)$$

Lemma 7.6. *Let $x_0, y \in X_0$ such that $[x, y[\subset \Gamma_{x_0}$. Then,*

$$\Lambda([x, y[) = \tau_{\Lambda^{\mathcal{A}}([x_0,y[)}(x_0) - \tau_{\Lambda^{\mathcal{A}}([x_0,x[)}(x_0).$$

Proof. By Lemma 7.4 and Proposition 4.12 we have

$$\begin{aligned} \Lambda([x, y[) &= \Lambda([x_0, y[) - \Lambda([x_0, x[) \\ &= \Psi(x_0, y) - \Psi(x_0, x) \\ &= \tau_{\Psi_{\mathcal{A}}(x_0,y)}(x_0) - \tau_{\Psi_{\mathcal{A}}(x_0,x)}(x_0) \\ &= \tau_{\Lambda^{\mathcal{A}}([x_0,y[)}(x_0) - \tau_{\Lambda^{\mathcal{A}}([x_0,x[)}(x_0). \end{aligned}$$

□

Next, we denote by $\varphi_{x_0} := \tau_{\Psi_{\mathcal{A}}(x_0,.)}(x_0)$ and we set by $\mu_{\varphi_{x_0}}$ the measure defined on the set $\mathcal{C} = \{[x,y[\subset \Gamma_{x_0}\}$ by

$$\mu_{\varphi_{x_0}}([x,y[) = \tau_{\Psi_{\mathcal{A}}(x_0,y)}(x_0) - \tau_{\Psi_{\mathcal{A}}(x_0,x)}(x_0).$$

Theorem 7.7. *There exists a unique positive measure $\tilde{\mu}_{\varphi_{x_0}}$ on the Borel σ-algebra of Γ_{x_0} called the Stieltjes measure on Γ_{x_0} such that*

$$\tilde{\mu}_{\varphi_{x_0}} = \mu_{\varphi_{x_0}} \text{ on } \mathcal{C}.$$

Proof. See [11]. □

Next, we consider the particular case when $\mathcal{A}$ is a time contraction.

7.1 Contraction of time

In this section, we will consider a particular case where $\mathcal{A}$ is a contraction of time. Note that if $t \le \tau_t(x)$ for all $t \ge 0$, then it is obvious that $\Psi_{\mathcal{A}}(x,y) \le \tau_{\Psi_{\mathcal{A}}(x,y)} = \Psi(x,y)$. Hence, we can give the following result:

Proposition 7.8. *Let $\mathcal{A}$ be an additive functional which is a contraction of time. Then*

$$t \le \tau_t(x)$$

for all $t \in [0, \rho_{\mathcal{A}}(x))$ and $x \in X$.

Proof. Let $\varepsilon > 0$, then $t + \varepsilon > t \ge A_t(x)$ implies that $t \le \tau_{t+\varepsilon}(x)$ for all $\varepsilon > 0$. Using the fact that $t \to \tau_t(x)$ is right continuous and letting $\varepsilon \to 0$, we get $t \le \tau_t(x)$. □

We shall prove the following result:

Theorem 7.9. *If $\mathcal{A}$ is a contraction, then there exists a $\mathcal{B}_0^*$-measurable function f such that $\Lambda^{\mathcal{A}} = f\Lambda$.*

Proof. Since $\Psi(x,y) = \tau_{\Psi_{\mathcal{A}}(x,y)}(x) \ge \Psi_{\mathcal{A}}(x,y)$, we get that

$$\Lambda([x,y[) \ge \Lambda^{\mathcal{A}}([x,y[) \text{ for every } [x,y[\subset X_0.$$

Then, for every $x \in X_0$ there exists a measurable function f_x on Γ_x, $0 \le f_x \le 1$ such that $\Lambda^{\mathcal{A}} = f_x\Lambda$ on Γ_x.

First, let $M = \bigcup\limits_n \Gamma_{x_n}$. We define a function f_M as follows

$$f_M = f_{x_1} \text{ on } \Gamma_{x_1}$$

and for $n > 1$,

$$f_M = f_{x_n} \text{ on } \Gamma_{x_n} \setminus (\bigcup_{i<n} \Gamma_{x_i}).$$

Now, we can set

$$X_0 = \bigcup_{M \in \mathcal{B}_0^\sigma} M$$

and we denote by $\mathcal{F}(M)$ the set of all measurable functions f defined on M. For $f, g \in \mathcal{F}(M)$, we will denote by $f \sim g$ if $f = g$ Λ a.e. in M and by

$$\overline{f_M} = \{g \in \mathcal{F}(M) : f_M \sim g\}.$$

Next, we shall prove that if $M \subset M'$, then $\overline{f_{M'}} = \overline{f_M}$ on M. Indeed, for every measurable set $B \subset M$, we have

$$\int_B f_{M'} d\Lambda = \Lambda^{\mathcal{A}}(B) = \int_B f_M d\Lambda$$

which yields that $f_{M'} = f_M$ Λ a.e. on M.

Finally, we conclude the proof by setting $f := \overline{f_M}$ on each M. □

8. Construction of additive functionals

Definition 8.1. Let U be a measurable subset of X_0 and let $s \in \mathcal{E}$. The function R_s^U given on X_0 by

$$R_s^U := \inf\{t \in \mathcal{E} : t \geq s \text{ on } U\}$$

is called the reduced function or reduit (réduite) of s on U with respect to $\mathcal{E}$. The excessive regularization

$$\widehat{R_s^U} := \sup_{\alpha > 0} \alpha V_\alpha(R_s^U) = R_s^U \; \Lambda \; a.e.$$

is called the balayage of s on U.

The above functions were introduced and intensively studied in classical potential theory by M.Brelot ([12, 14]). The reduit of the constant 1 on U is called the equilibrium potential of U. For standard reference books which contain more details about the notion of reduit, we can cite [8, 5, 16, 32]. See also [30] and in the case of ordered sets see [9, 10].

In this section, we consider two transient semidynamical systems Φ and Φ' having the same cofinal point ω. Given a measurable subset U of X, we define

$$^{\Phi}R_1^U = \inf\{s \in \mathcal{E}_\Phi : s \geq 1 \text{ on } U\}$$

and

$${}^{\Phi'}R_1^U = \inf\{s \in \mathcal{E}_{\Phi'} : s \geq 1 \text{ on } U\}.$$

Next, we shall prove the following result:

Theorem 8.2. *The following assumptions are equivalent.*

(1) *For any $x, y \in X_0$ we have $x \underset{\Phi}{\leq} y \Leftrightarrow x \underset{\Phi'}{\leq} y$,*

(2) *there exists a strict CAF $\mathcal{A}$ such that $\Phi' = \Phi_{\mathcal{A}}$,*

(3) $\mathcal{E}_\Phi = \mathcal{E}_{\Phi'}$,

(4) ${}^{\Phi}R_1^U = {}^{\Phi'}R_1^U$ *for every Borel subset U of X_0.*

Proof. Since a role of order is central in the proof, it is commode to prove (2) $\Leftrightarrow$ (1) $\Leftrightarrow$ (3) and (1) $\Leftrightarrow$ (4).

First, we prove that (1) $\Leftrightarrow$ (2). Indeed, (2) $\Rightarrow$ (1) is obvious by Proposition 4.13.

Conversely, let's fix $x \in X_0$. Then, for every $y \in \Gamma_x^\Phi = \Gamma_x^{\Phi'}$ we have

$$y = \Phi(t', x) = \Phi'(t, x)$$

for some $t \in [0, \rho'(x))$ and $t' \in [0, \rho(x))$. Set

$$\tau_t(x) = \Psi(x, \Phi'(t, x)).$$

Hence, $\Phi(\tau_t(x), x) = \Phi(t', x)$. Moreover it is obvious that $\tau_0(x) = \Psi(x, \Phi'(0, x)) = \Psi(x, x) = 0$.

Let $t, s \geq 0$, we have

$$\begin{aligned}\Phi(\tau_t(x) + \tau_s(\Phi(\tau_t(x), x), x)) &= \Phi(\tau_s(\Phi(\tau_t(x), x)), (\Phi(\tau_t(x), x))) \\ &= \Phi(\tau_s(\Phi'(t, x)), \Phi'(t, x)) \\ &= \Phi'(s, \Phi'(t, x)) \\ &= \Phi'(s + t, x) = \Phi(\tau_{t+s}(x), x).\end{aligned}$$

Consequently, since Φ is transient we obtain

$$\tau_{t+s}(x) = \tau_t(x) + \tau_s(\Phi(\tau_t(x), x)).$$

Let us set

$$A_t(x) = \Psi'(x, \Phi(t, x)).$$

Since for every $s, t > 0$, we have

$$x \underset{\Phi}{\leq} \Phi(t,x) \underset{\Phi}{\leq} \Phi(t+s,x)$$

we get

$$x \underset{\Phi'}{\leq} \Phi(t,x) \underset{\Phi'}{\leq} \Phi(t+s,x)$$

and therefore we obtain that

$$\begin{aligned} A_{s+t}(x) &= \Psi'(x, \Phi(s+t,x)) \\ &= \Psi'(x, \Phi(t,x)) + \Psi'(\Phi(t,x), \Phi(t+s,x)) \\ &= \Psi'(x, \Phi(t,x)) + \Psi'(\Phi(t,x), \Phi(s, \Phi(t,x))) \\ &= A_t(x) + A_s(\Phi(t,x)). \end{aligned}$$

Using Proposition 2.6, Theorem 6.2 and the fact that

$$A_t(x) = \Psi'(x, \Phi(t,x))$$

we deduce that $\mathcal{A} := (A_t)_{t \geq 0}$ is a strict CAF.
Moreover, it is obvious that

$$\Phi(t,x) = \Phi'(A_t(x), x).$$

Since

$$\Phi'(A_{\tau_t(x)}(x), x) = \Phi(\tau_t(x), x) = \Phi'(t,x)$$

we get $A_{\tau_t(x)}(x) = t$ and similarly $\tau_{A_t(x)}(x) = t$.

On the other hand, for $\varepsilon > 0$ we have

$$A_{\tau_t(x)+\varepsilon}(x) = \Psi'(x, \Phi(\tau_t(x)+\varepsilon, x)) > A_{\tau_t(x)}(x).$$

Consequently, we get

$$\inf\{u : A_u(x) > t\} = \inf\{u : A_u(x) > A_{\tau_t(x)}(x)\} = \tau_t(x).$$

Next, we prove that (1) $\Leftrightarrow$ (3). Indeed, since Φ and Φ' are transient, then by a result in [2]

$$x \underset{\Phi}{\leq} y \Leftrightarrow s(y) \leq s(x)$$

for every $s \in \mathcal{E}_\Phi$. Hence, condition (3) implies that

$$x \underset{\Phi}{\leq} y \Leftrightarrow x \underset{\Phi'}{\leq} y$$

and (3) $\Rightarrow$ (1) is proved.

Conversely, let $s \in \mathcal{E}_\Phi$, then s is decreasing with respect to "$\underset{\Phi}{\leq}$" and therefore with respect to "$\underset{\Phi'}{\leq}$". It is known by Theorem 6.1 that $\mathcal{T}_\Phi = \mathcal{T}_{\Phi'}$, so that s will be continuous with respect to the topology $\mathcal{T}_{\Phi'}$ and therefore $s \in \mathcal{E}'_\Phi$.

Finally, we shall prove that (1) $\Leftrightarrow$ (4). We start with (1) $\Rightarrow$ (4). Let U be a Borel subset of X_0. By a result in [2], ${}^\Phi R_1^U = 1_{\widetilde{U}^\Phi}$ where $\widetilde{U}^\Phi = \{x \in X_0 : \exists y \in U, x \underset{\Phi}{\leq} y\}$. Since $\widetilde{U}^\Phi = \widetilde{U}^{\Phi'}$, we get ${}^\Phi R_1^U = {}^{\Phi'} R_1^U$.

Conversely, let $x \underset{\Phi}{\leq} y$, thus $x \in \widetilde{\{y\}}^\Phi$ and ${}^\Phi R_1^{\{y\}}(x) = 1_{\widetilde{\{y\}}^\Phi}(x) = 1$. Since

$$ {}^\Phi R_1^{\{y\}} = {}^{\Phi'} R_1^{\{y\}} = 1_{\widetilde{\{y\}}^{\Phi'}} $$

we get $x \in \{\widetilde{y}\}^{\Phi'}$ and $x \underset{\Phi'}{\leq} y$ which concludes the proof. □

Remark 8.3. An analogous of the assertion (3) $\Rightarrow$ (2) is given by R. M. Blumenthal, R. K. Getoor and H. P. McKean (cf. [6, 7]) on the Markov processes having identical hitting distributions which is equivalent with Markov processes having identical excessive functions. An extension is given by P. J. Fitzsimmons, R. K. Getoor and M. J. Sharpe in [18].

References

[1] Bezzarga M.: Coexcessive functions and duality for semi-dynamical systems, *Romanian Journal of Pure and Applied Mathematics* **XLII** (1997) 15–30.

[2] Bezzarga M., Bucur Gh.: Théorie du potentiel pour les systèmes semi-dynamiques, *Rev. Roumaine Math. Pures Appl.* **39** (1994) 439–456.

[3] Bezzarga M., Bucur Gh.: Duality for semi-dynamical systems, In "*Potential Theory–ICPT94 (Kouty, 1994)*", 275–286, de Gruyter, Berlin, 1996.

[4] Bhatia N. P., Hajek O.: *Local semi-dynamical systems*, Lecture Notes in Math. 90. Springer Verlag, Berlin, 1969.

[5] Blumenthal R. M., Getoor R. K.: *Markov Processes and Potential Theory.* Academic Press. New York and London, 1968.

[6] Blumenthal R. M., Getoor R. K., McKean. H. P.: Markov Processes with identical hitting distributions, *Illinois J. Math.* **6** (1962) 402–420.

[7] Blumenthal. R. M., Getoor R. K., McKean. H. P.: A supplement to Markov Processes with identical hitting distributions, *Illinois J. Math.* **7** (1963) 540–542.

[8] Boboc N., Bucur G.H., Cornea. A.: *Order and Convexity in Potential Theory,* Lecture Notes in Math. 853, Springer Verlag, Berlin, 1981.

[9] Boboc N., Bucur G. H.: *Potential Theory on Ordered sets I. Rev. Roum. Math. Pures Appl.* **43** (1998) 277–298.

[10] Boboc N., Bucur G. H.: Potential Theory on Ordered Sets II, *Rev. Roum. Math. Pures Appl.* **43** (1998) 685–720.

[11] Bourbaki N.: *Éléments de mathématique. Intégration*, Hermann, Paris, 1965.

[12] Brelot M.: Fonctions sous-harmoniques et balayage, *Acad. Roy. Belgique. Bull. Cl. Sci.* **24** (1938) 301, 321.

[13] Brelot M.: Sur les ensembles effilés, *Bull. Sci. Math. (2)* **68** (1944) 12–36.

[14] Brelot M.: Minorantes sous-harmoniques, extrémales et capacités, *J. Math. Pures Appl. (9)* **24** (1945) 1–32.

[15] Dellacherie C., Meyer P.-A.: *Probabilités et Potentiel*, V-VIII, Hermann, Paris, 1980.

[16] Dellacherie C., Meyer, P.-A.: *Probabilités et potentiel*, XII-XVI, Hermann, Paris, 1987.

[17] Fitzsimmons P. J., Getoor R. K.: Revuz measures and time changes, *Math. Z.* **199** (1988) 233-256.

[18] Fitzsimmons P. J., Getoor R. K., Sharpe M. J.: The Blumenthal-Getoor-McKean theorem revisited, In "*Seminar on Stochastic Processes, 1989 (San Diego, CA, 1989)*", 35–57, Progr. Probab. **18**, Birkhäuser Boston, Boston, MA, 1990.

[19] Getoor R. K.: Transience and recurrence of Markov processes, In "*Séminaire de Probabilité, XIV (Paris, 1978-1979)*", 397–409, Lecture Notes in Math. **784**, Springer Verlag, Berlin, 1980. 397-409.

[20] Glover J.: *Identifying Markov processes up to time change.* Seminar on Stochastic processes, Birkhäuser, 171-194, 1982,1983.

[21] Glover J.: Discontinuous time changes of semiregenerative processes and balayage theorems. *Z. Wahrsch. Verw. Gebiete* **65** (1983) 145-–160.

[22] Hajek O.: *Dynamical systems in the plane.* Academic Press, London-New York, (1968).

[23] Hmissi M.: Semi-groupes déterministes, In "*Séminaire de théorie du potentiel, Paris, No.9*", Lecture Notes in Mathematics, 1393, 135–144, Springer-Verlag. Berlin, 1989.

[24] Hmissi M.: Cônes de potentiels stables par produit et systmes semi-dynamiques, *Exposition. Math.* **7** (1989) 265-–273.

[25] Hmissi M.: Sur l'équation fonctionnelle des cocycles d'un système semi-dynamique, In "*European Conference on Iteration Theory (Batschuns, 1989)*", World Sci. Publishing, River Edge, NJ, 1991, 149–156.

[26] Hmissi M.: Sur les solutions globales de l'quation des cocycles, *Aequationes Math.* **45** (1993) 195-–206.

[27] Hmissi M.: Sur les systèmes dynamiques instables, In "Proceedings from the International Conference on Potential Theory (Amersfoort, 1991)", *Potential Anal.* **3** (1994) 145–152.

[28] Itô K., McKean H. P.: *Diffusion Processes and their Sample Paths*, Springer, Berlin, Heidelberg, New York and Tokyo, 1965.

[29] Koopmann B. O.: Hamiltonian systems and transformation in Hilbert spaces, *Proc. Nat. Acad. Sci. USA* 17-5, (1931) 315–318.

[30] Mokobodzki G.: Structure des Cônes de Potentiels, *Séminaire Bourbaki*, 372 (1968/69).

[31] Saperstone S. H.: *Semidynamical systems in infinite dimensional space*, App. Math. Sciences 37. Springer Verlag (1981).

[32] Sharpe M.: *General Theory of Markov Processes.* Pure and Applied Mathematics, vol. 133, Academic Press. Boston, San Diego, New York, Berkley, London, Sydney, Tokyo, Toronto (1988).

INFINITE-DIMENSIONAL LAGRANGE PROBLEM AND APPLICATION TO STOCHASTIC PROCESSES

Alexey V. Uglanov
Yaroslavl State University, Math. Dept., 150000, Russia
uglanov@uniyar.ac.ru

Abstract We consider an infinite-dimensional analog of classical Lagrange problem. The main result of the work is the necessary conditions of the extremum. This result is new (the theory of surface integration in abstract locally convex space is the basic instrument of the investigation, therefore the indicated problem was not and just could not be investigated before). All our considerations and formulas are dimensional-invariant, and in conformity to finite-dimensional case they turn into classical ones.

Keywords: Lagrange problem, stochastic process

1. General notations

(Ω, Σ) an abstract measurable space.

$M(\Omega, Z)$ the set of all Z-valued measures (i.e., countably additive functions $\Sigma_\Omega \to Z$) where Z is a locally convex topological vector space (LCS).

$M(\Omega)$ $= M(\Omega, \mathbb{R})$.

$L(Y, Z)$ the set of all linear continuous mappings of a LCS Y into a LCS Z. $L(Y, Z)$ is assumed to be equipped with the uniform convergence topology on the bounded subsets of Y.

$(\cdot)^*$ the conjugate of $(\cdot)$. If Z is a LCS then $Z^* = L(X, \mathbb{R})$. If $A \in L(Y, Z)$ then $A^*: Z^* \to Y^*$ is a mapping conjugate to A.

f' the derivative of the function f. If f is a function of points (of LCS Z) then the derivative is understood in Fréchet sense (= derivative over the system of bounded subsets of Z). The definition of the sets function derivative will be given in section 2.

X a Hausdorff locally convex topological vector space.

X_0 a normed space, linearly and continuously enclosed into X.

$M^1(V)$ the totality of all Radon measures in $M(V)$, differentiable with respect to the space X_0 ($V \subset X$).

$D_a f$ the derivative of a function f with respect to the direction a (i.e. $D_a f = (a, f')$).

S. Albeverio et al. (eds.),
Proceedings of the International Conference on Stochastic Analysis and Applications, 239–248.

General assumptions. The vector spaces are real; unless otherwise stated the subsets of topological spaces are provided with the induced topology. Any topological space T is considered to be measurable, with a Borel σ-algebra Σ_T.

Section 2 contains various auxiliary concepts and results. Some results are new, for example, theorem 2.8. Note that results of an auxiliary nature have an independent interest too. Section 3 is the basic one. It is devoted to Lagrange problem.

2. Auxiliary concepts and results

Definition 2.1. The measure $\mu \in M(X)$ is called *differentiable* (with respect to the space X_0), if for any $h \in X_0$, $C \in \Sigma_X$ the limit

$$\lim_{t \to 0} \frac{1}{t}[\mu(C + th) - \mu(C)] \stackrel{\text{def}}{=} D_h\mu(C)$$

exists.

Further $G \subset X$ is an open subset, the space $M(G)$ is supposed to be endowed with the norm $\|\mu\| = \mathsf{var}\mu$.

Definition 2.2. The measure $\mu \in M(G)$ is called *differentiable* if there exists a differentiable measure $\bar{\mu} \in M(X)$, such that $\bar{\mu} \equiv \mu$ on G.

If $\mu \in M^1(G)$ then for any $h \in X_0$ the function $D_h\mu\colon \Sigma_G \to \mathbb{R}$, $C \mapsto D_h\bar{\mu}(C)$ is countably additive (Nikodym theorem about the limit of a sequence of numerical measures), i.e., $D_h\mu \in M(G)$. Therefore derivatives of higher order can be defined inductively.

As is known (see, for instance, [1, Chapter 4, §2]) if X_0 is complete and $\mu \in M^1(G)$ then the mapping $\mu'\colon X_0 \mapsto M(G)$, $h \mapsto D_h\mu$ is linear and continuous. Now the Pettis theorem makes the following definition correct.

Definition 2.3. Let X_0 be a reflexive space and $\mu \in M^1(G)$. The measure $\mu' \in M(G, X_0^*)$ defined by the equality

$$(\mu'(C), h) = D_h\mu(C) \quad (\forall\, h \in X_0,\, C \in \Sigma_G),$$

is called *derivative* of the measure μ.

Definition 2.4. A set $Q \in \Sigma_X$ is called X_0-smooth hypersurface (of C^1 class) if for any point $x \in Q$ there exist an open neighborhood $U(x) \ni x$ and a function $f\colon U(x) \to \mathbb{R}$ which is continuously differentiable along X_0 and satisfies to conditions $f'(x) \neq 0$ and $Q \bigcap U(x) = f^{-1}(0)$.

Notations. Q is a X_0-smooth hypersurface in G, $M_{\mathrm{lf}}(Q)$ is a set of all locally finite Borel measures on Q, $V \subset G$ is an open set, $F\colon V \to \mathbb{R}$

is a function which has a continuous derivative $F' \colon V \to X_0^*$ along the space X_0, $Q_t = \{x \in V \mid F(x) = t\}$. We assume that Q_t is a part of some X_0-smooth hypersurface for any $t \in \mathbb{R}$.

Theorem 2.5. *There exists a linear mapping*

$$M^1(G) \to M_{\mathrm{lf}}(Q), \quad \mu \mapsto \mu_Q$$

which has the following properties:

1. *if $\mu \equiv 0$ in a neighborhood of Q, then $\mu_Q \equiv 0$;*
2. *if $\mu \geq 0$ then $\mu_Q \geq 0$;*
3. *if $V = G$ then $(F\mu)_Q = F\mu_Q$;*
4. *if $|\mu|\{x \in V \mid F'(x) = 0\} = 0$ ($|\mu|$ is the total variation of measure μ) and the function $\varphi \colon V \to \mathbb{R}$ is μ-integrable then*

$$\int_V \varphi \mathrm{d}\mu = \int_{-\infty}^{\infty} \int_{Q_t} \varphi \|F\|^{-1} \mathrm{d}\mu_{Q_t} \mathrm{d}t$$

(iterated integration formula);

5. *if $\mu \geq 0$, $\mu(V) = 1$ then, under the assumptions of the previous subsection, for the conditional mean value $\mathsf{E}(\varphi|F) : V \to \mathbb{R}$ the equality*

$$\mathsf{E}(\varphi|F)(x) = \left[\int_{Q_{F(x)}} \|F'\|^{-1} d\mu_{Q_{F(x)}}\right]^{-1} \left[\int_{Q_{F(x)}} \varphi \|F'\|^{-1} d\mu_{Q_{F(x)}}\right]$$

is correct;

6. *under the natural assumptions concerning the boundary ∂V and the function $u \colon V \cup \partial V \to \mathbb{R}$ the equality*

$$\int_V D_a u \mathrm{d}\mu + \int_V u \, dD_a \mu = \int_{\partial V} u(n, a) \mathrm{d}\mu_{\partial V}$$

is correct (integration by parts formula; here $a \in X_0$, $n \in X_0^*$ is an outer unit normal to the surface ∂V with respect to V);

7. *if X_0 is a Hilbert space then under the natural assumptions concerning the boundary ∂V and function $b \colon V \cup \partial V \to X_0$ the Gauss–Ostrogradskii formula*

$$\int_V \operatorname{div} b \mathrm{d}\mu + \int_V b \, \mathrm{d}\mu' = \int_{\partial V} (b, n) \mathrm{d}\mu_{\partial V} \tag{2.1}$$

is correct (here $\operatorname{div} b(x) := \operatorname{Tr} b'(x)$*, with* b' *the derivative along the space* X_0 *of the function* b*,* $\int_V b d\mu'$ *is a defined in a special way integral of vector function* b *with respect to the* X_0^**-valued measure* μ'*, for such integrals see [2]);*

8. *if* X_0 *is the same as in 7) then both Green formulas are correct (the formulas are not written here because they require additional concepts and explanations);*

9. *if* $X = X_0 = \mathbb{R}^m$ *and* μ *is a Lebesgue measure then* μ_Q *is a classical (geometric) surface measure on* Q*.*

Proof. See [3]. □

Remarks. 1. Measure μ_Q is called a *surface measure* on Q.

2. In fact the measure μ_Q is built constructively.

3. The surface Q can possess singular points (roughly speaking, the set of singular points must be a surface of codimension 2).

The following should be noted. Besides general-theoretical significance (the contribution to the infinite-dimensional analysis and the direction, that had been named by Kolmogorov as 'Nonlinear probability theory' (Kolmogorov used this term in oral talk; he looked like the study of probabilities on nonlinear algebraic and topological structures), the surface integration theory in LCS has applications in such fields as the theory of infinite-dimensional distributions and differential equations, stochastic processes, approximation of functions of an infinite-dimensional argument, calculus of variations on infinite-dimensional spaces and others. See [3].

If $\mu \in M(\Omega, X)$ and $S \in L(X, Y)$ then the (induced) measure $S\mu \in M(\Omega, Y)$, $A \mapsto S(\mu(A))$ is well defined. It is possible to get acquainted in [4] with the concept of absolutely summing operator.

Later on: X is a Banach space; X_0 is a Hilbert space; $M(\Omega, [0, \infty])$ is the space of all nonnegative (not necessary finite) measures on Ω; $|\nu| \in M(\Omega, [0, \infty])$ is the total variation of the measure $\nu \in M(\Omega, Y)$ (Y is a Banach space); $MB(\Omega, Y) = \{\nu \in M(\Omega, Y) : |\nu|(\Omega) < \infty\}$. Note that we have a triplet $X^* \subset X_0 \subset X$.

Lemma 2.6. *If* $\mu \in M(\Omega, X)$ *and* $S \in L(X, Y)$ *is an absolutely summing operator then* $S\mu \in MB(\Omega, Y)$*.*

Proof. See [3, Chapter 1, section 2]. □

If $\mu \in M^1(X)$ then the measure $\mu' \in M(X, X_0)$ is defined correctly by virtue of X_0 reflexivity (see above). Let $\nu \in M(\Omega, X)$, $\rho \in M(\Omega)$. We remind that the ρ-integrable in Bochner sense function $\frac{d\nu}{d\rho} \colon \Omega \to X$ is called a Radon–Nikodym derivative of the measure ν with respect to the the measure ρ if for any $A \in \Sigma$ the equality $\nu(A) = \int_A \frac{d\nu}{d\rho}\, d\rho$ is valid. We suppose below that the embedding operator $I \colon X_0 \to X$ is an absolutely summing operator.

Lemma 2.7. *If $\mu \in M^1(X)$ then the Radon–Nikodym derivative $\frac{dI\mu'}{d\mu}$ exists.*

Proof. See [3, Chapter 3, section 6]. □

Remark. The function $\frac{dI\mu'}{d\mu}$ is called a *logarithmic gradient* of the measure μ. The existence of a logarithmic gradient has been shown under other conditions in [5, 6]. As is known to the author the results of works [5, 6, 3] are the first general nature results about logarithmic gradient existence (this gradient plays an important role in a great number of problems of infinite-dimensional analysis – for example, see [7, 8] and below theorem 2.8; its existence could be only assumed in various thematic works).

Let $V \in \Sigma_X$ and the set ∂V is an X_0-smooth hypersurface. The space $C^1(\overline{V}, X^*)$ appears in the following theorem; the definition of this space is given in section 4.

Theorem 2.8. *Let $\mu \in M^1(V)$, $b \in C^1(\overline{V}, X^*)$. If $\|b\|_{X_0} \colon \partial V \to \mathbb{R}$ is $\mu_{\partial V}$-integrable, then*

$$\int_V \left[\operatorname{div} b + \left(b, \frac{dI\mu'}{d\mu}\right)\right] d\mu = \int_{\partial V} (b, n) d\mu_{\partial V} \tag{2.2}$$

(Gauss–Ostrogradskii formula).

Proof. Follows from equality (2.1) and lemma 2.7. □

Remarks. 1. The Gauss–Ostrogradskii formula was either announced or proved in [2, 3]. But in these works the formula contained either vector integrals [2] or some limit expressions [3]. The variant (2.2) is most effective.

2. There were found quite acceptable purely geometric restrictions to the set V which guarantee the validity of (2.2) without any conditions of surface integral existence (these conditions often appeared to be hardly verifiable); the condition of $\mu_{\partial V}$-integrability of the function $\partial V \to \mathbb{R}$, $x \mapsto 1$ can be omitted for such sets in the main theorem of this paper. See below, theorem 3.2.

3. Lagrange problem

Later on: X_1, X_2 are Banach spaces; B is a Fréchet space, $X = X_1 \times X_2$; X_0 is a Hilbert space, linearly, continuously and compactly embedded into X; $V \subset X$ is an open set with X_0-smooth boundary; $C^m(V, Z)$ is the set of all Z-valued (Z is a LCS) functions on V whose Fréchet derivatives of orders $0, \ldots, m$ are continuous and bounded on V; $C^m(\overline{V}, Z)$ is the collection of functions in $C^m(V, Z)$ whose derivatives of the orders $0, \ldots, m-1$ extend continuously to the closure $\overline{V}$; $C_0^m(V, Z)$ is the collection of functions in $C^m(\overline{V}, Z)$, vanishing on the boundary ∂V; $C_1^m(V, Z), C_1^m(\overline{V}, Z)$ are the same as $C^m(V, Z), C^m(\overline{V}, Z)$, but differentiability is understood along the space X_1; $C = C^0$; $\mu \in M^1(V)$; $\lambda = \frac{dI\mu'}{\mathrm{d}\mu}$ and we assume that $\lambda \in C(V, \mathbb{R})$. We remind that the embedding operator $I\colon X_0 \to X$ is an absolutely summing operator.

For $p \in C^*(V, X_1)$ we denote $\operatorname{Div} p \in [C_0^1(V, \mathbb{R})]^*$ defined by:

$$(\operatorname{Div} p, h) = -(p, h') - (p, \lambda h).$$

The generalized function $\operatorname{Div} p$ may be called as a generalized divergence of generalized function p. The following statement shows that in regular case the generalized function $\operatorname{Div} p$ is an ordinary function, coinciding with an ordinary divergence of the vector function p. Note, that we have a triplet $X_1^* \subset X_0 \subset X_1$.

Lemma 3.1. *If $p \in C_1^1(V, X_1^*)$, so that for any $g \in C(V, X_1)$, $(p, g) = \int_V p\, g \mathrm{d}\mu$, then the following equalities are valid:*

$$(\operatorname{Div} p, h) = \int_V h \operatorname{div} p \,\mathrm{d}\mu = \int_V [(p, h') + h(p, \lambda)]\mathrm{d}\mu. \tag{3.1}$$

Proof. The second equality follows from (2.2) (the surface integral is absent because $h|_{\partial V} = 0$). The first equality follows from the second one. □

Let $\Phi \in C^2(V \times \mathbb{R} \times B, \mathbb{R})$; $\varphi \in C(V \times \mathbb{R} \times B, X_1^*)$; $Q \subset \partial V$ is a closed set; $f \in C(Q, \mathbb{R})$. For $y \in C_1^1(\overline{V}, \mathbb{R})$, $u \in C(V, B)$ let

$$J(y, u) = \int_V \Phi(x, y(x), u(x))\mathrm{d}\mu(x).$$

We have considered the following Lagrange problem

$$J(y, u) \to \text{extr}, \quad y'(x) = \varphi(x, y(x), u(x)), \quad y|_Q = f. \tag{3.2}$$

(Here y is a phase trajectory, u is a control. $Q = \partial V$ or $Q = \emptyset$ as a rule.) Let us note some special cases of the problem (3.2).

1. $X_2 = 0$, $B = X^*$, $\varphi = u$. Then (3.2) is a variational problem (the author had met such problem while optimizing some queueing system, see [9]). Here we considered the case of moving domain V; the analogies of fundamental results of classical finite-dimensional theory were deduced (the basic formula for variation of functional, necessary conditions of extremum, Noether theorem, Hamilton system. See [3]).

2. $\dim X_1 = 1$, $V = V_1 \times V_2$ $(V_i \subset X_i)$, $\mu = \mu_1 \times \mu_2$, where μ_1 is a measure on V_1, μ_2 is a probability on V_2. In this case (3.2) is a problem of control of stochastic process y (X_1 is a time-space, V_2 is a probabilistic space).

We add one remark to this case. Let the (ordinary) differential equation $y' = \varphi(x, y, u)$ be replaced by the stochastic differential equation $\mathrm{d}y(x_1) = \varphi(x, y(x), u(x))\mathrm{d}x_1 + \psi(x, y(x), u(x))\mathrm{d}w(x_1)$, where ψ is a suitable real-valued function, $w = w(x_1) \in \mathbb{R}$ is the Wiener process (or, more generally, a martingale). We have investigated problem (3.2) for this case too and some results have been received.

3. $\dim X_1 = \infty$, μ_1, μ_2, μ are the same as in case 2. Then (3.2) is a problem of control of stochastic process y with "infinite-dimensional time" (X_1 is a "time-space", V_2 is a probabilistic space).

4. X_1, μ_2 are the same as in cases 2 or 3, μ_1 is a positive normalized measure, concentrated in the point $a \in V_1$, $\mu = \mu_1 \times \mu_2$. Then (3.2) is a problem of minimization (maximization) of mathematical expectation of the random value $\Phi((a, x_2), y(a, x_2), u(a, x_2))$.

The following result appears to be central in the work. We remind that a Borel measure ν on topological space T is called topologically regular, if $|\nu|(A) > 0$ for any open set $A \subset T$. For understanding the following equalities (3.3), (3.4) let us recall, that we have a triplet $X_1^* \subset X_0 \subset X_1$.

Theorem 3.2. *Let (y, u) be a solution of problem (3.2). Let the function $\partial V \to \mathbb{R}$, $x \mapsto 1$ be $\mu_{\partial V}$-integrable. Finally let for any $x \in V$ the set $[\frac{\partial \varphi}{\partial u}(x, y(x), u(x))](B)$ be closed in in the space X_1^*. Then there is a vector $p \in C^*(V, X_1)$ such that the following equalities*

$$\frac{\partial \Phi}{\partial u} - \left(\frac{\partial \varphi}{\partial u}\right)^* p = 0 \tag{3.3}$$

$$\frac{\partial \Phi}{\partial y} - \operatorname{Div} p - (p, \lambda) - \frac{\partial \varphi}{\partial y} p = 0 \tag{3.4}$$

are valid respectively in the spaces $C^*(V, B)$ and $\left[C_0^1(V, \mathbb{R})\right]^*$.

If in addition the norm in X is a continuously differential function (outside zero), $p \in C_1^1(\overline{V}, X_1^*)$ and the measures μ, $\mu_{\partial V}$ are topologically regular respectively on the sets V, ∂V then equalities (3.3), (3.4) hold *identically with respect to* $x \in V$ *((3.3) in* B^**, (3.4) in* $\mathbb{R}$*). In addition the equality* $(p, n) = 0$ *holds identically with respect to* $x \in \partial V \setminus Q$*.*

Sketch of proof. Equalities (3.3), (3.4) are the standard Euler–Lagrange equations for smooth extremal problem (3.2) ($\frac{\partial \Phi}{\partial u} \in C^*(V, B)$ is understood in natural meaning: $\forall g \in C(V, B)$ $\left(\frac{\partial \Phi}{\partial u}, g\right) = \int_V \frac{\partial \Phi}{\partial u} g \mathrm{d}\mu$; the situation with $\frac{\partial \Phi}{\partial y} \in [C_0^1(V, \mathbb{R})]^*$ is analogous). Using equalities (2.2), (3.1) one can prove the second part of the theorem by standard methods. □

Remarks. 1. If $X = \mathbb{R}^m$, $B = \mathbb{R}^l$, μ the Lebesgue measure, then $\lambda = 0$ and (3.3), (3.4) are the classical Euler–Lagrange equations.

2. Simple sufficient conditions for the measure μ, that guarantee topological regularity of the measure $\mu_{\partial V}$ were obtained. In particular, these conditions are realized if μ is a nondegenerated Gaussian measure.

3. The condition of X-norm differentiability is absolutely necessary for our proof. In the general case Eq. (3.4) is absolutely useless: as is known there exist separable Banach spaces X such that $C_0^1(V, \mathbb{R}) = 0$ for each bounded subset $V \subset X$, see [10]. However an analog of theorem's second part may be proved for a general space X but the proof is completely different and much more difficult.

For variational problems (see special case 1 above) our results are much more effective than Theorem 3.2. We shall give here two examples of such results for a fixed domain V. Namely, let us consider the functional

$$K\colon C^1(V, \mathbb{R}) \to \mathbb{R}, \qquad y \mapsto \int_V \Phi(x, y(x), y'(x)) \mathrm{d}\mu(x).$$

Theorem 3.3. *Let measures* μ*,* $\mu_{\partial V}$ *be topologically regular respectively on the sets* V*,* ∂V*, the set* V *complies with some more conditions of geometric type. In order the function* $y \in C^2(\overline{V}, \mathbb{R})$ *to be the point of local extremum of the functional* K *the identity*

$$\frac{\partial \Phi}{\partial y} - \left(\lambda, \frac{\partial \Phi}{\partial y'}\right) - \operatorname{div} \frac{\partial \Phi}{\partial y'} = 0 \tag{3.5}$$

on V *and the identity*

$$\left(\frac{\partial \Phi}{\partial y'}, n\right) = 0$$

on ∂V *are needed.*

Proof. See [3, Chapter 3, section 8]. □

Theorem 3.4. *Let a function $f \in C(\partial V, \mathbb{R})$ be fixed and the premises of Theorem 3.3 hold. In order the function $y \in C^2(\overline{V}, \mathbb{R})$ to be the point of local extremum of the functional K with condition $y|_{\partial V} = f$ the identity* (3.5) *is needed.*

Proof. See [3]. □

So, the indicated variational problems are reduced to boundary value problems for the second order differential equation (Euler equation).

Remark. The variational problems on a Banach space were considered in the works [8, 5], but these works concern only the domain V, coinciding with the whole space. It appears to be principal: the domain boundary is empty, that is why the authors could manage without any attraction of surface integrals. But the case of a general domain can be hardly considered without such attraction.

Acknowledgments

This research was carried out with the support of the Russian Foundation for Basic Research (grant 01-01-00701).

References

[1] Daletskii Yu. L., Fomin S. V.: *Measures and differential equations in infinite dimensional spaces*, Nauka, Moscow, 1983; English transl.: Kluwer Academic Publishers, Dordrecht, 1991.

[2] Uglanov A. V.: Vector integrals, *Dokl. Acad. Nauk Russia* **373** (2000) 737–740; English transl.: *Russian Acad. of Sci. Dokl.* (2000).

[3] Uglanov A.V.: *Integration on infinite-dimensional surfaces and its applications*, Kluwer Academic Publishers, Dordrecht, 2000.

[4] Pietsch A.: *Nukleare lokalkonvexe Räume*, Akademie-Verlag, Berlin, 1965.

[5] Smollyanov O. G., Weizsäcker H.: Noether theorem for infinite-dimensional variational problems, *Docl. Akad. Nauk Russia*, **369** (1999) 158–162; English transl.: *Russian Acad. of Sci. Dokl.* (1999).

[6] Albeverio S., Daletskii A. Yu., Kondratyev Yu. G.: Stochastic analysis on product manifolds: Dirichlet operators on differential forms, *J. Funct. Anal.* **176** (19??) 280–316.

[7] Norin N. V.: Stochastic integrals and differentiable measures, *Teor. Ver. i Prim.* **32** (1987) 114–124; English transl.: *Theory Prob. Appl.* **32** (1987).

[8] Daletskii Yu. L., Steblovskaya V. R.: On infinite-dimensional variational problems, *Stochastic Analysis and Applications* **14** (1966) 47–71.

[9] Uglanov A.V., Filatova L.Yu.: On optimal organization of queueing systems with finite sources of requirements, In "*Seminari di Probabilita e Statistica Matematica*", Univ. di Cassino Publ. (Italy), 1998, 64–95.

[10] Bonic R., Frumpton J.: Differentiable functions on certain Banach spaces, *Bull. Amer. Math. Soc.* **71** (1965) 393–395.

ON PORTFOLIO SEPARATION IN THE MERTON PROBLEM WITH BANKRUPTCY OR DEFAULT

Nils Chr. Framstad
Skogvollveien 40, N-0580 Oslo, Norway
ncf@math.uio.no

Abstract When portfolio choice in discontinuous asset price models is considered, the possibility of a jump to 0 is often overlooked. We solve the Merton problem for a (modified) HARA agent in a generalized geometric Lévy market where the market coefficients can change simultaneously with the jumps in the prices. We show that two fund separation does only to some extent carry over, as agents with same exponent and different intertemporal trade-offs may no longer have the same mutual fund in the presence of such possible changes.

Keywords: Merton problem, bankruptcy and default, geometric Lévy motion, portfolio separation, incomplete markets.
Mathematics Subject Classification (2000): 91B28, 60G51, 93E20.
JEL classification: G11, D81, D52.

1. Introduction

This paper considers the Merton optimal consumption-portfolio problem, which is well known in the case with Brownian driving noise. When direct utility is of hyperbolic absolute risk aversion (HARA) type, the qualitative results carry over even if one admits other distributions than the Gaussian (see [1] and [5]). When generalizing from the continuous Brownian case to the discontinuous case where the assets are driven by Lévy processes, a feature is often overlooked, namely the fact that an asset, hence an investment opportunity, may disappear at the same time as a sudden jump in its price. Indeed, the usual transformation via self-financing yields a wealth process (2.2) below, does implicitly assume that if an asset jumps to 0, it is immediately reborn with some nonzero value and its original dynamics. The reason is formula (2.1), where $v_i = \xi_i S_i$

S. Albeverio et al. (eds.),
Proceedings of the International Conference on Stochastic Analysis and Applications, 249–265.

is supposed to be a control variable taking arbitrary values, which is impossible if the price S_i is zero.

Two fund portfolio separation, or more generally k fund separation is, in economic terms, the property that the market can be replaced by k (strictly less than the number of distinct available investment opportunities) index portfolios, without welfare loss to the investor. The original case (Tobin [10]) was a single period model with multinormal returns, where the same two funds suffice for all[1] agents; one of these two can be taken to be the risk free investment opportunity, provided that such one exists[2], so-called *monetary separation* in which case the other index is frequently referred to as *the mutual fund* (in singularis.) Merton [7][3] shows the same in a geometric Brownian market, i.e. with multinormal *log*returns. Only a few probability laws other than the Gaussian (and symmetric stable) admit a similar result, see Ross [9] for a complete (modulo integrability conditions) characterization in the single period setting. A dual problem is obtained when we replace the multinormal return with an arbitrary law, and ask what utility functions admit separation into k funds which are *individual* but *independent of wealth;* mathematically, this is to say that the portfolio optimization problem associated to this utility function admits an optimal solution of the form

$$\zeta_1(y)\boldsymbol{g}_1 + \cdots + \zeta_k(y)\boldsymbol{g}_k$$

where the funds $\boldsymbol{g}_i$ are simply vectors independent of the wealth y. The classical reference (again for the single period model), Cass and Stiglitz [2], cite the exponential and power functions as the prototypical examples of utility functions admitting two fund monetary separation.

This paper will consider in a continuous-time market, where direct utility functions are concave modification of the HARA (translated power) functions[4] $c \mapsto \Delta(t) \cdot \frac{1-\gamma}{\gamma}\left(\frac{c+\beta(t)}{1-\gamma}\right)^{\gamma}$. We shall see that

- if prices obey a geometric Lévy process (constant coefficients), then for each given γ there exists a mutual fund $\boldsymbol{f}^*$, invariant wrt. β, Δ, wealth and time;
- more generally, if the coefficients are deterministic functions, then so will $\boldsymbol{f}^*$ be; if the coefficients are random, obeying a Markov point process independent of price movements, then $\boldsymbol{f}^*$ will be a stochastic process too, but still invariant wrt. Δ, β and wealth;
- if however the coefficients may jump at the same time as prices jump, the Δ-invariance breaks down; we give an explicit example to this. Consequently (up to degeneracies), no $\boldsymbol{f}$ will anymore

serve as a mutual fund for agents who at each time have the same risk aversion, but with different intertemporal trade-offs.

Bankruptcy is such a case where we expect simultaneous changes in prices and in dynamics; in such an event, stocks become worthless and at the same time the investment opportunity disappears. Also a credit event (default) on bonds may trigger restructuring and alter the law of the dynamics even in cases where the bond's value does not vanish completely.

2. The model

Assume as given a filtered probability space $(\Omega, \mathfrak{S}, \{\mathcal{F}_t\}, \mathrm{P})$ satisfying the usual conditions. We will implicitly assume all processes to be $\mathcal{F}_t$-predictable. The market prices will be driven by a stochastic differential equation, with the coefficients themselves being stochastic, determined by a discrete valued *market parameter process* Θ driven by a ($\mathcal{F}_t$-)Poisson random measure N only:

$$\mathrm{d}\Theta(s) = \int (\nu(s, \Theta(s^-), \varpi) - \Theta(s^-)) N(\mathrm{d}s, \mathrm{d}\varpi),$$

i.e. that a jump will bring the market dynamics from state Θ to a new state ν (we choose $\nu(s, \Theta(s^-))$ as the new parameter, not the increment, hence the form of the differential equation.) The market consists of a safe asset S_0, which by discounting can be assumed constant, and a vector $\boldsymbol{S} = (S_1, \ldots, S_n)^\mathsf{T}$ of risky assets following

$$\mathrm{d}\boldsymbol{S}(s) = \mathrm{Diag}(\boldsymbol{S}(s^-))\mathrm{d}\boldsymbol{\Lambda}(s)$$

where $\boldsymbol{\Lambda}$ is an independent increment process of the form

$$\mathrm{d}\boldsymbol{\Lambda}(s) = \boldsymbol{\mu}(s, \Theta(s))\mathrm{d}s + \mathrm{d}\boldsymbol{W}(s) + \int \boldsymbol{\eta}(s, \Theta(s), \varpi)\, \bar{N}(\mathrm{d}s, \mathrm{d}\varpi)$$

where

$$\bar{N}(\mathrm{d}s, \mathrm{d}\varpi) = N(\mathrm{d}s, \mathrm{d}\varpi) - \bar{\chi}\lambda(\mathrm{d}\varpi)\mathrm{d}s = N(\mathrm{d}s, \mathrm{d}\varpi) - \bar{\chi}\mathrm{E}[N(1, \mathrm{d}\varpi)]\mathrm{d}s$$

and $\bar{\chi} = \chi_{|\boldsymbol{\eta}| \leqslant 1}$. A jump in N may then cause a jump in either Θ or in $\boldsymbol{S}$ or in both. $\boldsymbol{W}$ is a ($\mathcal{F}_t$-)Wiener process with covariance matrix $R = R(s, \Theta(s))$.

As mentioned in the introduction, we shall allow bankruptcy, where an investment opportunity disappears. For technical convenience, we shall nevertheless leave the number of assets constant; a disappearing investment opportunity will get zero dynamics and may be identified with the safe asset after the loss has incurred.

The total wealth invested in the risky assets at time s at will then be

$$Y(s) = \xi_0(s)S_0(s) + \boldsymbol{\xi}^\mathsf{T}(s)\,\boldsymbol{S}(s),$$

where ξ_i is the number held of asset i. The mathematical definition of a *self financing portfolio*, motivated from a discrete time approximation, is that the number ξ_0 of monetary units is reserved to satisfy

$$\mathrm{d}Y(s) = \xi_0(s^-)\mathrm{d}S_0(s) + \boldsymbol{\xi}^\mathsf{T}(s^-)\mathrm{d}\boldsymbol{S}(s) = \boldsymbol{v}^\mathsf{T}(s^-)\mathrm{d}\boldsymbol{\Lambda}(s) \tag{sf}$$

(since $\mathrm{d}S_0 \equiv 0$) so that

$$v_i := \xi_i S_i \tag{2.1}$$

is the market value of the agent's holding in asset i. Formula (sf) has the interpretation that the instantaneous changes in wealth are solely due to changes in the market values $\boldsymbol{S}$. However, in our setting the portfolio shall also finance consumption at rate $c(s)$, so that the wealth process has the dynamics

$$\mathrm{d}Y(s) = \boldsymbol{v}^\mathsf{T}(s^-)\mathrm{d}\boldsymbol{\Lambda}(s) - c(s)\mathrm{d}s \tag{2.2}$$

(instead of the dynamics (sf)).

We shall assume that the jumps will a.s. not change the sign of Y; then each portfolio weight must be upper bounded if the corresponding asset may jump downwards and nonpositive (nonnegative) if the jumps are not lower (upper) bounded. By scaling with a constant, we may assume that no asset will jump by a factor (strictly) less than -1. For simplicity, we assume that the intensity of jumps in Θ, namely $\lambda(\{\varpi;\ \nu(t,\vartheta,\varpi) \neq \lambda\vartheta\})$, is (uniformly) bounded, a condition which should admit generalizations. We also assume, ad hoc, absence of arbitrage opportunities.

3. The preferences

The HARA family of utility functions wrt. y consists of the functions for which the Arrow-Pratt measure of absolute risk aversion is a hyperbola in y. These are the translated power functions

$$\frac{1-\gamma}{\gamma}\Big(\frac{y+\beta}{1-\gamma}\Big)^\gamma$$

where $0 \neq \gamma \neq 1$, and in addition the translated logarithm

$$\log(y+\beta) = \lim_{\gamma\to 0} \frac{1-\gamma}{\gamma}\left[\left(\frac{y+\beta}{1-\gamma}\right)^{\gamma} - 1\right]$$

to correspond to $\gamma = 0$. Linear utility ($\gamma = 1$) is a degenerate case, which we will return to in (3.1b) below. Strictly speaking, the HARA class has an additional positive scaling parameter – which without loss of generality can be omitted in our setting, as we are soon to introduce a discounting factor. We shall also soon do away with the β parameter, but we will need to specify the problem first.

For $\frac{y+\beta}{1-\gamma} < 0$, the utility function is not very interesting ($\gamma > 1$) or not possibly not even defined ($\gamma < 1$). To cope with this, we can modify the utility function[5] by putting it equal to a constant when $\frac{y+\beta}{1-\gamma} < 0$. Now if $\gamma < 1$, then this constant has to be minus infinity for the utility function to be concave; otherwise, some infinite position could (and will!) be optimal (we skip the details.) If $\gamma > 1$, the only constant extension which makes utility nondecreasing and concave, is to put utility equal to zero when $\frac{y+\beta}{1-\gamma} < 0$. We shall therefore choose to work with the utility functions

$$\Upsilon_{\gamma,\beta}(y) = \begin{cases} \frac{1-\gamma}{\gamma}\left(\frac{y+\beta}{1-\gamma}\right)^{\gamma} & \text{if } 0 \neq \gamma \neq 1 \text{ and } \frac{y+\beta}{1-\gamma} \geq 0, \\ \log(y+\beta) & \text{if } \gamma = 0 \text{ and } \frac{y+\beta}{1-\gamma} \geq 0, \\ -\infty & \text{if } \gamma < 1 \text{ and } \frac{y+\beta}{1-\gamma} < 0, \\ 0 & \text{if } \gamma > 1 \text{ and } \frac{y+\beta}{1-\gamma} < 0, \end{cases} \tag{3.1a}$$

with $\log 0 = -0^{\gamma} = -\infty$ if $\gamma < 0$. This truncated construction also permits

$$\Upsilon(y) = \max(0, y+\beta) \text{ if } \gamma = 1. \tag{3.1b}$$

The calculations will then be a simplified version of the case $\gamma > 1$, only a bit notationally inconvenient, and will be left to the reader. We note that the cases $\gamma \geq 1$ are not contained in [2] because they are not C^2 as assumed therein.

Let $\Delta \geq 0$ be the discount term. We will consider the finite horizon problem

$$\Phi(t,y,\vartheta) := \sup \mathrm{E}^{t,y,\vartheta}\left[\int_t^T \Delta(s,\Theta(s))\cdot\Upsilon_{\gamma,\beta(s)}(c(s))\mathrm{d}s + \bar{\Delta}(\Theta(T))\cdot\Upsilon_{\gamma,\bar{\beta}}(Y(T))\right] \tag{3.2}$$

where $\mathrm{E}^{t,y,\vartheta}$ denotes expectation with respect to the probability law $\mathrm{P}^{t,y,\vartheta}$ of the processes starting at $\Theta(t) = \vartheta$, $Y(t) = y$. We shall assume

$\bar{\Delta} > 0$; the limiting case $\bar{\Delta} = 0$ can be considered as lower bound on terminal wealth if $\gamma < 1$ (in which case one may also obtain the infinite horizon case by letting $T \to \infty$; see Proposition 4.1.) We assume that $\bar{\beta}$ is a constant and that $s \mapsto \beta$ is a deterministic function. Under this assumption, we can make the affine transformation

$$\begin{aligned}\tilde{Y}(s) &= Y(s) + \int_s^T \beta(s')\mathrm{d}s' + \bar{\beta}\\ \tilde{c}(s) &= c(s) + \beta(s).\end{aligned}$$

Inserting this into (3.2), we see that modulo a translate of y in the value function, we can without loss of generality – and we will – assume

$$\beta = \bar{\beta} = 0 \tag{3.3}$$

in the finite horizon case. It will then easily follow that the construction of (4.1) below holds with merely a translate of y in place of y, and that the mutual fund we will find in Section 4 does not depend on $\beta, \bar{\beta}$ nor on y. Again, we skip the details. We do however note that we also assume (3.3) in the infinite horizon case; then the assumption indeed has some substance, as it is easy to construct a β which will yield negative infinite value.

We shall also impose the restriction that $c(s) \geq 0$, which is essential when $\Delta(s) = 0$ (if not, one could exploit these points in time for negative consumption and gain arbitrary wealth.) We will have to assume further technical conditions which will be stated in Proposition 4.1; we may also for example wish to forbid short sale. A convenient class of conditions is given in (4.2).

The Θ-dependence in terminal utility seems natural, as the bequest function may represent future investment optimization problems (beyond time T). The Θ-dependence in running utility has not the same intuitive interpretation, but does not complicate the below calculations.

4. Finding the optimal strategy

Recall from (3.3) that we can and do assume $\beta = \bar{\beta} = 0$. If $\gamma > 1$, then for $y \geq 0$ one can use the zero portfolio, consume zero (which yields maximum direct utility) and still end up with $Y(T) \geq 0$ and maximum terminal utility. On the other hand, if $Y(t) < 0$, then there is no portfolio (none that satisfy the tameness conditions in Proposition 4.1 below) for which we can achieve maximal utility, and so $\Phi(t, y, \vartheta) = 0$ iff $y \geq 0$.

Arguing the same way for $\gamma < 1$, we arrive at the following property:

$$\Phi(t,0,\vartheta) = \begin{cases} 0 & \text{if } \gamma > 0 \\ -\infty & \text{if } \gamma \le 0 \end{cases}$$

$$\Phi(t,y,\vartheta) = \begin{cases} 0 & \text{if } \gamma \ge 1 \\ -\infty & \text{if } \gamma < 1 \end{cases} \quad \text{if } \frac{y}{1-\gamma} < 0.$$

Therefore, it is no restriction to assume

$$\boldsymbol{v} = \Big(\frac{y}{1-\gamma} \vee 0\Big)\boldsymbol{f} \tag{4.1}$$

if $\gamma \neq 1$ (and $\boldsymbol{v} = (y \vee 0)\boldsymbol{f}$ in the truncated linear case). If $\boldsymbol{f}$ is unrestricted, it will turn out that the optimal $\boldsymbol{f}^*$ does not depend on β nor y. This also holds if the following kind of restriction applies:

$$\boldsymbol{f} \in \Gamma \quad \text{for some set } \Gamma \text{ not depending on } y. \tag{4.2}$$

The requirement that Γ should not depend on y seems quite artificial in its general ad hoc formulation, but it covers forbidding short sale (i.e. Γ is the first orthant.) Furthermore any restriction on portfolio *weights* satisfy (4.2), and we may in particular forbid both short sale and borrowing; this property does however not transform back to a case with nonzero β or $\bar{\beta}$.

We will consider candidates $\phi = \phi_\epsilon$ for the value function:

$$\phi = \phi_\epsilon(t,y,\vartheta) = \begin{cases} D(t,\vartheta)\Upsilon_{\gamma,\epsilon}(y) + g(t,\vartheta)\cdot \chi_{\{\gamma=0\}} & \text{if } \frac{y}{1-\gamma} > 0 \\ 0 \text{ or } -\infty \text{ as above} & \text{if } \frac{y}{1-\gamma} \le 0 \end{cases} \tag{4.3}$$

where D is supposed nonnegative. The case $\frac{y}{1-\gamma} \le 0$ is already solved, so assume $\frac{y}{1-\gamma} > 0$.

In the following, the dot accent will denote $\frac{\partial}{\partial t}$. Put

$$V = V_\epsilon(s) = \phi_\epsilon(s, \Theta(s), Y(s)).$$

Using the feedback control $(c, (\frac{y+\epsilon}{1-\gamma} \vee 0)\boldsymbol{f})$ and expanding V using the Itô formula we get, for $\gamma \neq 0$,

$$
\begin{aligned}
&\Delta(s,\Theta) \cdot \Upsilon_{\gamma,0}(c)\mathrm{d}s + \mathrm{d}V \\
&= \Big[\Delta \frac{1-\gamma}{\gamma}\Big(\frac{c}{1-\gamma} \vee 0\Big)^{\gamma} - cV\frac{\gamma}{y+\epsilon}\Big]\mathrm{d}s + V\frac{\dot{D}(s,\Theta)}{D(s,\Theta)}\mathrm{d}s \\
&+ V\Big\{\frac{\gamma}{1-\gamma}\Big(\boldsymbol{f}^{\mathsf{T}}\boldsymbol{\mu} - \frac{1}{2}\boldsymbol{f}^{\mathsf{T}}R\boldsymbol{f}\Big) \\
&+ \int\Big[\frac{D(s,\nu(\Theta(s^-)))}{D(s,\Theta(s^-))}\Big[\Big(1 + \frac{\boldsymbol{f}^{\mathsf{T}}\boldsymbol{\eta}}{1-\gamma}\Big) \vee 0\Big]^{\gamma} - 1 - \bar{\chi}\frac{\gamma}{1-\gamma}\boldsymbol{f}^{\mathsf{T}}\boldsymbol{\eta}\Big]\mathrm{d}\lambda\Big\}\mathrm{d}s \\
&+ V\Big\{\frac{\gamma}{1-\gamma}\boldsymbol{f}^{\mathsf{T}}\mathrm{d}\boldsymbol{W} \\
&\qquad + \int\Big[\frac{D(s,\nu(\Theta(s^-)))}{D(s,\Theta(s^-))}\Big[\Big(1 + \frac{\boldsymbol{f}^{\mathsf{T}}\boldsymbol{\eta}}{1-\gamma}\Big) \vee 0\Big]^{\gamma} - 1\Big]\mathrm{d}\tilde{N}\Big\}
\end{aligned}
\tag{4.4}
$$

where $\mathrm{d}\tilde{N} = \mathrm{d}N - \mathrm{d}\lambda\,\mathrm{d}s$ and where if $\gamma < 1$, we only consider $\boldsymbol{f}$ such that

$$1 - \gamma + \boldsymbol{f}^{\mathsf{T}}\boldsymbol{\eta} \geq 0 \quad \lambda\text{-a.e} \tag{4.5}$$

as the others yield minus infinity. The c^* maximizing the right hand side of (4.4) with respect to c, satisfies

$$c^* = \Big(\frac{D}{\Delta}\Big)^{\frac{1}{\gamma-1}}(y+\epsilon)\chi_{\Delta} \tag{4.6}$$

where $\chi_{\Delta}(s) = \chi_{\Delta(s)>0}$, and thus

$$\Delta\frac{1-\gamma}{\gamma}\Big[\Big(\frac{c^*}{1-\gamma}\Big) \vee 0\Big]^{\gamma} - c^*V\frac{\gamma}{y+\epsilon} = V(1-\gamma)\Big(\frac{D}{\Delta}\Big)^{\frac{1}{\gamma-1}}\chi_{\Delta}.$$

Therefore, if D satisfies the inequality

$$
\begin{aligned}
0 &\geq \dot{D}(t,\vartheta)\frac{1-\gamma}{\gamma} + D(t,\vartheta)\frac{(1-\gamma)^2}{\gamma}\Big(\frac{D(t,\vartheta)}{\Delta(t,\vartheta)}\Big)^{\frac{1}{\gamma-1}} \cdot \chi_{\Delta} \\
&+ D(t,\vartheta)\Big(\boldsymbol{f}^{\mathsf{T}}\boldsymbol{\mu} - \frac{1}{2}\boldsymbol{f}^{\mathsf{T}}R\boldsymbol{f}\Big) \\
&+ \frac{1-\gamma}{\gamma}\int\Big\{D(t,\nu(\vartheta))\Big[\Big(1 + \frac{\boldsymbol{f}^{\mathsf{T}}\boldsymbol{\eta}}{1-\gamma}\Big) \vee 0\Big]^{\gamma} \\
&\qquad - D(t,\vartheta) - \bar{\chi}\frac{\gamma}{1-\gamma}D(t,\vartheta)\boldsymbol{f}^{\mathsf{T}}\boldsymbol{\eta}\Big\}\mathrm{d}\lambda
\end{aligned}
\tag{4.7}
$$

for all $\boldsymbol{f}$, then

$$\begin{aligned}&\Delta(s,\Theta)\cdot\Upsilon_{\gamma,0}(c)\mathrm{d}s+\mathrm{d}V\\&=V\Big[\int\Big\{\frac{D(s,\nu(\Theta(s^-)))}{D(s,\Theta(s^-))}\Big[\Big(1+\frac{\boldsymbol{f}^\top\boldsymbol{\eta}}{1-\gamma}\Big)\vee 0\Big]^\gamma-1\Big\}\mathrm{d}\tilde{N}\\&\qquad+\frac{\gamma}{1-\gamma}\boldsymbol{f}^\top\mathrm{d}\boldsymbol{W}\Big]+p\mathrm{d}t\end{aligned}\tag{4.8}$$

for some $p\geq 0$, where $p=0$ iff both $c=c^*$ and (4.7) holds with equality for the maximizing $\boldsymbol{f}^*$.

Similarly, for log utility we get an optimal consumption c^* given by (4.6) with $\gamma=0$, and the following for D, g:

$$0=\dot{D}(t,\vartheta)+\int(D(t,\nu(\vartheta))-D(t,\vartheta))\mathrm{d}\lambda+\Delta(t,\vartheta)\tag{4.9a}$$

$$\begin{aligned}0\geq\;&\dot{g}(t,\vartheta)+\int(g(t,\nu(\vartheta))-g(t,\vartheta))\mathrm{d}\lambda\\&+\Delta(t,\vartheta)\big[\log(\Delta(t,\vartheta))-\log(D(t,\vartheta))-1\big]\\&+D(t,\vartheta)\Big[\boldsymbol{f}^\top\boldsymbol{\mu}-\frac{1}{2}\boldsymbol{f}^\top R\boldsymbol{f}\Big]\\&+\int(D(t,\nu(\vartheta))\log(1+\boldsymbol{f}^\top\boldsymbol{\eta})-D(t,\vartheta)\bar{\chi}\boldsymbol{f}^\top\boldsymbol{\eta})\mathrm{d}\lambda\end{aligned}\tag{4.9b}$$

for all $\boldsymbol{f}$ such that $\boldsymbol{f}^\top\boldsymbol{\eta}\geq -1$ (here, $0\log 0:=0$). If (4.9) holds, then

$$\begin{aligned}&\Delta(s,\Theta)\log(c+\beta)\mathrm{d}s+\mathrm{d}V=p\,\mathrm{d}t+D(s,\Theta(s))\boldsymbol{f}^\top\mathrm{d}\boldsymbol{W}\\&\quad+\int\Big[D(s,\nu(\Theta(s^-)))\log(1+\boldsymbol{f}^\top\boldsymbol{\eta})\\&\qquad\qquad+g(s,\nu(\Theta(s^-)))-g(s,\Theta(s^-))\Big]\mathrm{d}\tilde{N}\end{aligned}\tag{4.10}$$

for some $p\geq 0$, where again $p=0$ iff both $c=c^*$ and (4.9b) holds with equality for the maximizing $\boldsymbol{f}^*$.

We may now proceed to prove that the solution suggested is optimal under certain conditions. The case $\gamma=1$ will follow as $\gamma>1$ with minor modifications:

Proposition 4.1. *Consider problem* (3.2), $T=\infty$ *permitted if* $\bar{\Delta}=0$, *where the supremum is taken over all the consumption-investment strategies* $(c,\boldsymbol{v})=\{(c(s),(\frac{Y(s)}{1-\gamma}\vee 0)\boldsymbol{f}(s))\}_{s\in[t,T]}$ *admitting unique weak solution to* (2.2), *satisfying* $c\geq 0$ *and with* $\boldsymbol{f}$ *so regular that* (4.8) *(resp.* (4.10)*)* *are submartingales, and such that*

$$\boldsymbol{f} = \mathbf{0} \text{ whenever } Y(s) \text{ is less than some lower bound;}$$
$$\text{above this lower bound, } \boldsymbol{f} \text{ satisfies (4.2) and is bounded.} \tag{4.11}$$

Assume furthermore that there exist bounded D satisfying (4.7) (if $\gamma \neq 0$), or D, g satisfying (4.9) (if $\gamma = 0$), with $D \geq 0$ and the terminal condition

$$\begin{aligned} &D(T,\,\cdot\,) \geq \bar{\Delta}, \quad g(T,\,\cdot\,) \geq 0 && \text{if } T < \infty \\ &\liminf_{m\to\infty} \mathrm{E}[V_\epsilon(\inf\{s \geq t; |V_\epsilon(s)| \geq m\})] \geq 0 && \text{if } T = \infty. \end{aligned} \tag{4.12}$$

Then $\Phi \leq \phi_0$ (given by (4.3).)

Suppose in addition that c^* is given by (4.6), and that (4.7) (if $\gamma \neq 0$) or (4.9b) (if $\gamma = 0$) holds with equality and that (4.8) (resp. (4.10)) are indeed martingales if we insert the $\boldsymbol{f}^* = \boldsymbol{f}^*(\vartheta)$ maximizing the right hand side, and that formula (4.12) holds with equality.

Then $\Phi = \phi_0$ and attained by $(c^*, (\frac{Y}{1-\gamma} \vee 0)\boldsymbol{f}^*)$, at worst modulo integrability conditions for the particular case $\gamma = 0$.

Proof. Consider the process $V_\epsilon(s)$. For $\gamma \leq 0$ let $\epsilon > 0$ be given and note that for $y < 0$, then c and $\boldsymbol{f}$ are both zero; For $\gamma > 0$, use $\epsilon = 0$. Thus if we start at $y > 0$, $V_\epsilon(s)$ is lower bounded if $\gamma < 1$. Let $\tau_m = T \wedge \inf\{s \geq t; |V_\epsilon(s)| \geq m\}$. By the boundedness of $\boldsymbol{f}$ and D,

$$V_\epsilon(t) \geq \mathrm{E}\left[V_\epsilon(\tau_m) + \int_t^{\tau_m} \Delta \cdot \Upsilon_{\gamma,0}(c)\mathrm{d}s\right] \tag{4.13}$$

Let $m \to \infty$. By lower boundedness and Fatou's lemma, we get $V_\epsilon(t) \geq \Phi$ and thus $V_0(t) \geq \Phi$ if $\gamma < 1$. The same conclusion holds for $\gamma > 1$; though V is not necessarily lower bounded, (4.11) and the submartingale assumption suffice. The infinite horizon case follows by taking the limit.

To prove optimality, consider $(c^*, (\frac{Y(s)}{1-\gamma} \vee 0)\boldsymbol{f}^*)$. The case of log utility is easy in the finite horizon case, so consider the case $\gamma \neq 0$. Notice that

$$\Delta \cdot \Upsilon(c^*) = \chi_\Delta \Big(\frac{D}{\Delta}\Big)^{\frac{1}{\gamma-1}} V \tag{4.14}$$

and therefore, V is a geometric process, since $\boldsymbol{f}^*$ depends only on V through D. Solving the equation (4.7) for V, we get

$$V(T) \cdot \exp\Big\{\int_t^T \chi_\Delta \Big(\frac{D(s,\Theta(s))}{\Delta(s,\Theta(s))}\Big)^{\frac{1}{\gamma-1}} \mathrm{d}s\Big\} = V(t) \cdot X(T)$$

where $X(t) = 1$ and where

$$\mathrm{d}X = X_- \cdot \left\{ \int_t^T \frac{\gamma}{1-\gamma} \boldsymbol{f}^{*\mathsf{T}} \,\mathrm{d}\boldsymbol{W} + \int \left(\frac{D(s,\nu(\Theta(s^-)))}{D(s,\Theta(s^-))} \left[\left(1 + \frac{\boldsymbol{f}^{*\mathsf{T}} \boldsymbol{\eta}}{1-\gamma}\right) \vee 0 \right]^\gamma - 1 \right) \mathrm{d}\tilde{N} \right\}$$

is a martingale by assumption. Differentiating, we get that the right hand side of (4.8) – which now holds with equality – has zero expectation. It follows that $V_0(t) = \Phi$. The same holds in the infinite horizon case for $0 \neq \gamma < 1$; each term on the right hand side of (4.13) will converge monotonically. In the log case, however, we will possibly have both positive and negative parts and one would presumably need extra information on the decay (in time) of Δ and its dependence on ϑ. □

Remark 4.2. Obviously the boundedness assumptions on D, g and $\boldsymbol{f}$ may be weakened. Note also that the optimal $\boldsymbol{f}^*$ does not satisfy (4.11) for $\gamma > 1$. The conditions specified on $\boldsymbol{f}$ should therefore not be interpreted as "admissibility". For the sake of brevity, such "tameness conditions" (which have the economic interpretation of excluding "doubling strategies") are not specified completely herein, but note that in view of [3] they cannot just be dropped.

Corollary 4.3. *The vector $\boldsymbol{f}^*$ is a mutual fund for all agents with $(\Delta, \bar{\Delta})$ common (up to the obvious positive multiplicative constant). If Θ is a.s. constant, then $\boldsymbol{f}^*$ does not depend on Δ, i.e. it is a mutual fund for all agents with (modified) discounted HARA utility. This property holds more generally, namely if Θ and $\boldsymbol{\Lambda}$ have no simultaneous jumps, i.e. if*

$$\boldsymbol{\eta}(s,\vartheta,\varpi) \cdot (\nu(s,\vartheta,\varpi)) - \theta) = \mathbf{0} \qquad \lambda\text{-a.s., all } s \text{ and } \theta. \tag{4.15}$$

If this holds, then the $\mathrm{d}\lambda$ *part of* (4.4) *split into a part not containing $\boldsymbol{f}$ and a part not containing D, and the maximization wrt. $\boldsymbol{f}$ will depend only on γ, $\boldsymbol{\mu}$, R, $\boldsymbol{\eta}$ and λ. Similar holds for* log *utility, cf.* (4.9b).

If (4.15) fails, then there exist counterexamples where the agents do not share $D(s,\nu(\vartheta))/D(s,\vartheta)$ and thus not $\boldsymbol{f}^*$ either; we will prove this at the end of the section, as it will utilize calculations from the following example, which is a special case where (4.7) becomes a family of Bernoulli equations to be solved inductively. The example generalizes [1] and [5]:

Example 4.4. Consider the infinite horizon case with $0 \neq \gamma < 1$. Assume that in addition to the safe investment opportunity, we have n risky securities; assets $1, \ldots n_0$ are always accessible, and assets $n_0 + 1, \ldots, n$

represent eternal bonds with continuously capitalized interest, each being driven by one Poisson jump source N_i of which everything else is independent, and each bond will become inaccessible at first jump in N_i, whose intensity we denote λ_i. By scaling the drift term, we can without loss of generality assume that a bond jumps to 0; also, it is more convenient working with $\mathrm{d}N_i$ instead of $\mathrm{d}\bar{N}_i = \mathrm{d}N_i - \lambda_i\bar{\chi}\mathrm{d}t$. The bond then gets an adjusted drift term $\tilde{\mu}_i$ which has to be ≥ 1 in order to avoid arbitrage. The market parameter can now be represented as $\boldsymbol{\Theta}(s) = (\Theta_{n_0+1}(s), \ldots, \Theta_n(s))$, where Θ_i is one if the bond is still accessible, zero if not:

$$\mathrm{d}\Theta_i(s) = -\Theta_i(s^-)\mathrm{d}N_i(s), \quad \Theta_i(t) = 1$$

where each N_i has Lévy measure $\lambda_i > 0$. Consider (4.7) with equality. For the first n_0 assets, the maximization with respect to the vector $\hat{\boldsymbol{f}} := (f_1, \ldots, f_{n_0})^\mathsf{T}$, is independent of D:

$$\begin{aligned}
&\sup_{\hat{\boldsymbol{f}}}\Big\{\hat{\boldsymbol{f}}^\mathsf{T}\boldsymbol{\mu} - \frac{1}{2}\hat{\boldsymbol{f}}^\mathsf{T}\hat{R}\hat{\boldsymbol{f}} \\
&\qquad + \frac{1-\gamma}{\gamma}\int\Big(((1+\frac{\hat{\boldsymbol{f}}^\mathsf{T}\hat{\boldsymbol{\eta}}}{1-\gamma})\vee 0)^\gamma - 1 - \bar{\chi}\frac{\gamma}{1-\gamma}\hat{\boldsymbol{f}}^\mathsf{T}\hat{\boldsymbol{\eta}}\Big)\mathrm{d}\lambda\Big\} \\
&=: F(\mathbf{0})\frac{1-\gamma}{\gamma}
\end{aligned} \tag{4.16}$$

where the "ˆ" accents on the vectors and matrix denote the obvious truncation of the dimension. Again, if $\gamma < 1$ the supremum is only taken among the $\hat{\boldsymbol{f}}$ which satisfy $\frac{\hat{\boldsymbol{f}}^\mathsf{T}\hat{\boldsymbol{\eta}}}{1-\gamma} \geq -1$ λ-a.e. The last $n - n_0$ assets are each independent of everything else, so the portfolio optimization can be carried out for each $i > n_0$. We write $\boldsymbol{\vartheta} \succcurlyeq \boldsymbol{\epsilon}$ if $\boldsymbol{\epsilon}$ and $\boldsymbol{\vartheta} - \boldsymbol{\epsilon}$ are also vectors of zeros and ones (i.e., the latter has no negative components and all assets accessible with $\boldsymbol{\vartheta}$ are accessible with $\boldsymbol{\epsilon}$.) Writing $\boldsymbol{e}_i$ for the ith unit vector, the maximization with respect to f_i becomes:

$$\sup_{f_i}\Big\{D(t,\boldsymbol{\vartheta})f_i\tilde{\mu}_i + \frac{1-\gamma}{\gamma}\lambda_i\Big(D(t,\boldsymbol{\vartheta}-\boldsymbol{e}_i)\Big[\Big(1-\frac{f_i}{1-\gamma}\Big)\vee 0\Big]^\gamma - D(t,\boldsymbol{\vartheta})\Big)\Big\}$$

and the optimal f_i is given by

$$f_i^* = (1-\gamma)\Big[1 - \Big(\frac{\lambda_i D(t,\boldsymbol{\vartheta}-\boldsymbol{e}_i)}{\tilde{\mu}_i D(t,\boldsymbol{\vartheta})}\Big)^{\frac{1}{1-\gamma}}\Big]\vartheta_i. \tag{4.17}$$

Inserting into (4.7), we get the following Bernoulli equation for $D(t,\boldsymbol{\vartheta})$:

$$0 = \frac{F(\boldsymbol{\vartheta})}{1-\gamma} + \frac{\dot{D}(t,\boldsymbol{\vartheta})}{D(t,\boldsymbol{\vartheta})} + (1-\gamma)\Big[\big(\Delta(t,\boldsymbol{\vartheta})\big)^{\frac{1}{1-\gamma}}\chi_\Delta + \sum_{i>n_0}\big(\frac{\lambda_i}{\tilde{\mu}_i}D(t,\boldsymbol{\vartheta}-\boldsymbol{e}_i)\big)^{\frac{1}{1-\gamma}}\cdot\tilde{\mu}_i\vartheta_i\Big]\big(D(t,\boldsymbol{\vartheta})\big)^{-\frac{1}{1-\gamma}}$$

with

$$F(\boldsymbol{\vartheta}) := F(\mathbf{0}) + \sum_{i>n_0}(\gamma\tilde{\mu}_i - \lambda_i)\vartheta_i. \tag{4.18}$$

The solution is given (with abuse of notation if $\vartheta_i = 0$) by

$$D(t,\boldsymbol{\vartheta})\mathrm{e}^{F(\boldsymbol{\vartheta})t} = \left(A(\boldsymbol{\vartheta}) - \int^t \mathrm{e}^{\frac{F(\boldsymbol{\vartheta})}{1-\gamma}\cdot s}\cdot\Big[\big(\Delta(s,\boldsymbol{\vartheta})\big)^{\frac{1}{1-\gamma}}\chi_\Delta + \sum_{i>n_0}\Big(\frac{\lambda_i}{\tilde{\mu}_i}D(s,\boldsymbol{\vartheta}-\boldsymbol{e}_i)\Big)^{\frac{1}{1-\gamma}}\cdot\tilde{\mu}_i\vartheta_i\Big]\mathrm{d}s\right)^{1-\gamma},$$

to be solved inductively starting from $\boldsymbol{\vartheta} = \mathbf{0}$. In the special infinite horizon case

$$\Delta(t,\boldsymbol{\vartheta}) = H(\boldsymbol{\vartheta})\mathrm{e}^{-\delta t}, \qquad (\delta > 0 \text{ and independent of } \boldsymbol{\vartheta})$$

we find an easy generalization of the version of the Merton problem treated in [5]. We can take $A = 0$, and by induction, it follows that if $\delta > \max_{\boldsymbol{\epsilon}\preccurlyeq\boldsymbol{\vartheta}} F(\boldsymbol{\epsilon})$, then

$$D(t,\boldsymbol{\vartheta}) = K(\boldsymbol{\vartheta})\mathrm{e}^{-\delta t} \tag{4.19a}$$

with K inductively given by

$$K(\boldsymbol{\vartheta}) = \left(\frac{1-\gamma}{\delta - F(\boldsymbol{\vartheta})}\Big[\big(H(\boldsymbol{\vartheta})\big)^{\frac{1}{1-\gamma}} + \sum_{i>n_0}\Big(\frac{\lambda_i}{\tilde{\mu}_i}K(\boldsymbol{\vartheta}-\boldsymbol{e}_i)\Big)^{\frac{1}{1-\gamma}}\cdot\tilde{\mu}_i\vartheta_i\Big]\right)^{1-\gamma}. \tag{4.19b}$$

To prove optimality, note that by (4.14) and (4.10) with equality, it suffices that Δ/D is bounded away from 0 to conclude that $\mathrm{E}[Z(\tau_m)]$ tends to 0. However, in our case the fraction is constant. For $\gamma \in (0,1)$ it also follows, by solving the problem for a sequence $\delta_n \searrow F(\boldsymbol{\vartheta})$ that the value function is $+\infty$ for $\delta \le F$ (which is impossible for $\gamma < 0$, as $F < 0$ as well). Finally, for $\boldsymbol{\vartheta} = \mathbf{0}$ we have weakened the sufficient conditions for finiteness given in [5, formulae (24) – (26)] to an inequality which is also necessary.

To this end, we give the counterexample promised following the Corollary. The presence of market dynamics changes may cause agents with different intertemporal trade-off coefficients Δ to require different individual funds, even if they do not depend on $\boldsymbol{\vartheta}$ directly:

Proposition 4.5. *If two agents $j = 1, 2$ have the same Υ (both with $\beta = 0$) but Δ's given by $\Delta_j(t, \boldsymbol{\vartheta}) = \Delta_j(t) = \mathrm{e}^{-\delta_j t}$, then there is not in general a mutual fund $\boldsymbol{f}$ common to both.*

Proof. We give a counterexample for the infinite horizon case; it is not hard to find a similar one if $T < \infty$ either. Assume we are in the setting of the above Example with $n_0 = 1$, $n = 2$ so that there are only two risky assets, one of which a bond with default. For the other, the optimal f_1^* will be determined by (4.16) and independent of δ, so the optimal portfolio $\boldsymbol{f} = (f_1^*, f_2^*)^\mathsf{T}$ is then a mutual fund common to both agents iff f_2^* does not depend on δ; we shall see that such δ-dependence will indeed occur.

$\boldsymbol{\vartheta} = \vartheta$ will now be 1 iff the bond is still accessible, and then 0 for all eternity. It is easy to see that neither $F(0)$ nor $F(1)$ will depend on δ, so we skip those calculations, merely noting that we will have finite value if only δ is chosen big enough. From (4.19) (with $H = 1$) we have

$$K(0) = \left(\frac{1-\gamma}{\delta - F(0)}\right)^{1-\gamma} \quad \text{and}$$

$$K(1) = \left(\frac{1-\gamma}{\delta - F(1)}\Big[1 + \tilde{b}_2\big(\frac{\lambda_i}{\tilde{\mu}_i}K(0)\big)^{\frac{1}{1-\gamma}}\Big]\right)^{1-\gamma}. \tag{4.20}$$

We then remark that $D(t, 0)/D(t, 1) = K(1)/K(0)$. Plug this into (4.17) and see that $\boldsymbol{f}^*$ does indeed depend on δ. □

We conclude that *condition* (4.15), *which is violated by bankruptcy risk, is significant to the portfolio separation property.*

5. Appendix

In a single period model it is easy to see (cf. [2]) that only a special class of utility functions will admit separation irrespectively of the probability law of the returns. The same holds true in the continuous time setting; the author has not succeeded in finding a (counter)proof in the literature, and chooses to give one for the sake of completeness:

Proposition 5.1. *There are finite horizon problems of the form*

$$\Phi(y) = \sup_{\boldsymbol{v}} J^{(\boldsymbol{v})}(y) = \sup_{\boldsymbol{v}} \mathrm{E}^y[\Upsilon(Y(1))]$$

(without intermediate consumption,) with $\Upsilon \in \mathrm{C}^1$ *concave, and such that the optimal* $\boldsymbol{v}^*$ *does not separate into two funds independent of wealth.*

Proof. We shall see that there will be counterexamples of the following form: Let $\hat{\Upsilon}(y)$ be HARA, and consider a strictly increasing, concave and C^1 modification Υ such that

$$\Upsilon(y) \begin{cases} = \hat{\Upsilon}, & \text{if } y \leq \bar{y} \ (> 0) \\ < \hat{\Upsilon}, & \text{if } y > \bar{y}. \end{cases}$$

To prove the claim, it suffices – and simplifies notation – to prove the case $\bar{y} = 1$, $\beta = 0$, $\gamma = 1/2$ (so $\hat{\Upsilon}(y) = \sqrt{2y}$.) Let the market consist of two (nonnegative) assets,

$$\mathrm{d}X_i(t) = bX_i(t^-)(\mathrm{d}t - \mathrm{d}N_i(t))$$

where $\mathrm{E}[N_i(1)] = \lambda_i \in (0,1) \ni b$ such that X_i are submartingales. We assume the assets are not identical in law, i.e. $\lambda_1 \neq \lambda_2$. Let $\hat{\Phi}$ and $\hat{J}^{(\boldsymbol{v})}$ correspond to the limiting case $\Upsilon \nearrow \hat{\Upsilon}$; then we have the form $\hat{\Phi}(y) = K\sqrt{2y}$ and the (unique) optimal control is given by

$$b\hat{v}_i = (1 - \lambda_i^2)y. \tag{5.1}$$

Now if

$$y \leq y_0 := \exp\{-(2 - \lambda_1^2 - \lambda_2^2)\}, \tag{5.2}$$

we will have $Y(1) \leq 1$ and therefore for all choices of Υ as above, $\Phi(y) = \hat{\Phi}(y)$ for $y \leq y_0$, and with optimal control $\boldsymbol{v}^* = \hat{\boldsymbol{v}}$. On the other hand we clearly have

$$J^{(\hat{\boldsymbol{v}})}(y) < \hat{J}^{(\hat{\boldsymbol{v}})}(y) \quad \text{and thus} \quad \Phi(y) < \hat{\Phi}(y) \quad \text{for } y > y_0. \tag{5.3}$$

We now prove that Φ is differentiable at y_0, which by concavity (which by linearity of the dynamics is inherited from Υ) will imply continuous differentiability up to some $y_1 > y_0$. Consider the control $\hat{\boldsymbol{v}}$, and let $y = y_0 + \epsilon$ with $0 < \epsilon \leq b$ so that if there is at least one jump then $Y(1) \leq 1$. Let π be the probability that no jump occurs before time 1. Then

$$\begin{aligned} J^{(\hat{\boldsymbol{w}})}(y_0+\epsilon) - \hat{\Phi}(y_0+\epsilon) = \pi \cdot (\Upsilon((y_0+\epsilon)\mathrm{e}^{2-\lambda_1^2-\lambda_2^2}) \\ - \hat{\Upsilon}((y_0+\epsilon)\mathrm{e}^{2-\lambda_1^2-\lambda_2^2})) + (1-\pi) \cdot 0 \end{aligned}$$

Divide by ϵ and let $\epsilon \searrow 0$, we see that the right hand side tends to a constant times $\Upsilon'(1) - \hat{\Upsilon}'(1)$ which is 0 (the argument is 1 by the definition of y_0 (5.2)). So Φ is differentiable at y_0 as claimed. Therefore, since there is no restriction to assume nonnegative positions, the HJB equation holds for $y > y_0$ small enough; v_i then satisfies the first order condition

$$\Phi'(y) = \lambda_i \Phi'(y - bv_i) = \lambda_i K \cdot (2(y - bv_i))^{-1/2}$$

which implies that

$$\frac{v_2}{v_1} = \frac{y(\Phi'(y))^2 - (K\lambda_2)^2}{y(\Phi'(y))^2 - (K\lambda_1)^2} \tag{5.4}$$

Assume for contradiction that there is two fund separation; then by (5.1), the bank may be taken as one of the two funds, hence v_2/v_1 is constant, which by (5.4) implies that for $y > y_0$ small enough, $y(\Phi'(y))^2$ is constant and therefore by continuous differentiability, $\Phi(y) = K\sqrt{2y} = \hat{\Phi}(y)$ on some nonempty interval (y_0, y_2), contradicting (5.3). □

Acknowledgments

The author gratefully acknowledges financial support from the Research Council of Norway and NorFA, and the hospitality of Stockholm School of Economics and the University of Kansas where parts of the work were carried out. The paper has benefited from comments from Jan Ube and an anonymous referee.

Notes

1. Originally, greed and risk aversion was assumed; in [6] or more generally [4] it is shown (continuous time, though the argument applies to discrete time too) that only a mild preference of more to less is essential.

2. if not, one can take the safest possible portfolio.

3. see also Merton's erratum, [8].

4. the exponential case is as least as easy.

5. The reader may check that we may perform the calculations with the HARA utility unmodified for $\gamma > 1$.

References

[1] Aase K. K.: Optimum portfolio diversification in a general continuous-time model, *Stochastic Process. Appl.* **18** (1984) 81–98.

[2] Cass D., Stiglitz J.: The Structure of Investor Preferences and Asset Returns, and Separability in Portfolio Selection: A Contribution to the Pure Theory of Mutual Funds, *J. Econom. Theory* **2** (1970) 122–160.

[3] Dudley, R. M.: Wiener functionals as Itô integrals, *Ann. Probability* **5** (1977) 140–141.

[4] Framstad N. C.: Portfolio separation without stochastic calculus (almost), Preprint Pure Mathematics 10/2001, University of Oslo, Oslo, 2001, `http://www.math.uio.no/eprint/pure_math/2001/10-01.html`

[5] Framstad N. C., Øksendal B., Sulem A.: Optimal consumption and portfolio in a jump diffusion market, In *Workshop on Mathematical Finance*, A. Shiryaev, A. Sulem, eds., INRIA, Paris, 1998, 9–20.

[6] Khanna A., Kulldorff M.: A generalization of the mutual fund theorem, *Finance Stoch.* **3** (1999) 167–185.

[7] Merton R. C.: Optimum consumption and portfolio rules in a continuous-time model, *J. Econom. Theory* **3** (1971) 373–413.

[8] Merton, R. C.: Erratum: "Optimum consumption and portfolio rules in a continuous-time model" (*J. Econom. Theory* **3** (1971) 373–413), *J. Econom. Theory* **6** (1973) 213–214.

[9] Ross, S.A.: Mutual fund separation in financial theory—the separating distributions, *J. Econom. Theory* 17 (1978) 254–286.

[10] Tobin, J.: Liquidity preference as behavior toward risk, *Rev. Econom. Stud.* **27** (1958) 65–86.

SQUARE OF WHITE NOISE UNITARY EVOLUTIONS ON BOSON FOCK SPACE

Luigi Accardi
Centro Vito Volterra, Universitá di Roma Tor Vergata
Via di Tor Vergata, 00133 Roma, Italy
volterra@volterra.mat.uniroma2.it

Andreas Boukas
Department of Mathematics, American College of Greece
Aghia Paraskevi, Athens 15342, Greece
gxk-personnel@ath.forthnet.gr

Abstract With the help of *Mathematica* we deduce an explicit formula for bringing to normal order the product of two normally ordered monomials in the generators of the Lie algebra of $SL(2,\mathbb{R})$. We use this formula to prove the Itô multiplication table for the stochastic differentials of the universal enveloping algebra of the renormalized square of white noise defined on Boson Fock space. Using this Itô table we derive unitarity conditions for processes satisfying quantum stochastic differential equations driven by this noise. From these conditions we deduce in particular that a quantum stochastic differential involving only the three basic integrators (see (2.7)-(2.9) below) and dt cannot have a unitary solution. Computer algorithms for checking these conditions, for computing the product of stochastic differentials, and for iterating the differential of the square of white noise analogue of the Poisson-Weyl operator are also provided.

Keywords: Primary 81S25 ; Secondary 81S05

1. Introduction

The time-evolution of a quantum mechanical observable X (i.e a self-adjoint operator on the wave function Hilbert space $\mathcal{H}$) is described by a new observable $j_t(X) = U^*(t)\, X\, U(t)$ where $U(t) = \mathrm{e}^{-\mathrm{i}tH}$ is a unitary operator and H is a self-adjoint operator on the wave function Hilbert space.

S. Albeverio et al. (eds.),
Proceedings of the International Conference on Stochastic Analysis and Applications, 267–301.

If the evolution is not disturbed by noise, the operator processes $U = \{U(t) \,/\, t \geq 0\}$ and $j = \{j_t(X) \mid t \geq 0\}$ satisfy a deterministic differential equations in the "system" Hilbert space $\mathcal{H}$. In the case when the system is affected by quantum noise, described in terms of operators acting on a "noise" Fock space Γ and satisfying certain commutation relations, the equations for U and j are replaced by stochastic differential equations driven by that noise (cf. e.g. [22]), interpreted as stochastic differential equations in the tensor product $\mathcal{H} \otimes \Gamma$ and viewed as the Heisenberg picture of the Schrödinger equation in the presence of noise or as a quantum probabilistic analogue of the Langevin equation.

It is therefore important to be able to determine for specific quantum noises which processes U satisfying a quantum stochastic differential equation can be used to describe the time–evolution of an observable i.e. to decide when the solution of the equation satisfied by U is unitary.

Stochastic differential equations were introduced by Itô in [21] in the classical case and by Hudson and Parthasarathy in [20] in the quantum case.

In the early 1990's the development of the stochastic limit of quantum theory (cf. [13]) led Accardi, Lu and Volovich to a new approach to both types of equations: the white noise approach to stochastic calculus

This new approach made clear that both classical and quantum stochastic calculus were, from the white noise point of view, the calculus related to the first power of standard quantum white noise b_t, b_t^+, plus a single very special combination of the second power, i.e. $b_t^+ b_t$.

At this point the following question became completely clear from a mathematical point of view: *is it possible to develop a stochastic calculus for the higher powers of white noise*?

If the answer to this question turns out to be: *yes*, then it would be natural to call this new calculus "nonlinear stochastic calculus" because its restriction to the first powers of the white noise functionals (plus $b_t^+ b_t$) would produce the usual stochastic calculus.

The programme of developing a non-linear extension of classical and quantum stochastic calculus was first proposed in [14].

Starting from 1993 several papers were devoted to the attempt of realizing this program and several partial results were obtained. For reasons of space we will not describe these developments but refer to [14, 15].

The dificulty of the problem should not be surprising in view of the fact that it is equivalent to the long standing problem in quantum field theory of giving a meaning to the local powers of quantum fields (which are nothing but multiparameter white noises). The difficulty comes from the singularity of the problem: *white noises are operator valued distri-*

butions and their powers are ill defined. This problem has a counterpart also in classical probability where a number of authors (Nualart, Russo, ...) are now trying to extend stochastic calculus beyond the frame of semi-martingales. Both in quantum field theory and in white noise stochastic calculus a satisfactory solution of this problem, for arbitrary powers of the field, is still standing.

The first attempts to realize this program concretely in the simplest non linear situation,i. e. the square of white noise [2], [3], [4], [6], clearly indicated that the fundamental result of the theory, i.e. the formulation of the unitarity conditions, would be much deeper and complex than the corresponding result for linear white noise, proved in the early '80's by Hudson and Parthasarathy.

The first decisive step towards the realization of this programme was done, in 1999, in the paper [12] where Accardi, Lu and Volovich introduced in [12] a new renormalization technique which allowed to solve the problem in the case of second order noises, i.e second order powers of b_t and b_t^+ (that is why one speaks of "renormalized square of white noise" or RSWN). This includes in particular the square of classical white noise $w_t^2 = (b_t^+ + b_t)^2$ and already this goes beyond what can be obtained with the tools of classical or quantum probability theory.

This new renormalization technique was based on two new ideas:

(i) instead of the cut-off-and-take-limits procedure, used in the physical literature and instead of of renormalizing the Ito table as done in [14], one directly renormalizes the commutation relations themselves, and then looks for a Hilbert space representation of them. In fact the Ito table can, at least inprinciple, be derived from the commutation relations.

(ii) instead of subtracting infinite constants, use the identity $\delta(t)^2 = c\,\delta(t)$.

This allowed to give a constructive proof of the existence of the analogue of the Fock representation for the renormalized square of white noise.

Immediately after, Accardi and Skeide recognized, with considerable surprise, that the representation space of the RSWN coincided with the "finite difference Fock space" which was introduced 10 years before, in a completely different context, and starting from completely different motivation, by Feinsilver and Boukas in [17, 19].

This opened the way to the second breakthrough in this line of research, which was realized by Accardi, Franz and Skeide in [10]. In that paper the Lie algebra of the RSWN was identified to a current algebra over a certain extension of the Lie algebra of $SL(2;\mathbb{R})$ which is defined by the commutation relations (2.1). This allowed, using the Schürmann representation theorem for independent increment processes on *-bialgebras (cf. [23]), to concretely realize the RSWN basic integrators as simple sums of first order integrators, i.e of Hudson-Parthasarathy type (cf. (2.7)-(2.9)).

In other words: the renormalized square of white noise (RSWN) can be realized on a usual Fock space and the corresponding stochastic differentials can be expressed as sums of usual, first order, white noises acting on that space.

However, in this representation, the three basic integrators of the RSWN integrators (cf. (2.7)-(2.9)) have not a closed Itô table and this fact made it impossible to prove, in the paper [10], the main result of the theory, i.e. the unitarity condition. In fact the only thing one can get, from a naive application of the Hudson–Parthasarathy first order Itô table to the RSWN, is an infinite chain of coupled nonlinear operator equations on whose solution still now nothing is known.

This made clear that, for the solution of the problem, the simple application of the known formulae of first order stochastic calculus were not sufficient and new ideas and techniques were needed.

Such a situation is quite common in mathematics: abstract representation theorems are nice results and often quite useful, however they can rarely be used to get rid of the difficulties of a specific problem. For example it is well known that every Borel measure on a standard Borel space can be identified to the Lebesgue measure on the interval $[0, 1]$, but this is of little help in establishing the fine properties of the Wiener measure. In classical probability, where the huge literature on classical Levy processes is by no means a more or less routine application of known results on Brownian motion and Poisson processes.

The classical Levy-Khintchin theorem, of which the Schürmann representation theorem is a quantum generalization, shows that any infinitely divisible, stationary, independent increment process can be built up using as building blocks the standard Brownian motion and the standard Poisson processes with different intensities. But it is wel known that the stochastic calculus on such processes is by far not a simple corolary of the corresponding calculus for Wiener and Poisson and in fact many of the fundamental propertie of the these two processes (e.g. the chaos representatio property) do not hold in the genral case.

It is well known that every von Neumann algebra can be represented as a subalgebra of the algebra $\mathcal{B}(\mathcal{H})$ of all bounded operators on some Hilbert space, and this is surely a very useful tool in the theory of operator algebras. However the structure theory of von Neumann algebras is by far not reduced to the corresponding theory for $\mathcal{B}(\mathcal{H})$. Similarly the RSWN Ito algebra (as well as of the higher order renormalized powers of white noise, which now constitute the boundary of the present theory) can be represented as a proper sub–Ito algebra of the first order Ito algebra, and this possiblity surely represents a useful tool but, just as in the case of von Neumann algebras, new structure properties arise which make the theory more complex and prevent the possibility of a naive application of the first order theory.

For these reasons, in the three years after the paper [10], several alternative attacks to the unitarity problem were developed.

Accardi, Hida and Kuo calculated in [11] the Itô table of the RSWN directly, i.e without using the first order representation (2.7)-(2.9). This was possible at the cost of introducing the Hida derivative and its formal adjoint in the representation space of the RSWN. However, as remarked by Accardi, Boukas and Kuo in [8], the introduction of this operator makes the future increments of the basic integrators, linearly dependent on the past and this creates difficulties in the deduction of the unitarity conditions.

These partial results allowed at least to produce some concrete examples of stochastic equations driven by the RSWN and with unitary solutions (cf. [8]). However, although mathematically correct, these examples were artificial and the difficulty to produce them was an indication that a solution of the problem of obtaining an intuitively transparent, practically usable and, most of all, explicitly solvable unitarity condition for the RSWN was still far from the horizon.

In the present paper a new approach to the problem is presented which is based on a technique which could be called "normal ordering in the $sl(2,\mathbb{R})$–Lie algebra. This allows to give a finite set of conditions for unitarity (cf. Proposition 3.2). As an application we discuss in detail the unitary stochastic equation satisfied by the simplest Markovian cocycles (for the time-shift) in the RSWN space: the RSWN analogue of the Weyl-Poisson operator (see section 4 for explanation of this terminology). Since the Ito table for the three basic integrators of the RSWN is not closed and its closure requires the addition of infinitely many additional integrators, it is clear from the Hudson–Parthasarathy theory that a unitary stochastic differential equation for the RSWN cannot involve only the three basic integrators but must involve infinitely many integrators.

After completion of the present paper, and using the present results as starting point, we were able to find a closed simple form of the unitarity conditions (cf. [5]). Also after completion of the present paper, it was recognized that the classical processes generated by the Weyl-Poisson operators with respect to the vacuum vector can be identified with the three non standard (i.e neither Gaussian nor Poisson) arising in the Meixner classification theorem (cf. [1] for more details).

This result supports our conjecture that the difficulties one meets in dealing with general independent increment processes reflect their "nonlinear" character, in the sense that they are related to powers of quantum white noise.

For the first power (Gaussian and Poisson) this was discovered by Hudson and Parthasarathy, for the second power (the three remaining Meixner classes) this follows from the results of Accardi, Franz and Skeide. For higher powers this is still an open problem.

2. The Itô table for the renormalized SWN

The Lie algebra of SL(2; ℝ) is the three-dimensional simple Lie algebra with basis B^+, B^-, M satisfying the commutation relations

$$[B^-, B^+] = M,\ [M, B^+] = 2B^+,\ [M, B^-] = -2B^- \tag{2.1}$$

with involution

$$(B^-)^* = B^+,\ M^* = M \tag{2.2}$$

The map ρ^+ defined by

$$\rho^+(M)\, e_n = (2n+2)\, e_n \tag{2.3}$$

$$\rho^+(B^+)\, e_n = \sqrt{(n+1)(n+2)}\, e_{n+1} \tag{2.4}$$

$$\rho^+(B^-)\, e_n = \sqrt{n(n+1)}\, e_{n-1} \tag{2.5}$$

where e_n, $n = 0, 1, 2, \cdots$ is any orthonormal basis of l_2 (the space of square integrable complex sequences), defines a representation of the $SL(2; \mathbb{R})$ Lie algebra on l_2. In this paper we will only consider this representation. Index B^+, B^- and M by time $t \geq 0$ and replace M_t by $2ct + N_t$. The RSWN algebra is the current algebra over the Lie algebra of $SL(2; \mathbb{R})$ with generators B_t^+, B_t^-, N_t and commutation relations

$$[B_t^-, B_t^+] = 2ct + N_t,\ [N_t, B_t^+] = 2B_t^+,\ [N_t, B_t^-] = -2B_t^- \tag{2.6}$$

where $c > 0$ is a constant (coming from the renormalization $\delta^2(t) = c\,\delta(t)$ of [12]). These commutation relations are uniquely extended to the algebra of finite-valued step functions by the perscription that operators, whose test functions have disjoint support, commute. It was shown in [10] that the quantum stochastic differentials dB_t^+, dB_t^-, dM_t can be expressed in terms of the classic, first order white noise quantum stochastic differentials of Hudson and Parthasarathy , defined in terms of annihilation, creation, and conservation operators A_t, $A_t^\dagger$ and Λ_t respectively on the Boson Fock space $\Gamma(L^2(\mathbb{R}, l^2(\mathbb{N}))$, through

$$dM_t = d\Lambda_t(\rho^+(M)) + dt \tag{2.7}$$

$$dB_t^+ = d\Lambda_t(\rho^+(B^+)) + dA_t^\dagger(e_0) \tag{2.8}$$

$$dB_t^- = d\Lambda_t(\rho^+(B^-)) + dA_t(e_0). \tag{2.9}$$

The Itô multiplication table for the first order differentials is

$\cdot$	$dA_t^\dagger(u)$	$d\Lambda_t(F)$	$dA_t(u)$	dt
$dA_t^\dagger(v)$	0	0	0	0
$d\Lambda_t(G)$	$dA_t^\dagger(Gu)$	$d\Lambda_t(GF)$	0	0
$dA_t(v)$	$\langle v, u\rangle dt$	$dA_t(F^*v)$	0	0
dt	0	0	0	0

The corresponding Itô multiplication table for the RSWN quantum stochastic differentials dB_t^+, dB_t^- and dM_t is not closed. In order to discuss unitarity one should consider instead processes driven by time and the generalized square of white noise quantum stochastic differentials $d\Lambda_{n,k,l}(t)$, $dA_m(t)$ and $dA_m^\dagger(t)$, where $n, k, l, m = 0, 1, \ldots$, defined by

$$d\Lambda_{n,k,l}(t) = d\Lambda_t(\rho^+((B^+)^n M^k (B^-)^l)) \tag{2.10}$$

$$dA_m(t) = dA_t(e_m) \tag{2.11}$$

$$dA_m^\dagger(t) = dA_t^\dagger(e_m) \tag{2.12}$$

In the following we will prove that the smallest Itô algebra (i.e closed under Itô multiplication) containing the stochastic differentials dB_t^+, dB_t^- and dM_t is precisely the Itô algebra generated as a vector space by the stochastic differentials (2.10)-(2.12). The difficulty in the above proof comes from the Itô product of two $d\Lambda$-differentials. This gives rise to a monomial of the form $(B^+)^{n_1} M^{k_1} (B^-)^{l_1} (B^+)^{n_2} M^{k_2} (B^-)^{l_2}$ and we want to express t his as a linear combination of monomials of the form $(B^+)^n M^k (B^-)^l$. To achieve this goal we used the *Mathematica* program.

The following lemmata will be useful in obtaining the Itô table for the generalized RSWN stochastic differentials.

Lemma 2.1. *For all* $n, k, l, m = 0, 1, 2, \ldots$

$$(B^-)^n M^k = (M + 2n)^k (B^-)^n \tag{2.13}$$

$$M^n (B^+)^k = (B^+)^k (M + 2k)^n \tag{2.14}$$

$$(B^-)^n (B^+)^k = \tag{2.15}$$

$$\sum_{m=0}^{n} \binom{n}{m} (k)^{(n-m)} (B^+)^{k-n+m} (M + k + m - 1)^{(n-m)} (B^-)^m$$

$$\rho^+((B^+)^k)\, e_m = \frac{(m+1)_{k+1}}{\sqrt{(m+1)(m+k+1)}}\, e_{m+k} \tag{2.16}$$

$$\rho^+(M^k)\, e_m = (2m+2)^k\, e_m \tag{2.17}$$

$$\rho^+((B^-)^k)\, e_m = \begin{cases} \sqrt{\frac{m-k+1}{m+1}}(m+1)^{(k)}\, e_{m-k} & m \geq k \\ 0 & m < k, \end{cases} \tag{2.18}$$

and also

$$\rho^+((B^+)^n M^k (B^-)^l)\, e_m = \theta_{n,k,l,m}\, e_{n+m-l} \tag{2.19}$$

where

$$\theta_{n,k,l,m} = H(n+m-l) \sqrt{\frac{m-l+n+1}{m+1}} \times$$
$$\times\, 2^k (m-l+1)_n (m+1)^{(l)} (m-l+1)^k$$

and

$$H(x) = \begin{cases} 1 & \textit{if } x \geq 0 \\ 0 & \textit{if } x < 0 \end{cases}$$

is the Heaviside function,

$$0^0 = 1, \ (B^+)^n = (B^-)^n = N^n = 0, \ \textit{for } n < 0$$

and "factorial powers" are defined by

$$x^{(n)} = x(x-1)\cdots(x-n+1)$$
$$(x)_n = x(x+1)\cdots(x+n-1)$$
$$(x)_0 = x^{(0)} = 1.$$

Proof. The proof follows from (2.1) and (2.3)-(2.5) with the use of mathematical induction. We will only give the proof for (2.13) and (2.16). The proof of the rest is similar, with (2.19) following from (2.16)-(2.18) and the fact that ρ^+ is a homomorphism.

If one or both of n,k is zero then (2.13) is obviously true. So assume $n,k\geq 1$. We will first show that for all $n\geq 1$

$$(B^-)^n M = (M+2n)(B^-)^n.$$

For $n=1$ the above reduces to (2.1) and is therefore true. Assume it to be true for $n=n_0$. Then for $n=n_0+1$

$$\begin{aligned}(B^-)^{n_0+1}M &= B^-(B^-)^{n_0}M = (B^-)(M+2n_0)(B^-)^{n_0}\\ &= B^-M(B^-)^{n_0}+2n_0(B^-)^{n_0+1}\\ &= (M+2)B^-(B^-)^{n_0}+2n_0(B^-)^{n_0+1}\\ &= (M+2(n_0+1))(B^-)^{n_0+1}.\end{aligned}$$

We will now show that for each n and all $k\geq 1$

$$(B^-)^n M^k = (M+2n)^k(B^-)^n$$

For $k=1$ it was just proved. Assume it to be true for $k=k_0$. Then for $k=k_0+1$

$$\begin{aligned}(B^-)^n M^{k_0+1} &= (B^-)^n M M^{k_0} = (M+2n)(B^-)^n M^{k_0}\\ &= (M+2n)(M+2n)^{k_0}(B^-)^n\\ &= (M+2n)^{k_0+1}(B^-)^n.\end{aligned}$$

Turning to (2.16), for $k=0$ it reduces to $\rho^+(I)=I$ which is true. For $k=1$ it reduces to the definition of $\rho^+(B^+)$. Assume it to be true for $k=k_0$. Then for $k=k_0+1$

$$\begin{aligned}&\rho^+((B^+)^{k_0+1})\,e_m = \rho^+((B^+)^{k_0})\rho^+(B^+)\,e_m\\ &= \rho^+((B^+)^{k_0})\sqrt{(m+1)(m+2)}\,e_{m+1}\\ &= \sqrt{(m+1)(m+2)}\frac{(m+2)_{k_0+1}}{\sqrt{(m+2)(m+2+k_0)}}\,e_{m+k_0+1}\\ &= \sqrt{\frac{m+1}{m+2+k_0}}(m+2)_{k_0+1}\,e_{m+k_0+1}\\ &= \sqrt{\frac{m+1}{m+2+k_0}}(m+2)(m+3)\cdots(m+2+k_0)\,e_{m+k_0+1}\\ &= \sqrt{m+1}\,(m+2)(m+3)\cdots(m+2+k_0-1)\times\\ &\qquad\qquad\times\sqrt{m+2+k_0}\,e_{m+k_0+1}\\ &= \frac{1}{\sqrt{(m+1)(m+(k_0+1)+1)}}(m+1)_{(k_0+1)+1}\,e_{m+k_0+1}. \qquad\square\end{aligned}$$

Lemma 2.2 (The SWN Multiplication Law). *For $\alpha,\beta,\gamma,a,b,c \in \{0,1,2,\dots\}$*

$$(B^+)^\alpha M^\beta (B^-)^\gamma (B^+)^a M^b (B^-)^c$$
$$= \sum_{\lambda=0}^{\gamma}\sum_{\rho=0}^{\gamma-\lambda}\sum_{\sigma=0}^{\gamma-\lambda-\rho}\sum_{\omega=0}^{\beta}\sum_{\epsilon=0}^{b} c_{\beta,\gamma,a,b}^{\lambda,\rho,\sigma,\omega,\epsilon}\,(B^+)^{a+\alpha-\gamma+\lambda} M^{\omega+\sigma+\epsilon}(B^-)^{\lambda+c}$$

where

$$c_{\beta,\gamma,a,b}^{\lambda,\rho,\sigma,\omega,\epsilon} = \binom{\gamma}{\lambda}\binom{\gamma-\lambda}{\rho}\binom{\beta}{\omega}\binom{b}{\epsilon} 2^{\beta+b-\omega-\epsilon} S_{\gamma-\lambda-\rho,\sigma}\, a^{(\gamma-\lambda)}$$
$$(a+\lambda-1)^{(\rho)}(a-\gamma+\lambda)^{\beta-\omega}\lambda^{b-\epsilon}.$$

Here $S_{\gamma-\lambda-\rho,\sigma}$ are the "Stirling numbers of the first kind" and $0^0 = 1$.

Proof. Recalling the binomial theorem for factorial powers of two commuting variables x, y and the connection between factorial and ordinary powers through the "Stirling numbers of the first kind" $S_{n,k}$, namely

$$(x+y)^{(n)} = \sum_{k=0}^{n}\binom{n}{k} x^{(n-k)} y^{(k)}$$
$$x^{(n)} = \sum_{k=0}^{n} S_{n,k} x^k$$

the result follows using (2.13)-(2.15) to commute powers of B^+, B^- and M. In more detail, repeated use of Lemma 2.1 yields

$$(B^+)^\alpha M^\beta (B^-)^\gamma (B^+)^a M^b (B^-)^c$$
$$= (B^+)^\alpha M^\beta \sum_{\lambda=0}^{\gamma}\binom{\gamma}{\lambda}(a)^{(\gamma-\lambda)}(B^+)^{a-\gamma+\lambda}$$
$$(M+a+\lambda-1)^{(\gamma-\lambda)}(B^-)^\lambda M^b (B^-)^c$$
$$= \sum_{\lambda=0}^{\gamma}\binom{\gamma}{\lambda}(a)^{(\gamma-\lambda)}(B^+)^\alpha M^\beta (B^+)^{a-\gamma+\lambda}$$
$$(M+a+\lambda-1)^{(\gamma-\lambda)}(B^-)^\lambda M^b (B^-)^c$$
$$= \sum_{\lambda=0}^{\gamma}\binom{\gamma}{\lambda}(a)^{(\gamma-\lambda)}(B^+)^{\alpha+a-\gamma+\lambda}(M+2(a-\gamma+\lambda))^\beta$$
$$(M+a+\lambda-1)^{(\gamma-\lambda)}(B^-)^\lambda M^b (B^-)^c$$

$$= \sum_{\lambda=0}^{\gamma} \binom{\gamma}{\lambda} (a)^{(\gamma-\lambda)} (B^+)^{\alpha+a-\gamma+\lambda} (M + 2(a-\gamma+\lambda))^{\beta}$$

$$(M + a + \lambda - 1)^{(\gamma-\lambda)} (M + 2\lambda)^b (B^-)^{\lambda+c}$$

$$= \sum_{\lambda=0}^{\gamma} \binom{\gamma}{\lambda} (a)^{(\gamma-\lambda)} (B^+)^{\alpha+a-\gamma+\lambda} \sum_{\omega=0}^{\beta} \binom{\beta}{\omega} M^{\omega} 2^{\beta-\omega} (a-\gamma+\lambda)^{\beta-\omega}$$

$$\sum_{\rho=0}^{\gamma-\lambda} \binom{\gamma-\lambda}{\rho} (a+\lambda-1)^{(\rho)} M^{(\gamma-\lambda-\rho)} \sum_{\epsilon=0}^{b} \binom{b}{\epsilon} M^{\epsilon} 2^{b-\epsilon} \lambda^{b-\epsilon} (B^-)^{\lambda+c}$$

$$= \sum_{\lambda=0}^{\gamma} \sum_{\omega=0}^{\beta} \sum_{\rho=0}^{\gamma-\lambda} \sum_{\epsilon=0}^{b} \binom{\gamma}{\lambda} (a)^{(\gamma-\lambda)} \binom{\beta}{\omega} 2^{\beta-\omega} (a-\gamma+\lambda)^{\beta-\omega}$$

$$\binom{\gamma-\lambda}{\rho} (a+\lambda-1)^{(\rho)} \binom{b}{\epsilon} 2^{b-\epsilon} \lambda^{b-\epsilon} M^{\omega} M^{(\gamma-\lambda-\rho)} M^{\epsilon} (B^-)^{\lambda+c}$$

and the result follows using the Stirling numbers to expand $M^{(\gamma-\lambda-\rho)}$ in terms of ordinary powers. □

Remark 2.3. By Lemma 2.2 the highest power of B^+, M, B appearing in the formula for

$$(B^+)^{\alpha} M^{\beta} (B^-)^{\gamma} (B^+)^{a} M^{b} (B^-)^{c}$$

is $a + \alpha$, $\beta + b + \gamma$, $\gamma + c$ respectively.

Proposition 2.4 (The SWN Itô Table). *If* $\alpha, \beta, \gamma, a, b, c \in \{0, 1, \dots\}$

$$d\Lambda_{\alpha,\beta,\gamma}(t)\, d\Lambda_{a,b,c}(t) = \sum c_{\beta,\gamma,a,b}^{\lambda,\rho,\sigma,\omega,\epsilon}\, d\Lambda_{a+\alpha-\gamma+\lambda,\omega+\sigma+\epsilon,\lambda+c}(t) \quad (2.20)$$

$$d\Lambda_{\alpha,\beta,\gamma}(t)\, dA_n^{\dagger}(t) = \theta_{\alpha,\beta,\gamma,n}\, dA_{\alpha+n-\gamma}^{\dagger}(t) \quad (2.21)$$

$$dA_m(t)\, d\Lambda_{a,b,c}(t) = \theta_{c,b,a,m}\, dA_{c+m-a}(t) \quad (2.22)$$

$$dA_m(t)\, dA_n^{\dagger}(t) = \delta_{m,n}\, dt \quad (2.23)$$

where

$$\sum = \sum_{\lambda=0}^{\gamma} \sum_{\rho=0}^{\gamma-\lambda} \sum_{\sigma=0}^{\gamma-\lambda-\rho} \sum_{\omega=0}^{\beta} \sum_{\epsilon=0}^{b}$$

All other products are equal to zero.

Proof. We will only prove (2.20). The proof of (2.21), (2.22) and (2.23) is similar. By (2.10), the Itô table following (2.9), and Lemma 2.2

$$\begin{aligned}
&d\Lambda_{\alpha,\beta,\gamma}(t)\, d\Lambda_{a,b,c}(t) \\
&= d\Lambda_t(\rho^+((B^+)^\alpha M^\beta (B^-)^\gamma))\, d\Lambda_t(\rho^+((B^+)^a M^b (B^-)^c)) \\
&= d\Lambda_t(\rho^+((B^+)^\alpha M^\beta (B^-)^\gamma)\, \rho^+((B^+)^a M^b (B^-)^c)) \\
&= d\Lambda_t(\rho^+((B^+)^\alpha M^\beta (B^-)^\gamma (B^+)^a M^b (B^-)^c)) \\
&= d\Lambda_t(\rho^+(\textstyle\sum c^{\lambda,\rho,\sigma,\omega,\epsilon}_{\beta,\gamma,a,b}\, (B^+)^{a+\alpha-\gamma+\lambda} M^{\omega+\sigma+\epsilon} (B^-)^{\lambda+c})) \\
&= \textstyle\sum c^{\lambda,\rho,\sigma,\omega,\epsilon}_{\beta,\gamma,a,b}\, d\Lambda_t(\rho^+((B^+)^{a+\alpha-\gamma+\lambda} M^{\omega+\sigma+\epsilon} (B^-)^{\lambda+c})) \\
&= \textstyle\sum c^{\lambda,\rho,\sigma,\omega,\epsilon}_{\beta,\gamma,a,b}\, d\Lambda_{a+\alpha-\gamma+\lambda,\omega+\sigma+\epsilon,\lambda+c}(t)
\end{aligned}$$

□

3. The Unitarity Conditions

Consider the quantum stochastic differential equation, with constant coefficients acting on a system Hilbert space H_0,

$$dU(t) = \Big(A\, dt + \sum_{n,k,l=0}^{+\infty} B_{n,k,l}\, d\Lambda_{n,k,l}(t) + \sum_{m=0}^{+\infty} C_m\, dA_m(t) + \sum_{m=0}^{+\infty} D_m\, dA^\dagger_m(t)\Big)\, U(t),\;\; U(0) = I,\; 0 \le t \le t_0 < +\infty \tag{3.1}$$

interpreted as an $H_0 \otimes \Gamma$ (where Γ denotes Boson Fock space) quantum stochastic differential equation with infinite degrees of freedom in a way similar to that of [22], with adjoint

$$dU^*(t) = U^*(t)\Big(A^*\, dt + \sum_{n,k,l=0}^{+\infty} B^*_{n,k,l}\, d\Lambda_{l,k,n}(t) + \sum_{m=0}^{+\infty} C^*_m\, dA^\dagger_m(t) + \sum_{m=0}^{+\infty} D^*_m\, dA_m(t)\Big),\;\; U^*(0) = I,\; 0 \le t \le t_0 < +\infty \tag{3.2}$$

From first order stochastic calculus it is known (see [22]) that, under certain summability conditions on its coefficients, equation (3.1) admits a unique solution. The precise estimates, involving the specific structure of the integrators considered here, will appear elsewhere.

Proposition 3.1 (Necessary and sufficient unitarity conditions). *The solution $U = \{U(t)\,/\,t \ge 0\}$ of (3.1) is unitary, i.e $U(t)\, U^*(t) = U^*(t)\, U(t) = I$ for each $t \ge 0$, if and only if the coefficient operators* *satisfy*

$$A + A^* + \sum_{m=0}^{+\infty} D_m^* D_m = 0 \tag{3.3}$$

$$A + A^* + \sum_{m=0}^{+\infty} C_m C_m^* = 0, \tag{3.4}$$

for each $m = 0, 1, 2, \ldots$

$$C_m + D_m^* + \sum_{n,l=0}^{+\infty} D_{m+n-l}^* \sum_{k=0}^{+\infty} \theta_{l,k,n,m+n-l} B_{n,k,l} = 0 \tag{3.5}$$

$$C_m + D_m^* + \sum_{n,l=0}^{+\infty} C_{m+l-n} \sum_{k=0}^{+\infty} \theta_{n,k,l,m+l-n} B_{n,k,l}^* = 0, \tag{3.6}$$

and for each $n, k, l = 0, 1, 2, \ldots$

$$B_{n,k,l} + B_{l,k,n}^* + \sum_{\alpha,\beta,\gamma,a,b,c=0}^{n,k,\min(k,l),l,k,n} B_{\alpha,\beta,\gamma} B_{a,b,c}^* \, g_{\alpha,a,\beta,\gamma,c,b}^{n,k,l} = 0 \tag{3.7}$$

$$B_{n,k,l} + B_{l,k,n}^* + \sum_{\alpha,\beta,\gamma,a,b,c=0}^{\min(k,l),k,n,n,k,l} B_{\alpha,\beta,\gamma}^* B_{a,b,c} \, g_{\gamma,c,\beta,\alpha,a,b}^{n,k,l} = 0 \tag{3.8}$$

where, with δ *denoting Kronecker's delta,*

$$g_{x,y,z,X,Y,Z}^{n,k,l} = \sum_{\lambda=0}^{X} \sum_{\rho=0}^{X-\lambda} \sum_{\sigma=0}^{X-\lambda-\rho} \sum_{\omega=0}^{z} \sum_{\epsilon=0}^{Z} \delta_{x+Y-X+\lambda,n} \, \delta_{\omega+\sigma+\epsilon,k} \, \delta_{\lambda+y,l} \, c_{z,X,Y,Z}^{\lambda,\rho,\sigma,\omega,\epsilon}$$

and

$$\sum_{\alpha,\beta,\gamma,a,b,c=0}^{n,k,\min(k,l),l,k,n}$$

means that α *ranges from* 0 *to* n, β *ranges from* 0 *to* k *e.t.c with a similar interpretation for*

$$\sum_{\alpha,\beta,\gamma,a,b,c=0}^{\min(k,l),k,n,n,k,l}$$

Proof. In the theory of quantum stochastic differential equations, one obtains sufficient unitarity conditions for stochastic evolutions driven by quantum noise by starting with the definition of unitarity

$$U(t)\,U^*(t) = U^*(t)\,U(t) = I\,,\; U(0) = U^*(0) = I$$

which is equivalent to

$$d\,(U(t)U^*(t)) = dU(t)\,U^*(t) + U(t)\,dU^*(t) + dU(t)\,dU^*(t) = 0$$

and

$$d\,(U^*(t)U(t)) = dU^*(t)\,U(t) + U^*(t)\,dU(t) + dU^*(t)\,dU(t) = 0$$

replacing $dU(t)$ and $dU^*(t)$ by (in the SWN case) (3.1) and (3.2), using the Itô multiplication rule of Proposition 2.4 to multiply the stochastic differentials, and then equating coefficients of the time and noise differentials to zero. In the SWN case this method yields (3.3)-(3.8) as sufficient conditions for the unitarity of U. In view of the linear independence of the generalized SWN stochastic differentials conditions (3.3)-(3.8) are also necessary for the unitarity of U. In more detail, by (3.1) and (3.2),

$$d\,(U(t)U^*(t)) = 0$$

implies

$$\begin{aligned}
&\Big[A\,dt + \sum_{n,k,l=0}^{+\infty} B_{n,k,l}\,d\Lambda_{n,k,l}(t) + \sum_{m=0}^{+\infty} C_m\,dA_m(t)\\
&\qquad\qquad + \sum_{m=0}^{+\infty} D_m\,dA_m^\dagger(t)\Big]U(t)U^*(t)\\
&+U(t)U^*(t)\Big[A^*\,dt + \sum_{n,k,l=0}^{+\infty} B^*_{n,k,l}\,d\Lambda_{l,k,n}(t)\\
&\qquad\qquad + \sum_{m=0}^{+\infty} C^*_m\,dA^\dagger_m(t) + \sum_{m=0}^{+\infty} D^*_m\,dA_m(t)\Big]\\
&+\Big[A\,dt + \sum_{n,k,l=0}^{+\infty} B_{n,k,l}\,d\Lambda_{n,k,l}(t) + \sum_{m=0}^{+\infty} C_m\,dA_m(t)\\
&+\sum_{m=0}^{+\infty} D_m\,dA^\dagger_m(t)\Big]U(t)U^*(t)\Big[A^*\,dt + \sum_{n,k,l=0}^{+\infty} B^*_{n,k,l}\,d\Lambda_{l,k,n}(t)\\
&\qquad\qquad + \sum_{m=0}^{+\infty} C^*_m\,dA^\dagger_m(t) + \sum_{m=0}^{+\infty} D^*_m\,dA_m(t)\Big] = 0
\end{aligned}$$

which using $U(t)\,U^*(t) = I$ can be written as

$$\begin{aligned}
&(A + A^*)dt \\
&+ \sum_{n,k,l=0}^{+\infty} (B_{n,k,l} + B^*_{l,k,n})\, d\Lambda_{n,k,l}(t) \\
&+ \sum_{m=0}^{+\infty} (C_m + D^*_m)\, dA_m(t) + \sum_{m=0}^{+\infty} (D_m + C^*_m)\, dA^\dagger_m(t) \\
&+ \sum_{n,k,l=0}^{+\infty} \sum_{n_0,k_0,l_0=0}^{+\infty} B_{n,k,l} B^*_{n_0,k_0,l_0} hid\Lambda_{n,k,l}(t)\, d\Lambda_{l_0,k_0,n_0}(t) \\
&+ \sum_{m,n,k,l=0}^{+\infty} B_{n,k,l} C^*_m\, d\Lambda_{n,k,l}(t)\, dA^\dagger_m(t) \\
&+ \sum_{m=0}^{+\infty} \sum_{m_0=0}^{+\infty} C_m C^*_{m_0}\, dA_m(t)\, dA^\dagger_{m_0}(t) \\
&+ \sum_{m,n,k,l=0}^{+\infty} C_m B^*_{l,k,n}\, dA_m(t)\, d\Lambda_{n,k,l}(t) = 0
\end{aligned}$$

which by Proposition 2.4 implies

$$\begin{aligned}
&\left(A + A^* + \sum_{m=0}^{+\infty} C_m C^*_m \right) dt \\
&+ \sum_{n,k,l=0}^{+\infty} (B_{n,k,l} + B^*_{l,k,n})\, d\Lambda_{n,k,l}(t) \\
&+ \sum_{\substack{\alpha,\beta,\gamma=0 \\ a,b,c=0}}^{+\infty} B_{\alpha,\beta,\gamma} B^*_{a,b,c} \sum_{\lambda,\rho,\sigma,\omega,\epsilon} c^{\lambda,\rho,\sigma,\omega,\epsilon}_{\beta,\gamma,c,b}\, d\Lambda_{\alpha+c-\gamma+\lambda,\omega+\sigma+\epsilon,\lambda+a}(t) \\
&\quad + \sum_{m=0}^{+\infty} (C_m + D^*_m)\, dA_m(t) \\
&\quad + \sum_{m,n,k,l=0}^{+\infty} C_m B^*_{l,k,n} \theta_{l,k,n,m}\, dA_{l+m-n}(t) \\
&\quad + \sum_{m=0}^{+\infty} (D_m + C^*_m)\, dA^\dagger_m(t) \\
&\quad + \sum_{m,n,k,l=0}^{+\infty} B_{n,k,l} C^*_m \theta_{n,k,l,m}\, dA^\dagger_{n+m-l}(t) = 0
\end{aligned}$$

and by reindexing we obtain

$$\left(A + A^* + \sum_{m=0}^{+\infty} C_m C_m^*\right) dt + \sum_{n,k,l=0}^{+\infty} (B_{n,k,l} + B_{l,k,n}^*$$

$$+ \sum_{\alpha,\beta,\gamma,a,b,c=0}^{+\infty} \sum_{\lambda,\rho,\sigma,\omega,\epsilon} c_{\beta,\gamma,c,b}^{\lambda,\rho,\sigma,\omega,\epsilon} B_{\alpha,\beta,\gamma} B_{a,b,c}^*) \, d\Lambda_{n,k,l}(t)$$

$$+ \sum_{m=0}^{+\infty} (C_m + D_m^* + \sum_{n,k,l=0}^{+\infty} C_{m+n-l} B_{l,k,n}^* \theta_{l,k,n,m+n-l}) \, dA_m(t)$$

$$+ \sum_{m=0}^{+\infty} (D_m + C_m^* + \sum_{n,k,l=0}^{+\infty} B_{n,k,l} C_{m+l-n}^* \theta_{n,k,l,m+l-n}) \, dA_m^\dagger(t) = 0$$

where $\sum_{\lambda,\rho,\sigma,\omega,\epsilon}$ is over all $\lambda, \rho, \sigma, \omega, \epsilon$ such that

$$\alpha + c - \gamma + \lambda = n$$
$$\omega + \sigma + \epsilon = k$$
$$\lambda + a = l$$

In view of Remark 2.3 we may replace $\sum_{\alpha,\beta,\gamma,a,b,c=0}^{+\infty}$ by the finite sum appearing in (3.7). By equating coefficients to zero we obtain (3.4), (3.6) and its adjoint, and (3.7). Similarly, starting with

$$d(U^*(t)U(t)) = 0$$

we obtain (3.3), (3.5) and its adjoint, and (3.8). □

Proposition 3.2 (Matrix form of the unitarity conditions). *Unitarity conditions* (3.3)-(3.8) *can be put in the matrix form*

$$A + A^* + D^\dagger D = 0 \tag{3.9}$$

$$A + A^* + CC^\dagger = 0 \tag{3.10}$$

$$C + D^\dagger + \theta \Delta^\dagger \hat{B} = 0 \tag{3.11}$$

$$C + D^\dagger + \theta \Gamma \hat{E} = 0 \tag{3.12}$$

$$\tilde{B} + \tilde{E} + \bar{B} G \hat{E} = 0 \tag{3.13}$$

$$\tilde{B} + \tilde{E} + \bar{E} G \hat{B} = 0 \tag{3.14}$$

where (in standard vector and matrix notation, denoting $\delta(x_0, x_1, \ldots)$ *the diagonal matrix with main diagonal* $x_0, x_1, \ldots$*, denoting dual operator by using the superscript* $*$*, transpose by* $\top$ *and conjugate transpose* *by* †)

$$C=(C_0,C_1,\dots), \qquad D=(D_0,D_1,\dots)^\top,$$
$$\hat{B}=\delta(B,B,\dots), \qquad \bar{B}=\delta(B^\top,B^\top,\dots),$$
$$\Delta=\delta(\Delta_0,\Delta_1,\dots), \qquad \hat{E}=\delta(E,E,\dots),$$
$$\Gamma=\delta(\Gamma_0,\Gamma_1,\dots), \qquad \theta=(\theta_0,\theta_1,\dots),$$
$$B=(B_0,B_1,\dots)^\top, \qquad E=(E_0,E_1,\dots)^\top,$$
$$\tilde{B}=\delta(B_0,B_1,\dots), \qquad \tilde{E}=\delta(E_0,E_1,\dots),$$

$$G=\delta(g^{0,0,0},g^{0,1,0},\dots,g^{0,0,1},\\ g^{0,1,1},\dots,g^{1,0,0},g^{1,1,0},\dots,g^{1,0,1},\dots,g^{1,1,1},\dots)$$

and for $n,k,l,m,a,b,c,\alpha,\beta,\gamma\in\{0,1,2,\dots\}$

$$\Delta_m=\delta(D_0(m),D_1(m),\dots),$$

$$\Gamma_m=\delta(C_0(m),C_1(m),\dots),$$

$$C_n(m)=\delta(C_{m+n},C_{m+n-1},C_{m+n-2},\dots),$$

$$D_n(m)=\delta(D_{m+n},D_{m+n-1},D_{m+n-2}\cdots),$$

$$\theta_m=(\theta_0(m),\theta_1(m),\dots),$$

$$E_n=(E_{n,0},E_{n,1},\dots)^\top, \quad B_n=(B_{n,0},B_{n,1},\dots)^\top,$$

$$\theta_n(m)=(\theta_{0,n}(m),\theta_{1,n}(m),\dots),$$

$$\theta_{l,n}(m)=(\theta_{l,0,n,m+n-l},\theta_{l,1,n,m+n-l},\dots),$$

$$B_{n,l}=(B_{n,0,l},B_{n,1,l},\dots)^\top, \quad E_{n,l}=(E_{n,0,l},E_{n,1,l},\dots.)^\top,$$

$$E_{l,k,n}=B^*_{n,k,l},$$

$$g^{n,k,l}=(g_0^{n,k,l},g_1^{n,k,l},\dots), \quad g_c^{n,k,l}=(g_{0,c}^{n,k,l},g_{1,c}^{n,k,l},\dots),$$

$$g_{a,c}^{n,k,l}=(g_{a,c,0}^{n,k,l},g_{a,c,1}^{n,k,l},\dots)$$

$$g_{a,c,b}^{n,k,l}=(g_{0,a,c,b}^{n,k,l},g_{1,a,c,b}^{n,k,l},\dots)^\top, \quad g_{\alpha,a,c,b}^{n,k,l}=(g_{\alpha,a,0,c,b}^{n,k,l},g_{\alpha,a,1,c,b}^{n,k,l},\dots)^\top,$$

$$g_{\alpha,a,\gamma,c,b}^{n,k,l}=(g_{\alpha,a,0,\gamma,c,b}^{n,k,l},g_{\alpha,a,1,\gamma,c,b}^{n,k,l},\dots)^\top.$$

Proof. We will only show that (3.5) can be written as (3.11), the proof of the rest of (3.9)-(3.14) is similar. To that end, we notice that (3.5) can be written as

$$C_m + D_m^* + \sum_{n,l} D_{m+n-l}^* \theta_{l,n}(m) B_{n,l}$$
$$= C_m + D_m^* + \sum_n \theta_n(m) D_n^\dagger(m) B_n = C_m + D_m^* + \theta_m \Delta_m^\dagger B = 0$$

for all m, which implies (3.11). □

Corollary 3.3 (Compatibility of the unitarity conditions). *In order for the pairs* (3.9)-(3.10), (3.11)-(3.12), *and* (3.13)-(3.14) *to be compatible it is necessary that*

$$D^\dagger D = CC^\dagger \tag{3.15}$$
$$\Delta^\dagger \hat{B} = \Gamma \hat{E} \tag{3.16}$$
$$\bar{B} G \hat{E} = \bar{E} G \hat{B}. \tag{3.17}$$

Proof. The proof follows by a direct comparison of (3.9)-(3.10), (3.11)-(3.12), and (3.13)-(3.14). □

Corollary 3.4. *Let the coefficients* A_i, $i = 1, 2, 3, 4$ *of the quantum stochastic differential equation*

$$\begin{aligned} dU(t) &= (A_1\, dt + A_2\, dB_t^- + A_3\, dB_t^+ + A_4\, dM_t)U(t) \\ U(0) &= I \end{aligned} \tag{3.18}$$

be time independent bounded operators on the system space. Then conditions (3.3)-(3.8) *are satisfied if and only if, denoting real part by* Re,

$$\operatorname{Re} A_1 = A_2 = A_3 = A_4 = 0. \tag{3.19}$$

Therefore (3.18) *admits a unitary solution if and only if its coefficients satisfy* (3.19).

Proof. In view of (2.7)-(2.9), (3.18) can be written as

$$dU(t) = ((A_1 + A_4)\, dt + A_2\, dA_0(t) + A_3\, dA_0^\dagger(t) + A_2\, d\Lambda_{0,0,1}(t)$$
$$+ A_3\, d\Lambda_{1,0,0}(t) + A_4\, d\Lambda_{0,1,0}(t))U(t),\ U(0) = I$$

which is of the form of (3.1) with

$$A = A_1 + A_4,\ C_0 = A_2,\ D_0 = A_3,\ B_{0,0,1} = A_2,\ B_{1,0,0} = A_3,\ B_{0,1,0} = A_4$$

Attempting to satisfy (3.3)-(3.8) we find that for $(n,k,l) = (0,2,0)$, (3.7) implies

$$B_{0,1,0}B_{0,1,0}^* = A_4A_4^* = 0$$

i.e. $A_4 = 0$. Similarly, for $(n,k,l) = (1,0,1)$, (3.7) and (3.8) imply

$$A_3A_3^* = 0$$
$$A_2A_2^* = 0$$

i.e. $A_2 = A_3 = 0$. Finally by (3.3) $\operatorname{Re} A_1 = 0$. Thus (3.18) reduces to

$$dU(t) = iH\,dt\,U(t),\ U(0) = I$$

where H is self-adjoint, with solution $U(t) = e^{iHt}$. □

Corollary 3.5. *The quantum stochastic differential equation*

$$\begin{aligned} dU(t) &= (a_0\,dt + a_1\,dA_0(t) + a_2\,dA_0^\dagger(t) \\ &\qquad + a_3\,d\Lambda_{0,0,0,}(t) + a_4\,dB_t^- + a_5\,dB_t^+ + a_6\,dM_t)U(t) \\ U(0) &= I \end{aligned} \tag{3.20}$$

containing first and second order white noise terms with (as in Corollary 3.4) constant operator coefficients a_i, $i = 0,1,2,\ldots,6$, does not admit a unitary solution unless $a_4 = a_5 = a_6 = 0$ in which case it is reduced to a standard, Hudson-Parthasarathy type, first order white noise quantum stochastic differential equation (see [22]).

Proof. For $(n,k,l) = (1,0,1)$, (3.7) and (3.8) imply

$$a_5a_5^* = 0$$
$$a_4a_4^* = 0$$

i.e. $a_5 = a_4 = 0$. Similarly, for $(n,k,l) = (2,0,0)$, (3.7) implies

$$a_6a_6^* = 0$$

i.e. $a_6 = 0$. □

4. The SWN analogue of the Poisson-Weyl operator

Proposition 4.1. *Let $\lambda, k \in \mathbb{R}$ and $z \in \mathbb{C}$, with $|z|, |k|$ less than a sufficiently small positive number. We call "Poisson-Weyl operator" any operator of the form*

$$\begin{aligned} E(t) &= \lambda\,t + z\,B_t^- + \bar{z}\,B_t^+ + k\,M_t \\ &= (\lambda + k)\,t + z\,A_0(t) + \bar{z}\,A_0^\dagger(t) \\ &\qquad + z\,\Lambda_{0,0,1}(t) + \bar{z}\,\Lambda_{1,0,0}(t) + k\,\Lambda_{0,1,0}(t). \end{aligned} \tag{4.1}$$

This terminology is justified by the fact that if B_t^-, B_t^+, *and* M_t *are the usual first order integrators, then the process (4.1) is a classical Poisson process expressed in terms of Weyl operators. Consider* $U = \{U(t) = \mathrm{e}^{\mathrm{i}\, E(t)}, t \geq 0\}$ *where* $E(t)$ *is given by* (4.1). *Then* U *is a unitary process satisfying*

$$dU(t) = U(t)[\tau(\lambda, z, k)\, dt + \sum_{m=0}^{+\infty} [a_m(z,k)\, dA_m(t) + \bar{a}_m(z,k)\, dA_m^\dagger(t)] + \sum_{0<i+j+r<+\infty} l_{i,j,r}(z,k)\, d\Lambda_{i,j,r}(t)] \tag{4.2}$$

where the coefficients $\tau(\lambda, z, k)$, $a_m(z,k)$, $\bar{a}_m(z,k)$, *and* $l_{i,j,r}(z,k)$ *are given by*

$$\tau(\lambda, z, k) = i\lambda - |z|^2/2 + \sum_{n=3}^{+\infty} \frac{i^n}{n!} \sum_{\alpha=0}^{n-2} \Big[\sum_{\substack{j_1,\dots,j_{n-2} \in \{-1,0,1\} \\ j_1+\dots+j_{n-2}=0}} \prod_{\epsilon=1}^{n-2} \hat{\theta}_{\epsilon, j_\epsilon}(0) \Big](\alpha) |z|^{2(\alpha+1)} k^{n-2(\alpha+1)} \tag{4.3}$$

$$a_m(z,k) = iz + \sum_{n=2}^{+\infty} \frac{i^n}{n!} \sum_{\alpha=0}^{n-1} \Big[\sum_{\substack{j_1,\dots,j_{n-1} \in \{-1,0,1\} \\ m+j_1+\dots+j_{n-1}=0}} \prod_{\epsilon=1}^{n-1} \hat{\theta}_{\epsilon, j_\epsilon}(m) \Big](\alpha) |z|^{2\alpha} z^{m+1} k^{n-2\alpha-m-1} \tag{4.4}$$

$$\bar{a}_m(z,k) = \overline{a_m(z,k)} \quad \textit{(the complex conjugate of } a_m(z,k))$$

$$l_{v,j,r}(z,k) = \phi_{v,j,r}(z,k) \times \sum_{n=1}^{+\infty} i^n/n! \sum_{1} \cdots \sum_{n-1} \prod_{s=1}^{n-1} [1 - \delta_{\epsilon_s,1}(\delta_{q_s,-1} + \delta_{q_s,0})] \cdot \phi_{v+\gamma_{n-s}-r+\sum_{\lambda=1}^{n-s} q_\lambda, \beta_{n-s}, \gamma_{n-s}}(z,k) \cdot \hat{c}_{\beta_s, \gamma_s, \delta_{q_s,-1}, \delta_{q_s,0}}^{\gamma_{s-1}-\delta_{q_s,1}, \rho_s, \beta_{s-1}-\omega_s-\delta_{q_s,0}\epsilon_s, \omega_s, \delta_{q_s,0}\epsilon_s} \tag{4.5}$$

where for $\xi \in \{1, 2, \dots, n-1\}$

$$\sum_{\xi} = \sum_{\substack{q_\xi \in \{-1,0,1\},\ 0 \leq \beta_\xi \leq n-\xi \\ 0 \leq \gamma_\xi \leq n-\xi,\ 0 \leq \omega_\xi \leq \beta_\xi,\ 0 \leq \epsilon_\xi \leq 1 \\ 0 \leq \rho_\xi \leq \gamma_\xi - \gamma_{\xi-1} + \delta_{q_\xi,1}}}$$

with $\gamma_0 = r$ *and* $\beta_0 = j$, *and*

$$\hat{\theta}_{\epsilon,1}(m) = \theta_{0,0,1,m+1+\sum_{\lambda=1}^{\epsilon-1} j_\lambda} = \sqrt{(m+1+\sum_{\lambda=1}^{\epsilon-1} j_\lambda)(m+2+\sum_{\lambda=1}^{\epsilon-1} j_\lambda)} \tag{4.6}$$

$$\hat{\theta}_{\epsilon,-1}(m) = \theta_{1,0,0,m-1+\sum_{\lambda=1}^{\epsilon-1} j_\lambda} = \sqrt{(m-1+\sum_{\lambda=1}^{\epsilon-1} j_\lambda)(m+\sum_{\lambda=1}^{\epsilon-1} j_\lambda)} \tag{4.7}$$

$$\hat{\theta}_{\epsilon,0}(m) = \theta_{0,1,0,m+\sum_{\lambda=1}^{\epsilon-1} j_\lambda} = 2\Big(m+1+\sum_{\lambda=1}^{\epsilon-1} j_\lambda\Big) \tag{4.8}$$

$$\hat{c}^{\lambda,\rho,\sigma,\omega,\epsilon}_{\beta,\gamma,a,b} = \begin{cases} c^{\lambda,\rho,\sigma,\omega,\epsilon}_{\beta,\gamma,a,b} & \text{if } 0 \le \lambda \le \gamma, 0 \le \rho \le \gamma - \lambda, \\ & 0 \le \sigma \le \gamma - \lambda - \rho, \\ & 0 \le \omega \le \beta, 0 \le \epsilon \le b, \\ 0 & \text{otherwise} \end{cases} \tag{4.9}$$

$$\phi_{I,J,K}(z,k) = \begin{cases} \bar{z} & \text{if } I = 1,\ J = K = 0, \\ k & \text{if } J = 1,\ I = K = 0, \\ z & \text{if } K = 1,\ I = J = 0, \\ 0 & \text{otherwise} \end{cases} \tag{4.10}$$

δ *denotes Kronecker's delta,* $c^{\lambda,\rho,\sigma,\omega,\epsilon}_{\beta,\gamma,a,b}$ *is as in Lemma 2.2,* θ *is as in Lemma 2.1, and the dependence on* α *in (4.3) (resp. (4.4)) is in the sense that* α j_ϵ*'s are equal to 1,* α *(resp.* $\alpha + m$*)* j_ϵ*'s are equal to* -1*, and* $n - 2\alpha - 2$ *(resp.* $n - 2\alpha - m - 1$*)* j_ϵ*'s are equal to* 0.

Proof. Computing the differential of $U(t)$ we find

$$\begin{aligned} dU(t) &= d(\mathrm{e}^{\mathrm{i}\,E(t)}) \\ &= \mathrm{e}^{\mathrm{i}\,E(t+dt)} - \mathrm{e}^{\mathrm{i}\,E(t)} \\ &= \mathrm{e}^{\mathrm{i}\,(E(dt)+E(t))} - \mathrm{e}^{\mathrm{i}\,E(t)} \\ &= \mathrm{e}^{\mathrm{i}\,E(dt)}\,\mathrm{e}^{\mathrm{i}\,E(t)} - \mathrm{e}^{\mathrm{i}\,E(t)} \quad (E(dt) \text{ and } E(t) \text{ commute}) \\ &= \mathrm{e}^{\mathrm{i}\,E(t)}\,[\mathrm{e}^{\mathrm{i}\,dE(t)} - I] \\ &= U(t) \sum_{n=1}^{\infty} \frac{(i\,dE(t))^n}{n!}\,. \end{aligned}$$

By Proposition 2.4

$$dE(t)^n = \tau_n(\lambda, z, k)\, dt + \sum_{m=0}^{n-1} a_{m,n}(z, k)\, dA_m(t) + \sum_{m=0}^{n-1} \bar{a}_{m,n}(z, k)\, dA_m^\dagger(t) + \sum_{0<i+j+k\leq n} l_{i,j,k,n}(z, k)\, d\Lambda_{i,j,k}(t)$$

for some coefficients $\tau_n(\lambda, z, k)$, $a_{m,n}(z, k)$, $\bar{a}_{m,n}(z, k)$ and $l_{i,j,k,n}(z, k)$. We will obtain recursive relations satisfied by these coefficients and by iterating these recursions we will derive explicit formulas for each one of them.

Again by Proposition 2.4

$$\begin{aligned} dA_{m+1}(t)\, d\Lambda_{1,0,0}(t) &= \theta_{0,0,1,m+1}\, dA_m(t) \\ dA_m(t)\, d\Lambda_{0,1,0}(t) &= \theta_{0,1,0,m}\, dA_m(t) \\ dA_{m-1}(t)\, d\Lambda_{0,0,1}(t) &= \theta_{1,0,0,m-1}\, dA_m(t) \end{aligned}$$

Thus, based on the right multiplication by $dE(t)$ recursive scheme,

$$\begin{aligned} dE(t)^n &= dE(t)^{n-1}(t)\, dE(t) \\ &= dE(t)^{n-1}(t)((\lambda + k)\, dt + z\, dA_0(t) + \bar{z}\, dA_0^\dagger(t) \\ &\quad + z\, d\Lambda_{0,0,1}(t) + \bar{z}\, d\Lambda_{1,0,0}(t) + k\, d\Lambda_{0,1,0}(t)) \end{aligned}$$

we obtain

$$a_{m,n}(z, k) = \bar{z}\, \theta_{0,0,1,m+1} a_{m+1,n-1}(z, k) + k\, \theta_{0,1,0,m} a_{m,n-1}(z, k) + z\, \theta_{1,0,0,m-1} a_{m-1,n-1}(z, k) \tag{4.11}$$

with

$$a_{0,1}(z, k) = z. \tag{4.12}$$

We can write (4.12) as

$$\begin{aligned} a_{m,n}(z, k) &= \sum_{j_1 \in \{-1,0,1\}} c_{1,j_1} a_{m+j_1,n-1}(z, k) \\ &= \sum_{j_1,j_2 \in \{-1,0,1\}} c_{1,j_1} c_{2,j_2} a_{m+j_1+j_2,n-2}(z, k) \\ &\;\;\vdots \\ &= \sum_{j_1,\ldots,j_{n-1} \in \{-1,0,1\}} c_{1,j_1} c_{2,j_2} \cdots c_{n-1,j_{n-1}} a_{m+j_1+\cdots+j_{n-1},1}(z, k) \end{aligned} \tag{4.13}$$

where

$$c_{1,1} = \bar{z}\,\theta_{0,0,1,m+1},\ c_{1,0} = k\,\theta_{0,1,0,m},\ c_{1,-1} = z\,\theta_{1,0,0,m-1}$$
$$c_{2,1} = \bar{z}\,\theta_{0,0,1,m+j_1+1},\ c_{2,0} = k\,\theta_{0,1,0,m+j_1},\ c_{2,-1} = z\,\theta_{1,0,0,m+j_1-1}$$
$$\vdots$$
$$c_{n-1,1} = \bar{z}\,\theta_{0,0,1,m+1+\sum_{q=1}^{n-2} j_q},$$
$$c_{n-1,0} = k\,\theta_{0,1,0,m+\sum_{q=1}^{n-2} j_q},\ c_{n-1,-1} = z\,\theta_{1,0,0,m-1+\sum_{q=1}^{n-2} j_q}$$

In view of (4.13) we only keep $j_1, \ldots, j_{n-1}$ such that $j_1 + \cdots + j_{n-1} = -m$ and (4.14) becomes

$$a_{m,n}(z,k) = \sum_{\substack{j_1,\ldots,j_{n-1} \in \{-1,0,1\} \\ j_1+\cdots+j_{n-1}=-m}} \prod_{\epsilon=1}^{n-1} c_{\epsilon,j_\epsilon}\, z \tag{4.14}$$

where

$$c_{\epsilon,1} = \bar{z}\,\theta_{0,0,1,m+1+\sum_{q=1}^{\epsilon-1} j_q},$$
$$c_{\epsilon,0} = k\,\theta_{0,1,0,m+\sum_{q=1}^{\epsilon-1} j_q},\ c_{\epsilon-1,-1} = z\,\theta_{1,0,0,m-1+\sum_{q=1}^{\epsilon-1} j_q}$$

which can be written as

$$c_{\epsilon,1} = \bar{z}\,\hat{\theta}_{\epsilon,1}(m),\ c_{\epsilon,0} = k\,\hat{\theta}_{\epsilon,0}(m),\ c_{\epsilon,-1} = z\,\hat{\theta}_{\epsilon,-1}(m)$$

Suppose that among the $j_1, \ldots, j_{n-1}$ we have α, 1's, β 0's, and γ (-1)'s (corresponding to the "basic monomial" $\bar{z}^\alpha k^\beta z^\gamma z = \bar{z}^\alpha k^\beta z^{\gamma+1}$) where $\alpha \cdot 1 + \beta \cdot 0 + \gamma \cdot (-1) = -m$ i.e. $\gamma = \alpha + m$. Since $\alpha + \beta + \gamma + 1 = n$ it follows that $\beta = n - \alpha - \gamma - 1$ and the basic monomial becomes $\bar{z}^\alpha k^{n-2\alpha-m-1} z^{\alpha+m+1} = |z|^{2\alpha} k^{n-2\alpha-m-1} z^{m+1}$ with coefficient (for $n \geq 2$)

$$\sum_{\alpha=0}^{n-1} \Bigg[\sum_{\substack{j_1,\ldots,j_{n-1} \in \{-1,0,1\} \\ m+j_1+\cdots+j_{n-1}=0}} \prod_{\epsilon=1}^{n-1} \hat{\theta}_{\epsilon,j_\epsilon}(m) \Bigg] (\alpha)$$

Thus, for $n \geq 2$,

$$a_{m,n}(z,k) = \sum_{\alpha=0}^{n-1} \Bigg[\sum_{\substack{j_1,\ldots,j_{n-1} \in \{-1,0,1\} \\ m+j_1+\cdots+j_{n-1}=0}} \prod_{\epsilon=1}^{n-1} \hat{\theta}_{\epsilon,j_\epsilon}(m) \Bigg] (\alpha) |z|^{2\alpha} k^{n-2\alpha-m-1} z^{m+1}$$

while

$$a_{m,1}(z,k) = \delta_{m,0}\, z.$$

Thus

$$a_m(z,k) = \sum_{n=1}^{+\infty} a_{m,n}(z,k) i^n / n!$$

from which (4.4) follows.

Regarding the convergence of the above infinite series we notice that, since

$$\begin{aligned} |\hat{\theta}_{\epsilon,j_\epsilon}(m)| &\le 2(m+2+\sum_{q=1}^{\epsilon-1} 1) \\ &\le 2(m+\epsilon+1) \\ &\le 2(m+n), \end{aligned}$$

we have that

$$\begin{aligned} &\Big|\sum_{\alpha=0}^{n-1}\Big[\sum_{\substack{j_1,\ldots,j_{n-1}\in\{-1,0,1\}\\ m+j_1+\cdots+j_{n-1}=0}} \prod_{\epsilon=1}^{n-1} \hat{\theta}_{\epsilon,j_\epsilon}(m)\Big](\alpha)|z|^{2\alpha} k^{n-2\alpha-m-1} z^{m+1}\Big| \\ &\le \sum_{\alpha=0}^{n-1}\Big[\sum_{\substack{j_1,\ldots,j_{n-1}\in\{-1,0,1\}\\ m+j_1+\cdots+j_{n-1}=0}} \prod_{\epsilon=1}^{n-1} |\hat{\theta}_{\epsilon,j_\epsilon}(m)|\Big](\alpha)|z|^{2\alpha} |k|^{n-2\alpha-m-1} |z|^{m+1} \\ &\le \sum_{\alpha=0}^{n-1}\Big[\sum_{\substack{j_1,\ldots,j_{n-1}\in\{-1,0,1\}\\ m+j_1+\cdots+j_{n-1}=0}} \prod_{\epsilon=1}^{n-1} |\hat{\theta}_{\epsilon,j_\epsilon}(m)|\Big](\alpha) \max(|z|,|k|)^n \\ &\le n\, 3^{n-1} 2^{n-1} (m+n)^{n-1} \max(|z|,|k|)^n \end{aligned}$$

and so, for each m, the ratio test yields

$$\begin{aligned} &\frac{|a_{m,n+1}(z,k) i^{n+1}/(n+1)!|}{|a_{m,n}(z,k) i^n/n!|} \\ &= \frac{6}{n}\frac{(m+n+1)^n}{(m+n)^{n-1}} \max(|z|,|k|) \\ &= \frac{6(m+n)}{n}\big[(1+\frac{1}{m+n})^{m+n}\big]^{\frac{n}{m+n}} \max(|z|,|k|) \\ &\to 6\,e \max(|z|,|k|) < 1 \end{aligned}$$

as $n \to +\infty$, provided that $\max(|z|,|k|) < \frac{1}{6\,e}$. As for (4.3), letting $\tau_n(\lambda,z,k)$ denote the coefficient of dt in dE^n, for $n \ge 2$ Proposition 2.4 implies

$$\tau_n(\lambda,z,k) = a_{0,n-1}(z,k)\,\bar{z} \tag{4.15}$$

with

$$\tau_1(\lambda, z, k) = \lambda \tag{4.16}$$

and since

$$\tau(\lambda, z, k) = \sum_{n=1}^{+\infty} \tau_n(\lambda, z, k) i^n / n!$$

(4.3) follows from (4.4) which has already been proved.

Turning to (4.5), we notice that by Proposition 2.4

$$d\Lambda_{a,b,s}(t)\, dA_0^\dagger(t) = \theta_{a,b,s,0}\, dA_{a-s}^\dagger(t)$$

Letting $a - s = m$ the above becomes

$$d\Lambda_{s+m,b,s}(t)\, dA_0^\dagger(t) = \theta_{s+m,b,s,0}\, dA_m^\dagger(t)$$

from which we obtain the recursion

$$\bar{a}_{m,n}(z, k) = \bar{z} \sum_{\substack{s,b \in \{0,1,\dots,n-1\} \\ 0<2s+b+m \leq n-1}} l_{s+m,b,s,n-1}(z, k)\, \theta_{s+m,b,s,0} \tag{4.17}$$

Though computationally useful the above recursion does not reveal the fact that

$$\bar{a}_m(z, k) = \overline{a_m(z, k)}.$$

To establish that we proceed as in the proof of (4.4) but this time using the left multiplication by $dE(t)$ recursive scheme

$$\begin{aligned} dE(t)^n &= dE(t)\, dE(t)^{n-1}(t) \\ &= [(\lambda + k)\, dt + z\, dA_0(t) + \bar{z}\, dA_0^\dagger(t) + z\, d\Lambda_{0,0,1}(t) \\ &\quad + \bar{z}\, d\Lambda_{1,0,0}(t) + k\, d\Lambda_{0,1,0}(t)]\, dE(t)^{n-1}(t) \end{aligned}$$

along with the following consequences of Proposition 2.4

$$\begin{aligned} d\Lambda_{0,0,1}(t)\, dA_{m+1}^\dagger(t) &= \theta_{0,0,1,m+1}\, dA_m^\dagger(t) \\ d\Lambda_{0,1,0}(t)\, dA_m^\dagger(t) &= \theta_{0,1,0,m}\, dA_m^\dagger(t) \\ d\Lambda_{1,0,0}(t)\, dA_{m-1}^\dagger(t) &= \theta_{1,0,0,m-1}\, dA_m^\dagger(t) \end{aligned}$$

to obtain

$$\begin{aligned} \bar{a}_{m,n}(z, k) &= z\, \theta_{0,0,1,m+1} \bar{a}_{m+1,n-1}(z, k) \\ &\quad + k\, \theta_{0,1,0,m} \bar{a}_{m,n-1}(z, k) \\ &\quad + \bar{z}\, \theta_{1,0,0,m-1} \bar{a}_{m-1,n-1}(z, k) \end{aligned} \tag{4.18}$$

with

$$\bar{a}_{0,1}(z,k) = \bar{z} \tag{4.19}$$

which are the complex conjugates of (4.12) and (4.13) respectively.

To prove (4.6) we notice that in the notation of Proposition 2.4

$$\begin{aligned} d\Lambda_{\alpha,\beta,\gamma}(t)\, d\Lambda_{1,0,0}(t) &= \sum c^{\lambda,\rho,\sigma,\omega,\epsilon}_{\beta,\gamma,1,0}\, d\Lambda_{1+\alpha-\gamma+\lambda,\omega+\sigma+\epsilon,\lambda}(t) \\ d\Lambda_{\alpha,\beta,\gamma}(t)\, d\Lambda_{0,1,0}(t) &= \sum c^{\lambda,\rho,\sigma,\omega,\epsilon}_{\beta,\gamma,0,1}\, d\Lambda_{\alpha-\gamma+\lambda,\omega+\sigma+\epsilon,\lambda}(t) \\ d\Lambda_{\alpha,\beta,\gamma}(t)\, d\Lambda_{0,0,1}(t) &= \sum c^{\lambda,\rho,\sigma,\omega,\epsilon}_{\beta,\gamma,0,0}\, d\Lambda_{\alpha-\gamma+\lambda,\omega+\sigma+\epsilon,\lambda+1}(t). \end{aligned}$$

Thus

$$\begin{aligned} l_{i,j,r,n} &= \bar{z} \sum \hat{c}^{r,\rho,j-\omega,\omega,0}_{\beta,\gamma,1,0}\, l_{i+\gamma-r-1,\beta,\gamma,n-1} \\ &\quad + k \sum \hat{c}^{r,\rho,j-\omega-\epsilon,\omega,\epsilon}_{\beta,\gamma,0,1}\, l_{i+\gamma-r,\beta,\gamma,n-1} \\ &\quad + z \sum \hat{c}^{r-1,\rho,j-\omega,\omega,0}_{\beta,\gamma,0,0}\, l_{i+\gamma-r+1,\beta,\gamma,n-1} \\ &= \phi_{i,j,r}(z,k) \times \\ &\quad \times \sum_{1} \left[1 - \delta_{\epsilon_1,1}(\delta_{q_1,-1} + \delta_{q_1,0})\right] \\ &\qquad \cdot \hat{c}^{r-\delta_{q_1,1},\rho_1,j-\omega_1-\delta_{q_1,0}\epsilon_1,\omega_1,\delta_{q_1,0}\epsilon_1}_{\beta_1,\gamma_1,\delta_{q_1,-1},\delta_{q_1,0}} \\ &\qquad \cdot l_{i+\gamma_1-r+q_1,\beta_1,\gamma_1,n-1} \end{aligned} \tag{4.20}$$

with

$$l_{1,0,0,1} = \bar{z} \tag{4.21}$$

$$l_{0,1,0,1} = k \tag{4.22}$$

$$l_{0,0,1,1} = z \tag{4.23}$$

or equivalently

$$l_{i,j,r,1} = \phi_{i,j,r}. \tag{4.24}$$

Iterating (4.21), using (4.25) in the last step, we obtain

$$\begin{aligned} l_{i,j,r,n} &= \phi_{i,j,r}(z,k) \sum_1 \sum_2 \phi_{i+\gamma_1-r+q_1,\beta_1,\gamma_1}(z,k) \\ &\quad \cdot (1-\delta_{\epsilon_1,1}(\delta_{q_1,-1}+\delta_{q_1,0}))\,(1-\delta_{\epsilon_2,1}(\delta_{q_2,-1}+\delta_{q_2,0})) \\ &\quad \cdot \hat{c}^{\,r-\delta_{q_1,1},\rho_1,j-\omega_1-\delta_{q_1,0}\epsilon_1,\omega_1,\delta_{q_1,0}\epsilon_1}_{\beta_1,\gamma_1,\delta_{q_1,-1},\delta_{q_1,0}} \\ &\quad \cdot \hat{c}^{\,\gamma_1-\delta_{q_2,1},\rho_2,\beta_1-\omega_2-\delta_{q_2,0}\epsilon_2,\omega_2,\delta_{q_2,0}\epsilon_2}_{\beta_2,\gamma_2,\delta_{q_2,-1},\delta_{q_2,0}}\; l_{i+\gamma_2-r+q_1+q_2,\beta_2,\gamma_2,n-2} \\ &= \dots\dots\dots\dots\dots\dots\dots\dots\dots\dots\dots\dots \\ &= \phi_{i,j,r}(z,k) \sum_1 \cdots \sum_{n-1} \prod_{s=1}^{n-1} (1-\delta_{\epsilon_s,1}(\delta_{q_s,-1}+\delta_{q_s,0})) \\ &\quad \cdot \phi_{i+\gamma_{n-s}-r+\sum_{\lambda=1}^{n-s} q_\lambda,\beta_{n-s},\gamma_{n-s}}(z,k) \\ &\quad \cdot \hat{c}^{\,\gamma_{s-1}-\delta_{q_s,1},\rho_s,\beta_{s-1}-\omega_s-\delta_{q_s,0}\epsilon_s,\omega_s,\delta_{q_s,0}\epsilon_s}_{\beta_s,\gamma_s,\delta_{q_s,-1},\delta_{q_s,0}} \end{aligned}$$

from which we obtain (4.6). The convergence of the series in (4.6) is proved as before. □

Remark 4.2 (The first order Poisson-Weyl operator). Let

$$U(t) = e^{i\,E(t)}$$

where

$$E(t) = \lambda t + zA_0(t) + \bar{z}A_0^+(t) + k\Lambda_{0,0,0}(t)$$

with $\lambda, k \in \mathbb{R}$, $z \in \mathbb{C}$.

(a) If $k \neq 0$ then

$$\begin{aligned} dU(t) = U(t)\Big[\Big(i\,\lambda + \frac{|z|^2}{k^2} M\Big)dt + \Big(i\,z + \frac{z}{k} M\Big)dA_0(t) \\ + \Big(i\,\bar{z} + \frac{\bar{z}}{k} M\Big)dA_0^+(t) + (i\,k + M)d\Lambda_0(t)\Big] \end{aligned}$$

where

$$M = e^{i\,k} - 1 - i\,k$$

(b) If $k = 0$ then

$$dU(t) = U(t)[(i\,\lambda - \frac{|z|^2}{2})dt + i\,z dA_0(t) + i\,\bar{z} dA_0^+(t)]$$

Proof. The proof is similar to that of Proposition 4.1 using the fact that for $k \neq 0$ and $n \geq 2$

$$dE(t)^n = |z|^2 k^{n-2} dt + z k^{n-1} dA_0(t) + \bar{z} k^{n-1} dA_0^+(t) + k^n d\Lambda_{0,0,0}(t)$$

while for $k = 0$

$$dE(t)^2 = |z|^2\, dt$$

and for $n > 2$

$$dE(t)^n = 0. \qquad \square$$

The above two equations are of the Hudson-Parthasarathy form (see [22]), namely,

$$\mathrm{d}U(t) = U(t)[(\mathrm{i}\,H - \frac{1}{2}\,L^* L)\mathrm{d}t - L^* W \mathrm{d}A_0(t) + L\mathrm{d}A_0^\dagger(t) + (W - I)\mathrm{d}\Lambda_{0,0,0}(t)]$$

with

$$W = \mathrm{e}^{\mathrm{i}\,k} I$$

$$L = \frac{\bar{z}}{k}\,(\mathrm{e}^{\mathrm{i}\,k} - 1)I$$

$$H = \left(\lambda - \frac{|z|^2}{k} - \frac{\mathrm{i}}{2}\,\frac{|z|^2}{k^2}\,[\mathrm{e}^{\mathrm{i}\,k} - \mathrm{e}^{-\mathrm{i}\,k} - 1]\right) I$$

and

$$H = \lambda I, \quad L = \mathrm{i}\,\bar{z} I, \quad W = -I$$

respectively.

5. Appendix: *Mathematica* algorithms

Some of the results contained in this paper would have been very hard to obtain without the use of computer algorithms for symbolic calculations and noncommutative iterations. The computer algebra software that we used was *Mathematica 4* (see [24] for operational instuctions). A typical example of what the computer was helpful in deriving is the unitary quantum stochastic differential equation of Proposition 4.1.

Following are the algorithms that we used in order to derive, verify, or develop intuition for some of the results contained the previous sections.

Algorithm 5.1 (The Itô-table for the SWN differentials). This algorithm computes the, in general, noncommutative products of the generalized SWN stochastic differentials $d\Lambda_{n,k,l}(t), \mathrm{d}A_m(t)$ and $\mathrm{d}A^\dagger_m(t)$, where $n, k, l, m = 0, 1, \ldots$, and "time" dt. Each sentence corresponds to a new input. Inputs are separated by space.

```
p[x_, y_] = If[x == y == 0, 1, x^y]
```

$u[x\_, n\_] = \text{Product}[x - i + 1, \{i, 1, n\}]$

$v[x\_, n\_] = \text{Product}[x + i - 1, \{i, 1, n\}]$

$\theta[n\_, k\_, l\_, m\_] = \text{If}[n + m - 1 < 0, 0, \text{Sqrt}[(m - l + n + 1)/(m + 1)] 2\hat{}\, k\, v[m - l + 1, n]\, u[m + 1, l]\, p[m - l + 1, k]]$

$c[\beta\_, \gamma\_, a\_, b\_, \lambda\_, \rho\_, \sigma\_, \omega\_, \epsilon\_] = \text{Binomial}[\gamma, \lambda]\ \text{Binomial}[\gamma - \lambda, \rho]\ \text{Binomial}[\beta, \omega]\ \text{Binomial}[b, \epsilon]\ 2\hat{}(\beta + b - \omega - \epsilon)\ \text{StirlingS1}[\gamma - \lambda - \rho, \sigma]\, u[a, \gamma - \lambda]\, u[a + \lambda - 1, \rho]\, p[a - \gamma + \lambda, \beta - \omega]\, p[\lambda, b - \epsilon]$

$\text{NCM}[d\Lambda[\alpha\_, \beta\_, \gamma\_], d\Lambda[a\_, b\_, s\_]] = \text{Sum}[c[\beta, \gamma, a, b, \lambda, \rho, \sigma, \omega, \epsilon]\, d\Lambda[a + \alpha - \gamma + \lambda, \omega + \sigma + \epsilon, \lambda + s], \{\lambda, 0, \gamma\}, \{\rho, 0, \gamma - \lambda\}, \{\sigma, 0, \gamma - \lambda - \rho\}, \{\omega, 0, \beta\}, \{\epsilon, 0, b\}]$

$\text{NCM}[d\Lambda[a\_, b\_, c\_], dA^\dagger[m\_]] = \theta[a, b, c, m] dA^\dagger[a + m - c]$

$\text{NCM}[dA[m\_], d\Lambda[a\_, b\_, c\_]] = \theta[c, b, a, m] dA[c + m - a]$

$\text{NCM}[dA[m\_], dA^\dagger[n\_]] = \text{KroneckerDelta}[m, n] dt$

$\text{NCM}[dA[m\_], dA[n\_]] = 0$

$\text{NCM}[dA^\dagger[m\_], dA^\dagger[n\_]] = 0$

$\text{NCM}[dA^\dagger[m\_], dA[n\_]] = 0$

$\text{NCM}[dA^\dagger[m\_], d\Lambda[\alpha\_, \beta\_, \gamma\_]] = 0$

$\text{NCM}[d\Lambda[\alpha\_, \beta\_, \gamma\_], dA[m\_]] = 0$

$\text{NCM}[d\Lambda[\alpha\_, \beta\_, \gamma\_], dt] = 0$

$\text{NCM}[dt, d\Lambda[\alpha\_, \beta\_, \gamma\_]] = 0$

$\text{NCM}[dA[m\_], dt] = 0$

$\text{NCM}[dt, dA[m\_]] = 0$

$\text{NCM}[dA^\dagger[m\_], dt] = 0$

$\text{NCM}[dt, dA^\dagger[m\_]] = 0$

$\text{NCM}[dt, dt] = 0$

For example using the above algorithm to compute $d\Lambda_{4,1,2}(t)\, d\Lambda_{1,2,1}(t)$ we obtain

$$\begin{aligned} &\text{NCM}[d\Lambda[4,1,2], d\Lambda[1,2,1]] \\ &= 8d\Lambda[4,1,2] + 16d\Lambda[4,2,2] + 10d\Lambda[4,3,2] + 2d\Lambda[4,4,2] \\ &\quad + 32d\Lambda[5,0,3] + 32d\Lambda[5,1,3] + 10d\Lambda[5,2,3] + d\Lambda[5,3,3] \end{aligned}$$

while for $d\Lambda_{4,2,1}(t)\, dA_2^\dagger(t)$ we obtain

$$\text{NCM}[d\Lambda[4,2,1], dA^\dagger[2]] = 48\sqrt{5}\, dA^\dagger[5]$$

Algorithm 5.2 (Powers of the SWN Poisson-Weyl operator differential). To compute $dE(t)^n$, where $E(t)$ is the SWN Poisson-Weyl operator of Proposition 4.1 and $n = 2, 3, \ldots$ (the value of n must be supplied by the user), we use Algorithm 5.1 with the following commands attached to it:

$\mathrm{NCM}[0, x\_] = \mathrm{NCM}[x\_, 0] = 0$

$\mathrm{NCM}[(x\_\ y\_), z\_] = x\ \mathrm{NCM}[y, z]$

$\mathrm{NCM}[w\_\ d\Lambda[a\_, b\_, s\_], q\_\ d\Lambda[d\_, h\_, f\_]] = w\,q\ \mathrm{NCM}[d\Lambda[a, b, s], d\Lambda[d, h, f]]$

$\mathrm{NCM}[w\_\ d\Lambda[a\_, b\_, s\_], q\_\ dA^{\dagger}[m\_]] = w\,q\ \mathrm{NCM}[d\Lambda[a, b, s], dA^{\dagger}[m]]$

$\mathrm{NCM}[w\_\ d\Lambda[a\_, b\_, s\_], q\_\ dA[m\_]] = 0$

$\mathrm{NCM}[q\_\ dA^{\dagger}[m\_], w\_\ d\Lambda[a\_, b\_, s\_]] = 0$

$\mathrm{NCM}[q\_\ dA[m\_], w\_\ d\Lambda[a\_, b\_, s\_]] = q\,w\ \mathrm{NCM}[dA[m], d\Lambda[a, b, s]]$

$\mathrm{NCM}[w\_\ dA^{\dagger}[m\_], q\_\ dA^{\dagger}[r\_]] = 0$

$\mathrm{NCM}[q\_\ dA[r\_], w\_\ dA^{\dagger}[m\_]] = q\,w\ \mathrm{NCM}[dA[r], dA^{\dagger}[m]]$

$\mathrm{NCM}[q\_\ dA[r\_], w\_\ dA[m\_]] = q\,w\ \mathrm{NCM}[dA[r], dA[m]]$

$\mathrm{NCM}[q\_\ dA[r\_], w\_\ dt] = 0$

$\mathrm{NCM}[q\_\ dA^{\dagger}[r\_], w\_\ dt] = 0$

$\mathrm{NCM}[q\_\ d\Lambda[a\_, b\_, s\_], w\_\ dt] = 0$

$\mathrm{NCM}[dt\,w\_, q\_\ dA[r\_]] = 0$

$\mathrm{NCM}[dt\,w\_, q\_\ dA^{\dagger}[r\_]] = 0$

$\mathrm{NCM}[dt\,w\_, q\_\ d\Lambda[a\_, b\_, s\_]] = 0$

$\mathrm{NCM}[dt\,w\_, dt\,q\_] = 0$

$dE[1] = (\lambda + k)\,dt + z\,dA[0] + \bar{z}\,dA^{\dagger}[0] + \bar{z}\,d\Lambda[1, 0, 0] + k\,d\Lambda[0, 1, 0] + z\,d\Lambda[0, 0, 1]$

$n =$

Do[Print[StringForm["dE"]^i, StringForm[" = "], $dE[i]$ = Collect[Expand[MapAll[Distribute, NCM[$dE[i -$ $1], dE[1]$]]], $\{dt, dA[\_], dA^{\dagger}[\_], d\Lambda[\_, \_, \_]\}$]], $\{i, 2, n\}$]

For example, running the algorithm for $n = 2$ we obtain

$$\begin{aligned} dE^2(t) = {} & z\bar{z}\,dt + 2kz\,dA_0(t) + \sqrt{2}z^2\,dA_1(t) + 2k\bar{z}\,dA_0^{\dagger}(t) \\ & + \sqrt{2}\bar{z}^2\,dA_1^{\dagger}(t) + 2kz\,d\Lambda_{0,0,1}(t) + z^2\,d\Lambda_{0,0,2}(t) \\ & + 2kz\,d\Lambda_{0,1,1}(t) + k^2\,d\Lambda_{0,2,0}(t) + z\bar{z}\,d\Lambda_{0,1,0}(t) \\ & + 2k\bar{z}\,d\Lambda_{1,0,0}(t) + 2z\bar{z}\,d\Lambda_{1,0,1}(t) + 2k\bar{z}\,d\Lambda_{1,1,0}(t) \\ & + \bar{z}^2\,d\Lambda_{2,0,0}(t). \end{aligned}$$

Algorithm 5.3 (SWN Poisson-Weyl recursions). This algorithm uses recursions (4.12), (4.13), (4.16),(4.17),(4.19)-(4.24) of Proposition 4.1 to compute the coefficients of dt, $dA_m(t)$, $dA_m^{\dagger}(t)$, and $d\Lambda_{i,j,h}(t)$

in $dE(t)^n$, denoted respectively by $\tau[n\_]$, $\alpha[m\_, n\_]$, $\alpha^\dagger[m\_, n\_]$, and $f[i\_, j\_, h\_, n\_]$. It also computes the m-th partial sum $\sum_{n=1}^{m} dE^n/n!$ where the value of m must be provided by the user.

$p[x\_, y\_] = \text{If}[x == y == 0, 1, x\hat{\ }y]$

$\text{upperfact}[x\_, n\_] = \text{Product}[x - i + 1, \{i, 1, n\}]$

$\text{lowerfact}[x\_, n\_] = \text{Product}[x + i - 1, \{i, 1, n\}]$

$\theta[n\_, h\_, l\_, m\_] = \text{If}[n + m - l < 0, 0, \text{Sqrt}[(m - l + n + 1)/(m + 1)]$
$2\hat{\ }h\ \text{lowerfact}[m - l + 1, n]\ \text{upperfact}[m + 1, l]\ p[m - l + 1, h]]$

$c[\beta\_, \gamma\_, a\_, b\_, \lambda\_, \rho\_, \sigma\_, \omega\_, \epsilon\_] = \text{If}[0 \le \lambda \le \gamma \&\& 0 \le \rho \le \gamma - \lambda$
$\&\& 0 \le \sigma \le \gamma - \lambda - \rho \&\& 0 \le \omega \le \beta \&\& 0 \le \epsilon \le b, \text{Binomial}[\gamma, \lambda]\ \text{Binomial}[\gamma - \lambda, \rho]$
$\text{Binomial}[\beta, \omega]\ \text{Binomial}[b, \epsilon]\ 2\hat{\ }(\beta + b - \omega - \epsilon)\ \text{StirlingS1}[\gamma - \lambda - \rho, \sigma]\ \text{upperfact}[a, \gamma - \lambda]$
$\text{upperfact}[a + \lambda - 1, \rho]\ p[a - \gamma + \lambda, \beta - \omega]\ p[\lambda, b - \epsilon], 0]$

$\alpha[m\_, 0] = 0$

$\alpha[m\_, 1] = \text{If}[m == 0, z, 0]$

$\alpha[m\_, n\_] = \text{If}[m > n, 0, \text{Collect}[\bar{z}\ \text{Sqrt}[(m + 1)\,(m + 2)]\,\alpha[m + 1, n - 1] +$
$k\,2\,(m + 1)\,\alpha[m, n - 1] + z\ \text{Sqrt}[m\,(m + 1)]\,\alpha[m - 1, n - 1], \{z, \bar{z}, k\}]]$

$\tau[1] = \lambda + k$

$\tau[n\_] = \text{Collect}[\bar{z}\ \alpha[0, n - 1], \{\lambda + k, z, \bar{z}, k\}]$

$f[1, 0, 0, 1] = \bar{z}$

$f[0, 1, 0, 1] = k$

$f[0, 0, 1, 1] = z$

$f[i\_, j\_, h\_, n\_] = \text{If}[i + j + h > n || i + j + h == 0 || i < 0 || j < 0 || h < 0,$
$0, \bar{z}\ \text{Sum}[f[i + \gamma - h - 1, \beta, \gamma, n - 1]\ c[\beta, \gamma, 1, 0, h, \rho, j - \omega, \omega, 0],$
$\{\gamma, 0, n - 1\}, \{\beta, 0, n - 1\}, \{\omega, 0, \beta\}, \{\rho, 0, \gamma - h\}] +$
$k\ \text{Sum}[f[i + \gamma - h, \beta, \gamma, n - 1]\ c[\beta, \gamma, 0, 1, h, \rho, j - \omega - \epsilon, \omega, \epsilon],$
$\{\gamma, 0, n - 1\}, \{\beta, 0, n - 1\}, \{\omega, 0, \beta\}, \{\rho, 0, \gamma - h\}, \{\epsilon, 0, 1\}] +$
$z\ \text{Sum}[f[i + \gamma - h + 1, \beta, \gamma, n - 1]\ c[\beta, \gamma, 0, 0, h - 1, \rho, j - \omega, \omega, 0],$
$\{\gamma, 0, n - 1\}, \{\beta, 0, n - 1\}, \{\omega, 0, \beta\}, \{\rho, 0, \gamma - h + 1\}]]$

$\alpha^\dagger[m\_, 0] = 0$

$\alpha^\dagger[m\_, 1] = \text{If}[m == 0, \bar{z}, 0]$

$\alpha^\dagger[m\_, n\_] = \text{If}[m > n, 0, \text{Collect}[z\ \text{Sqrt}[(m + 1)\,(m + 2)]\,\alpha^\dagger[m + 1, n - 1] +$
$k\,2\,(m + 1)\,\alpha^\dagger[m, n - 1] + \bar{z}\ \text{Sqrt}[m\,(m + 1)]\,\alpha^\dagger[m - 1, n - 1], \{z, \bar{z}, k\}]]$

$M =$

$\text{partialsum}[M\_] = \text{Collect}[\text{Sum}[I\hat{\ }N\,/\,N!\,\tau[N]\,dt +$
$\text{Expand}[I\hat{\ }N\,/\,N!\ \text{Sum}[\alpha[m, N]\,dA[m] + \alpha^\dagger[m, N]\,dA^\dagger[m], \{m, 0, N - 1\}] + I\hat{\ }N\,/\,N!$
$\text{Sum}[f[i, j, h, N]\,d\Lambda[i, j, h], \{i, 0, N\}, \{j, 0, N\}, \{h, 0, N\}]], \{N, 1, M\}],$
$\{dt, dA[\_], dA^\dagger[\_], d\Lambda[\_,\_,\_]\}]$

For example using the above algorithm we obtain

$$\alpha[2,3] = 2\sqrt{3}\,z^3,$$
$$\alpha^\dagger[0,4] = 8\,k^3\bar{z} + 16\,kz\bar{z}^2,$$
$$\tau[8] = 64\,k^6 z\bar{z} + 1824\,k^4 z^2\bar{z}^2 + 2880\,k^2 z^3\bar{z}^3 + 272\,z^4\bar{z}^4,$$
$$f[1,0,1,3] = 12\,kz\bar{z},$$

$$\begin{aligned}\sum_{n=1}^{2} dE^n/n! = {} & i(k+\lambda)\,dt + (i-k)z\,dA[0] - \frac{1}{\sqrt{2}}z^2\,dA[1] \\ & + (i-k)\bar{z}\,dA^\dagger[0] - \frac{1}{\sqrt{2}}\bar{z}^2\,dA^\dagger[1] + (i-k)z\,d\Lambda[0,0,1] \\ & - \frac{1}{2}z^2\,d\Lambda[0,0,2] - kz\,d\Lambda[0,1,1] - \frac{1}{2}k^2\,d\Lambda[0,2,0] \\ & - z\bar{z}\,d\Lambda[1,0,1] - k\bar{z}\,d\Lambda[1,1,0] - \frac{1}{2}\bar{z}^2\,d\Lambda[2,0,0] \\ & + (i-k)\bar{z}\,d\Lambda[1,0,0] + (ik - \frac{1}{2}z\bar{z})\,d\Lambda[0,1,0].\end{aligned}$$

Algorithm 5.4 (Unitarity Conditions). This algorithm checks unitarity conditions (3.3)-(3.8) of Proposition 3.1 for specific coefficient processes. We present here the classical example of [22] where

$$A = iH - \frac{1}{2}L^*L, \quad C_0 = -L^*W, \quad D_0 = L, \quad B_{0,0,0} = W - 1,$$

with L self-adjoint and W unitary, and all other coefficients are zero. Part of the algorithm deals with coding self-adjointness and unitarity for L and W respectively. For different examples different or additional commands may be needed:

$\mathrm{NCM}[x\_, y\_] = \mathrm{NonCommutativeMultiply}[x, y]$

$\mathrm{NCM}[\mathrm{NCM}[a\_, b\_], \mathrm{NCM}[c\_, d\_]] = \mathrm{NCM}[a, b, c, d]$

$\mathrm{NCM}[0, 0] = 0$

$\mathrm{NCM}[a\_, 1] = \mathrm{NCM}[1, a\_] = a$

$\mathrm{NCM}[a\_, -1] = \mathrm{NCM}[-1, a\_] = -a$

$\mathrm{NCM}[(-1)a\_, (-1)b\_] = \mathrm{NCM}[a, b]$

$\mathrm{NCM}[0, a\_] = \mathrm{NCM}[a\_, 0] = 0$

$\mathrm{NCM}[(-1)a\_, b\_] = \mathrm{NCM}[a\_, (-1)b\_] = -\,\mathrm{NCM}[a, b]$

$A = IH - 1/2\ \mathrm{NCM}[L^*, L]$

$A^* = -IH - 1/2\ \mathrm{NCM}[L^*, L]$

$C[m\_] = \mathrm{If}[m == 0, -\,\mathrm{NCM}[L^*, W], 0]$

$C^*[m\_] = \mathrm{If}[m == 0, -\,\mathrm{NCM}[W^*, L], 0]$

$D[m\_] = \text{If}[m == 0, L, 0]$

$D^*[m\_] = \text{If}[m == 0, L^*, 0]$

$B[n\_, k\_, l\_] = \text{If}[n == k == l == 0, W - 1, 0]$

$B^*[n\_, k\_, l\_] = \text{If}[n == k == l == 0, W^* - 1, 0]$

$\text{NCM}[W^*, W] = \text{NCM}[W, W^*] = 1$

$p[x\_, y\_] = \text{If}[x == y == 0, 1, x\hat{\ }y]$

$u[x\_, n\_] = \text{Product}[x - i + 1, \{i, 1, n\}$

$v[x\_, n\_] = \text{Product}[x + i - 1, \{i, 1, n\}]$

$\theta[n\_, k\_, l\_, m\_] = \text{UnitStep}[n + m - 1]\ \text{Sqrt}[(m - l + n + 1)/(m + 1)]$
$2\hat{\ }k\, v[m - l + 1, n]\, u[m + 1, l]\, p[m - l + 1, k]$

$K1 = \text{Expand}[A + A^* + \text{Sum}[\text{MapAll}[\text{Distribute}, \text{NCM}[D^*[m], D[m]]], \{m, 0, 0\}]]$

If$[K1 == 0$, Print[StringForm["Condition (3.3) is satisfied"]],
Print[StringForm "Condition (3.3) is not satisfied"]]

$K2 = \text{Expand}[A + A^* + \text{Sum}[\text{MapAll}[\text{Distribute}, \text{NCM}[C[m], C^*[m]]], \{m, 0, 0\}]]$

If$[K2 == 0$, Print[StringForm["Condition (3.4) is satisfied"]],
Print[StringForm "Condition (3.4) is not satisfied"]]

$m = 0;$

$K3 = \text{Expand}[C[m] + D^*[m] + \text{Sum}[\text{MapAll}[\text{Distribute},$
$\text{NCM}[D^*[m + n - l], B[n, k, l]]\ \theta[l, k, m + n - l]], \{n, 0, 0\}, \{l, 0, 0\}, \{k, 0, 0\}]$

If$[K3 == 0$, Print[StringForm["Condition (3.5) is satisfied for $m =$ "], m],
Print[StringForm "Condition (3.5) is not satisfied for $m =$ "], m]

$c[\lambda\_, \rho\_, \sigma\_, \omega\_, \epsilon\_, \beta\_, \gamma\_, a\_, b\_] =$
$\text{Binomial}[\gamma, \lambda]\ \text{Binomial}[\gamma - \lambda, \rho]\ \text{Binomial}[\beta, \omega]\ \text{Binomial}[b, \epsilon]$
$2\hat{\ }(\beta + b - \omega - \epsilon)\ \text{StirlingS1}[\gamma - \lambda - \rho, \sigma]\, u[a, \gamma - \lambda]\, (a - \gamma + \lambda)\hat{\ }(\beta - \omega)\, \lambda\hat{\ }(b - \epsilon);$

$g[n\_, k\_, l\_, x\_, y\_, z\_, X\_, Y\_, Z\_] = \text{Sum}[\text{KroneckerDelta}[x + Y - X +$
$\lambda, n]\ \text{KroneckerDelta}[\omega + \sigma + \epsilon, k]\ \text{KroneckerDelta}[\lambda + y, l]$
$c[\lambda, \rho, \sigma, \omega, \epsilon, z, X, Y, Z], \{\lambda, 0, X\}, \{\rho, 0, X, Y - \lambda\}, \{\sigma, 0, X - \lambda - \rho\}, \{\omega, 0, z\}, \{\epsilon, 0, Z\}]$

$n = 1$

$k = 1$

$l = 1$

$K4 =$
$\text{Expand}[B[n, k, l] + B^*[l, k, n] + \text{Sum}[\text{MapAll}[\text{Distribute}, \text{NCM}[B[\alpha, \beta, \gamma], B^*[a, b, c]]$
$g[n, k, l, \alpha, \beta, \gamma, c, b]], \{\alpha, 0, n\}, \{\beta, 0, k\},$
$\{\gamma, 0, Min[k, l]\}, \{a, 0, l\}, \{b, 0, k\}, \{c, 0, n\}]]$

If$[K4 == 0$, Print[StringForm["Condition (3.7) is satisfied for $(n, k, l) =$
("], n, StringForm[", "], k, StringForm[", "], l, StringForm[")"]],
Print[StringForm["Condition (3.7) is not satisfied for $(n, k, l) =$
("], n, StringForm[", "], k, StringForm[", "], l, StringForm[")"]]]

$K5 = \text{Expand}[B[n,k,l] + B^*[l,k,n] + \text{Sum}[\text{MapAll}[\text{Distribute},$
$\text{NCM}[B^*[\alpha,\beta,\gamma], B[a,b,c]]$
$g[n,k,l,\gamma,c,\beta,\alpha,a,b]], \{\alpha, 0, \text{Min}[k,l]\}, \{\beta,0,k\}, \{\gamma,0,n\}, \{a,0,n\},$
$\{b,0,k\}, \{c,0,l\}]]$

If[$K5 == 0$, Print[StringForm["Condition (3.8) is satisfied for $(n,k,l) =$ ("], n, StringForm[", "], k, StringForm[", "], l, StringForm[")"]],
Print[StringForm["Condition (3.8) is not satisfied for $(n,k,l) =$ ("], n, StringForm[", "], k, StringForm[", "], l, StringForm[")"]]]

Acknowledgments

The second author wishes to express his gratitude to Professor Luigi Accardi for his support and guidance as well as for the hospitality of the Centro Vito Volterra of the Universitá di Roma Tor Vergata on several occasions over the past 15 years.

References

[1] Accardi L.: Meixner classes and the square of white noise, in "*Finite and infinite dimensional analysis in honor of Leonard Gross (New Orleans, LA, 2001)*", 1–13, *Contemp. Math.* **317**, Amer. Math. Soc., Providence, RI, 2003.

[2] Accardi L., Boukas A.: Unitarity conditions for stochastic differential equations driven by nonlinear quantum noise, *Random Oper. Stochastic Equations* **10**(1) (2002) 1–12.

[3] Accardi L., Boukas A.: Stochastic evolutions driven by non-linear white noise, *Probab. Math. Statist.* **22**(1) (2002) 141–154.

[4] Accardi L., Boukas A.: Stochastic evolutions driven by nonlinear quantum noise. II, *Russ. J. Math. Phys.* **8**(4) (2001) 401–413.

[5] Accardi L., Boukas A.: The unitarity conditions for the square of white noise, *Infin. Dimens. Anal. Quantum Probab. Relat. Top.* **6**(2) (2003) 197–222.

[6] Accardi L., Boukas A.: Unitarity conditions for the renormalized square of white noise, in "*Trends in Contemporary Infinite Dimensional Analysis and Quantum Probability, Natural and Mathematical Sciences Series* **3** 7–36, "Italian School of East Asian Studies, Kyoto, Japan (2000)", Volterra Center preprint 466.

[7] Accardi L., Boukas A.: The semi-martingale property of the square of white noise integrators, in "*Stochastic Partial Differential Equations and Applications (Trento, 2002)*", eds Da Prato G., Tubaro L., 1–19, *Lecture Notes in Pure and Appl. Math.* **227**, Marcel Dekker, Inc., New York, 2002.

[8] Accardi L., Boukas A., Kuo H.-H.: On the unitarity of stochastic evolutions driven by the square of white noise, *Infin. Dimens. Anal. Quantum Probab. Relat. Top.* **4**(4) (2001) 1–10.

[9] Accardi L., Fagnola F., Quaegebeur J.: A representation free quantum stochastic calculus, *J. Funct. Anal.* **104**(1) (1992) 149–197.

[10] Accardi L., Franz U., Skeide M.: Renormalized squares of white noise and other non-Gaussian noises as Lévy processes on real Lie algebras, *Commun. Math. Phys.* **228**(1) (2002) 123–150.

[11] Accardi L., Hida T., Kuo H.-H.: The Itô table of the square of white noise, *Infin. Dimens. Anal. Quantum Probab. Relat. Top.* **4** (2001) 267–275.

[12] Accardi L., Lu Y. G., Volovich I. V.: White noise approach to classical and quantum stochastic calculi, Lecture Notes of the Volterra International School of the same title, Trento, Italy, 1999, Volterra Center preprint 375.

[13] Accardi L., Lu Y. G., Volovich I. V.: *Quantum theory and its stochastic limit*, Springer-Verlag, Berlin, 2002.

[14] Accardi L., Lu Y. G., Volovich I. V.: Non-linear extensions of classical and quantum stochastic calculus and essentially infinite dimensional analysis, Lecture Notes in Statistics, Springer-Verlag, 128 (1998) 1–33.

[15] Accardi L., Obata N.: Towards a non-linear extension of stochastic calculus, in "*Quantum stochastic analysis and related fields (Kyoto, 1995)*", Obata N. (ed.), *RIMS Kokyuroku* **957** (1996) 1–15.

[16] Accardi L., Skeide M.: On the relation of the square of white noise and the finite difference algebra, *Infin. Dimens. Anal. Quantum Probab. Relat. Top.* **3** (2000) 185–189.

[17] Boukas A., "Quantum Stochastic Analysis: a non Brownian case", Ph.D Thesis, Southern Illinois University, 1998.

[18] Boukas A.: An example of a quantum exponential process, *Monatsh. Math.* **112**(3) (1991) 209–215.

[19] Feinsilver P. J.: Discrete analogues of the Heisenberg-Weyl algebra, *Monatsh. Math.* **104** (1987) 89–108.

[20] Hudson R. L., Parthasarathy K. R.: Quantum Ito's formula and stochastic evolutions, *Commun. Math. Phys.* **93** (1984) 301–323.

[21] Itô K.: *On stochastic differential equations*, Memoirs Amer. Math. Soc. No. 4, Amer. Math. Soc., New York, 1951.

[22] Parthasarathy K. R.: *An introduction to quantum stochastic calculus*, Birkhäuser Boston Inc., Boston, 1992.

[23] Schürmann M.: *White noise on bialgebras*, Lecture Notes in Math., vol. 1544, Springer-Verlag, Berlin, 1993.

[24] Wolfram S.: *The Mathematica Book*, 4th ed., Wolfram Media/Cambridge University Press, 1999.

NUMERICAL SOLUTION OF WICK-STOCHASTIC PARTIAL DIFFERENTIAL EQUATIONS

Thomas Gorm Theting
Department of Mathematical Sciences
Norwegian University of Science and Technology
N-7491 Trondheim, Norway
tgt@math.ntnu.no

Abstract We consider the numerical approximation of linear Wick-stochastic boundary value problems of elliptic and parabolic type. Numerical methods based on a Galerkin type of approximation are described and convergence results are reported. To illustrate the approach we consider three specific examples: The stochastic Poisson equation, the parabolic Wick-stochastic pressure equation, and a Wick version of the viscous Burgers' equation with stochastic source. The latter example is nonlinear and falls outside the scope of the theory, but the ideas still can be applied and numerical results are reported.

Keywords: Stochastic partial differential equations, Wick product, Finite element method, Stochastic simulation.
65M60 (60H15, 60H40, 65N30)

1. Introduction

During the last twenty years there has been an increasing interest in numerical solutions of stochastic partial differential equations. References include [2, 1, 7, 9, 11, 10, 14, 17, 23, 27, 28, 35] and many others. We have been concerned with a special class of such equations referred to as Wick-stochastic partial differential equations. In particular, we have worked on boundary value problems on the form

$$\begin{aligned} D_t u + L^{\diamond} u &= f, && \text{on } (0,T)\times\mathcal{D}, \\ u|_{\partial\mathcal{D}} &= 0, && \text{on } (0,T)\times\partial\mathcal{D}, \\ u|_{t=0} &= u_0, && \text{on } \mathcal{D}, \end{aligned} \tag{1.1}$$

S. Albeverio et al. (eds.),
Proceedings of the International Conference on Stochastic Analysis and Applications, 303–349.

and the corresponding stationary problem

$$\begin{aligned} L^\diamond u &= f, && \text{on } \mathcal{D}, \\ u|_{\partial\mathcal{D}} &= 0, && \text{on } \partial\mathcal{D}, \end{aligned} \tag{1.2}$$

where $\mathcal{D} \subset \mathbb{R}^d$ denotes a bounded smooth domain, the right-hand side f denotes a generalized stochastic variable (in a sense made clear below), and the stochastic differential operator $L^\diamond$ is given as

$$L^\diamond u = -\sum_{i,j=1}^{d} D_i(a^{ij} \diamond D_j u) + \sum_{i=1}^{d} b^i \diamond D_i u + c \diamond u. \tag{1.3}$$

Here all coefficients a^{ij}, b^i, c $(i,j = 1,\dots,d)$ are allowed to be generalized stochastic variables. The symbol $\diamond$ denotes the Wick-product, which may be viewed as a renormalization of the product of generalized variables [16, 22].

The exact solution of the above type of problems may be attempted using the strategy of Hermite-transform (or $\mathcal{S}$-transform) and results from classical PDE-theory, as has been done successfully for a range of problems, see for example [12, 19, 21, 20, 22]. Although this approach works in many cases, one reaches the same obstacle as for the exact solution of classical PDEs: Even though one can provide existence and uniqueness results, and sometimes even a formula for the solution, the actual numerical evaluation or simulation of the solution can be quite difficult. In the deterministic theory, this problem of numerical evaluation is often solved by approximating the solution using some numerical method. We may attempt the same approach for our stochastic equations. It is therefore interesting to find good numerical methods for the above type of stochastic boundary value problems. The papers [27, 28] develop such methods based on a Galerkin type approximation. Other references for this type of approximation include [1, 2, 17, 31, 30, 29, 32]. The main purpose of the present paper is to describe and exemplify the ideas from [27, 28] by considering some examples of the above type of stochastic boundary value problems. In addition, we wanted to investigate how this approach applies to the non-linear setting, and to point out some aspects of the implementation of the methods.

After giving some notation and definitions, we define and justify a variational formulation of both the stationary (1.2) and the time-dependent stochastic boundary value problem (1.1). We state conditions securing existence of a unique solution in both cases, and study a related sequence of variational problems over the Sobolev space $H_0^1(\mathcal{D})$. Next, we show how one can approximate the solution of the variational equations by applying a Galerkin type argument. This leads us to algorithms for the solution of both the stationary and the time-dependent problem. In the rest of the paper we apply these algorithms on some specific examples. The first two examples are concerned with the stochastic Poisson equation with Dirichlet boundary conditions on a bounded domain. More specific, we study the two cases where we let the right-hand side in the Poisson equation be smoothed white noise and singular white noise, respectively. This type of equation has been studied by many authors, see for example [22, 25, 33]. In particular, the approach in [22] is related to the work in this paper. Next, we consider a parabolic version of the Wick-stochastic pressure equation, also with Dirichlet boundary conditions on a bounded domain. This equation was introduced in [21] as a stochastic model for the pressure in a fluid flowing through a porous medium. The equation (in particular its stationary version) has been studied extensively, see for example [1, 3, 18, 21, 27, 28, 32, 31]. Finally, in the last part of the article, we consider a Wick version of the one-dimensional viscous Burgers' equation with stochastic source. This equation was introduced as a stochastic model for growth of interfaces of solids in [20] and it is of interest to find good numerical methods for its solution. Even though the convergence results in [27, 28] only are valid for linear Wick-stochastic equations, we have attempted to apply the same ideas to the non-linear Burgers' equation. In fact, it turns out that much of the methodology from [27, 28] can be used also in this case. In particular, we still find the chaos expansion through a sequence of deterministic problems, but with the first equation being non-linear. The resulting numerical method seems to behave stably, but whether or not the method converges to the correct solution is still an open problem.

Through this introduction we have underlined that the stochastic partial differential equations we consider are of Wick type. This is because we are concerned with problems where the stochastic noise appears multiplicatively (1.3). It should be noted that the methods described here work also when the coefficients of the differential operator $L^{\diamond}$ are not stochastic. In this case, the Wick products in (1.3) are just ordinary products, something which leads to simplifications in our approach. We will see an example of this type of simplification below, when we consider the stochastic Poisson equation.

We give an outline of the paper: In Section 2 we introduce notation and present a few necessary preliminary results. In Section 3 we study variational (or weak) formulations of (1.1) and (1.2) over a type of Sobolev-space valued stochastic Hilbert space, and give sufficient conditions for existence and uniqueness of a solution. In Section 5 and Section 6 we present numerical methods for the variational formulations of (1.2) and (1.1), respectively. We also report convergence results and discuss some aspects of the implementation of these methods. In the rest of the article we apply the methods to some specific examples. In Section 7 we study the stochastic Poisson equation driven by singular and smoothed white noise. In Section 8 we study the time-dependent Wick-stochastic pressure equation. Finally, in Section 9 we attempt to apply our numerical method to the one-dimensional viscous Wick-stochastic Burgers' equation with stochastic source.

2. Preliminaries

In this section we introduce notation and present some results needed in the later parts of this article. For the interested reader we recommend the introduction to white noise analysis given in [22] and [16].

Our basic probability triple is $(\mathcal{S}'(\mathbb{R}^d), \mathcal{B}(\mathcal{S}'(\mathbb{R}^d)), \mu)$ where as usual $\mathcal{S}'(\mathbb{R}^d)$ denotes the space of tempered distributions endowed with the weak–$*$ topology, $\mathcal{B}(\mathcal{S}'(\mathbb{R}^d))$ denotes the family of Borel subsets of $\mathcal{S}'(\mathbb{R}^d)$, and μ denotes the Gaussian white noise probability measure on $\mathcal{B}(\mathcal{S}'(\mathbb{R}^d))$ [22].

For $n \in \mathbb{N}_0, x \in \mathbb{R}$ define the Hermite polynomial

$$h_n(x) = (-1)^n e^{x^2/2} \frac{d^n}{dx^n}(e^{-x^2/2}), \tag{2.1}$$

and for $n \in \mathbb{N}$ define the Hermite functions

$$\xi_n(x) = \pi^{-1/4}((n-1)!)^{-1/2} e^{-x^2/2} h_{n-1}(\sqrt{2}x). \tag{2.2}$$

It is well known that $\xi_n \in \mathcal{S}(\mathbb{R})$, $\|\xi_n\|_\infty \le 1$ $(n \in \mathbb{N})$, and that $\{\xi_n : n \in \mathbb{N}\}$ constitutes an orthonormal basis in $L^2(\mathbb{R}, dx)$. Moreover, by forming tensor products of $\{\xi_n : n \in \mathbb{N}\}$ as described in [27], we may construct an orthonormal basis for $L^2(\mathbb{R}^d, dx)$, denoted $\{\eta_j : j \in \mathbb{N}\} \subset \mathcal{S}(\mathbb{R}^d)$.

Let $\mathcal{I}$ denote the set of all multi-indices $\alpha = (\alpha_1, \alpha_2, \cdots)$ with $\alpha_i \in \mathbb{N}_0$ $(i \in \mathbb{N})$ and only finitely many non-zero elements. We define the partial ordering $\alpha \preceq \beta$ $(\alpha \prec \beta)$ if $\alpha_i \le \beta_i$ $(\alpha_i < \beta_i)$ for all $i \in \mathbb{N}$. That is, $\alpha \preceq \gamma$ $(\alpha \prec \gamma)$ if and only if there exists $\beta \in \mathcal{I}$ such that $\alpha + \beta = \gamma$ $(\alpha + \beta = \gamma$ and $\beta \neq 0)$.

For each $\alpha \in \mathcal{I}$ we define the stochastic variable

$$H_\alpha(\omega) := \prod_{j=1}^{l(\alpha)} h_{\alpha_j}(\langle \omega, \eta_j \rangle). \tag{2.3}$$

where $l(\alpha)$ denotes the greatest index $i \in \mathbb{N}$ for which α_i is non-zero. In [16] the following result is proved.

Theorem 2.1. *The family* $\{H_\alpha : \alpha \in \mathcal{I}\}$ *constitutes an orthogonal basis for* $L^2(\mathcal{S}', \mathcal{B}(\mathcal{S}'), \mu)$*, and it holds* $E[H_\alpha H_\beta] = \alpha!\delta_{\alpha\beta}$.

Thus, every f in $L^2(\mathcal{S}', \mathcal{B}(\mathcal{S}'), \mu)$ has a unique expansion $f = \sum_{\alpha\in\mathcal{I}} f_\alpha H_\alpha$ where $f_\alpha \in \mathbb{R}$, and $||f||^2_{L^2(\mu)} = \sum_{\alpha\in\mathcal{I}} c_\alpha^2 \alpha!$. This expansion is often referred to as the Wiener-Itô chaos expansion. We will in the following adopt the notation f_α to denote the αth chaos coefficient of a function f.

Now we define a class of Hilbert space valued stochastic function spaces, which we use in the variational formulation of the boundary value problems studied below.

Definition 2.2. Given constants $\rho \in [-1, 1]$, $k \in \mathbb{R}$ and a separable Hilbert space H. Define

$$(\mathcal{S})^{\rho,k,H} := \{f = \sum_{\alpha\in\mathcal{I}} f_\alpha H_\alpha \ : \ f_\alpha \in H, \ \|f\|_{\rho,k,H} < \infty\}, \tag{2.4}$$

where the norm $\|\cdot\|_{\rho,k,H}$ is induced by the inner product

$$(f, g)_{\rho,k,H} := \sum_{\alpha\in\mathcal{I}} (f_\alpha, g_\alpha)_H (\alpha!)^{1+\rho} (2\mathbb{N})^{k\alpha}, \tag{2.5}$$

with weights $(2\mathbb{N})^{k\alpha} := \prod_{j=1}^{l(\alpha)} (2j)^{k\alpha_j}$.

Note that for negative ρ and k the sums in (2.4) are formal sums. That is, they do not necessarily converge in $L^1(\mathcal{S}', \mathcal{B}(\mathcal{S}'), \mu)$. The following result was essentially proved in [30].

Lemma 2.3. *The space* $(\mathcal{S})^{\rho,k,H}$ *is a separable Hilbert space isomorphic to* $H \otimes (\mathcal{S})^{\rho,k,\mathbb{R}}$. *If* $k \leq k'$ *then* $(\mathcal{S})^{\rho,k',H} \hookrightarrow (\mathcal{S})^{\rho,k,H}$ *(where* $\hookrightarrow$ *denotes existence of a continuous linear injection with dense image). Finally, if* V *is a Hilbert space such that* $V \hookrightarrow H$*, then* $(\mathcal{S})^{\rho,k,V} \hookrightarrow (\mathcal{S})^{\rho,k,H}$.

Remark 2.4. The above spaces can be used to define Hilbert-space valued versions of the Kondratiev test function and distribution spaces [8]. This will not be needed in the present article.

Example 2.5. Let $\phi \in \mathcal{S}(\mathbb{R}^d)$, assume $\mathcal{D} \subset \mathbb{R}^d$ bounded, and set $\phi_x(\cdot) := \phi(\cdot - x)$. Then the smoothed white noise process, defined by $W_\phi(x, \omega) := \langle\omega, \phi_x\rangle = \sum_{i=1}^{\infty} (\phi_x, \eta_i) H_{\epsilon_i}(\omega)$, $(x \in \mathcal{D})$, is in $(\mathcal{S})^{1,k,m,\mathcal{D}}$ for any $k \leq 0$ and any $m \in \mathbb{N}_0$. Moreover, the singular white noise process, defined by $W(x) := \sum_{i=1}^{\infty} \eta_i(x) H_{\epsilon_i}$, $(x \in \mathcal{D})$, belongs to $(\mathcal{S})^{1,k,1,\mathcal{D}}$ for any $k < -2/3$.

Let H' denote the dual space of H and let $\langle\cdot,\cdot\rangle$ denote the dual pairing. Then the space $(\mathcal{S})^{-\rho,-k,H'}$ may be regarded as the dual of $(\mathcal{S})^{\rho,k,H}$ under the pairing

$$\langle\langle F, f\rangle\rangle := \sum_{\alpha\in\mathcal{I}} \langle F_\alpha, f_\alpha\rangle \alpha! \tag{2.6}$$

This paring is well defined [28].

If H is one of the Sobolev spaces $H^m(\mathcal{D})$ or $H_0^m(\mathcal{D})$ ($m \in \mathbb{N}_0$ and $\mathcal{D}$ an open subset of $\mathbb{R}^d$), we denote the associated spaces by $(\mathcal{S})^{\rho,k,m,\mathcal{D}}$ and $(\mathcal{S})_0^{\rho,k,m,\mathcal{D}}$, respectively. We denote the corresponding inner product by $(\cdot,\cdot)_{\rho,k,m,\mathcal{D}}$, and we omit the domain $\mathcal{D}$ from the notation whenever it is clear from the context.

Remark 2.6. It is well known that $H_0^1(\mathcal{D})$ and $L^2(\mathcal{D})$ form separable Hilbert spaces, with $H_0^1(\mathcal{D})$ densely embedded in $L^2(\mathcal{D})$. From Lemma 2.3 it follows

$$(\mathcal{S})_0^{\rho,k,1} \hookrightarrow (\mathcal{S})^{\rho,k,0} \hookrightarrow (\mathcal{S})^{-\rho,-k,-1}, \tag{2.7}$$

where $\rho \in [-1,1], k \in \mathbb{R}$ and $H^{-1}(\mathcal{D})$ denotes the dual of $H_0^1(\mathcal{D})$ [6].

Definition 2.7. Let $f = \sum_\alpha f_\alpha H_\alpha$ be in $(\mathcal{S})^{\rho,k,m,\mathcal{D}}$ ($\rho \in [-1,1], k \in \mathbb{R}, m \in \mathbb{N}$) and let β be some multi-index in $\mathbb{N}_0^d$ with $|\beta| \leq m$. Then the derivative $D^\beta f$ is defined as

$$D^\beta f := \sum_\alpha D^\beta f_\alpha H_\alpha. \tag{2.8}$$

Here $D^\beta f_\alpha$ denotes the derivative in the usual weak sense.

Definition 2.8. For any formal sum $f = \sum_\alpha f_\alpha H_\alpha$, the zeroth order chaos coefficient f_0 is called the generalized expectation of f and denoted $E[f]$.

Remark 2.9. The above definition of expectation generalizes the ordinary expectation of $f \in L^p(\mathcal{S}'(\mathbb{R}^d), \mathcal{B}(\mathcal{S}'(\mathbb{R}^d)), \mu)$ ($p > 1$). Note that $E[f \diamond g] = E[f]\, E[g]$ for any given pair of (generalized) stochastic variables f and g, a property which does not hold for the ordinary product [22].

Definition 2.10. The Wick product $f \diamond g$ of two formal series $f = \sum_\alpha f_\alpha H_\alpha$, $g = \sum_\alpha g_\alpha H_\alpha$ is defined as

$$f \diamond g := \sum_{\alpha,\beta\in\mathcal{I}} f_\alpha g_\beta H_{\alpha+\beta}. \tag{2.9}$$

This product is associative, commutative and distributive. For a discussion on the modeling properties of the Wick product we refer to [22], here we only note that it may be viewed as a regularization of the product of two generalized stochastic variables. The reader should note that if one of the terms are deterministic, then the Wick product coincides with the ordinary point-wise product.

In order to controll the behaviour of the Wick product we define the Banach space

$$\begin{aligned}\mathcal{F}_l(\mathcal{D}) := \{f(x) = \sum_\alpha f_\alpha(x)H_\alpha : f_\alpha(x) \text{ measurable}, \\ \|f\|_{l,*} := \sup_{x\in\mathcal{D}}(\sum_\alpha |f_\alpha(x)|(2\mathbb{N})^{l\alpha}) < \infty\}.\end{aligned} \tag{2.10}$$

From [32] we have the following result.

Proposition 2.11. *Let $\mathcal{D} \subset \mathbb{R}^d$ be open and let $l, k \in \mathbb{R}$ such that $k \leq 2l$. Then the map $g \mapsto f \diamond g$ defines a continuous linear operator on $(\mathcal{S})^{-1,k,0,\mathcal{D}}$. Furthermore, we have*

$$\|f \diamond g\|_{-1,k,0} \leq \|f\|_{k/2,*}\|g\|_{-1,k,0} \leq \|f\|_{l,*}\|g\|_{-1,k,0}, \tag{2.11}$$

for all $g \in (\mathcal{S})^{-1,k,0,\mathcal{D}}$.

Example 2.12. The singular white noise process W_x from Example 2.5 belongs to $\mathcal{F}_l(\mathcal{D})$ for $l < -1$ and any open set $\mathcal{D}$. Moreover, the Wick exponential of (singular) white noise, defined by

$$\exp^\diamond(W(x)) = \sum_{n=0}^{\infty} \frac{1}{n!} W(x)^{\diamond n},$$

is an element in $\mathcal{F}_l(\mathcal{D})$ for $l < -1$ and any open set $\mathcal{D}$.

From [28] we have

Corollary 2.13. *It holds $\mathcal{F}_l(\mathcal{D}) \subset (\mathcal{S})^{-1,k,0,\mathcal{D}}$ for all $k \leq 2l$ if and only if $\mathcal{D}$ has finite measure.*

Later we will need the following result.

Corollary 2.14. *Let $\mathcal{D} \subset \mathbb{R}^d$ be open, bounded, with Lipschitz continuous boundary $\partial\mathcal{D}$, and suppose $u \in (\mathcal{S})^{-1,q,m,\mathcal{D}}$ with $m > d/2$. Then $u \diamond u$ is in $(\mathcal{S})^{-1,k,0,\mathcal{D}}$ for all $k < q - 1$.*

Proof. Note that by Proposition 2.11 and Corollary 2.13 it suffices to show that $(\mathcal{S})^{-1,q,m,\mathcal{D}}$ is a subset of $\mathcal{F}_l(\mathcal{D})$ for $2l < q - 1$. This follows by an application of the Sobolev's Embedding Theorem; In particular, if

$m > d/2$ then there exists a constant $c < \infty$ such that for all $g \in H^m(\mathcal{D})$ it holds

$$\sup_{x \in \mathcal{D}} |g(x)| \leq c\|g\|_m. \tag{2.12}$$

Now, suppose that u is in $(\mathcal{S})^{-1,q,m,\mathcal{D}}$, then the chaos coefficients u_α are in $H^m(\mathcal{D})$ $(\alpha \in \mathcal{I})$. Thus, using (2.12) together with Cauchy-Schwarz's inequality gives

$$\begin{aligned}
\|u\|_{l,*} &\leq \sum_{\alpha \in \mathcal{I}} \sup_{x \in \mathcal{D}} |u_\alpha(x)|(2\mathbb{N})^{l\alpha} \\
&\leq c \sum_{\alpha \in \mathcal{I}} \|u_\alpha\|_m (2\mathbb{N})^{l\alpha} \\
&\leq c\Big(\sum_{\alpha \in \mathcal{I}} (2\mathbb{N})^{(2l-q)\alpha}\Big)^{1/2} \|u\|_{-1,q,m} < \infty,
\end{aligned}$$

where the last inequality follows from the assumption $2l < q-1$ and the result: $\sum_{\alpha \in \mathcal{I}} (2\mathbb{N})^{k\alpha} < \infty$ if and only if $k < -1$ [22]. □

Remark 2.15. Corollary 2.14 tells us that by restricting to the smaller space $(\mathcal{S})^{-1,q,m,\mathcal{D}} \subset (\mathcal{S})^{-1,k,0,\mathcal{D}}$ $(k \leq q-1)$ we are able to control the behavior of the Wick product. This is useful when looking at non-linear equations like the Burgers' equation (cf. Section 9).

Remark 2.16. Using a simple induction argument it is possible to strengthen the result in Corollary 2.14 and show that $u \diamond u$ is in the smaller space $(\mathcal{S})^{-1,k,m,\mathcal{D}}$ $(k < q-1)$. This will not be needed in the following.

3. Variational formulations

In this section we introduce variational formulations of (1.1) and (1.2). The stationary problem (1.2) was studied in [27] and the time-dependent problem (1.1) in [28], and most of the result presented in this section we give without detailed proofs. We refer to [27, 28] for the details. Other references for such variational formulations for Wick-stochastic partial differential equations include [31, 29, 32].

In the following we assume that $\mathcal{D} \subset \mathbb{R}^d$ is an open, bounded domain with Lipschitz continuous boundary $\partial\mathcal{D}$. For notational convenience, we let V and H denote $(\mathcal{S})_0^{-1,k,1,\mathcal{D}}$ and $(\mathcal{S})^{-1,k,0,\mathcal{D}}$, respectively. We let $\|\cdot\|$, $|\cdot|$, and $\|\cdot\|_*$ denote the norms in V, H and V', respectively, and we let $(\cdot,\cdot)$ denote the inner product on H.

Remark 3.1. Note that this notation hides the fact the V and H depends on the parameter k.

From Remark 2.6 we know that V and H are separable, and by identifying H with its dual H' we have from (2.7) that $V \hookrightarrow H \cong H' \hookrightarrow V'$. One consequence of this relation is that that the pairing between V' and V given in (2.6) may be viewed as the continuous extension of the inner product on H, using the relation

$$\langle\langle \mathrm{Id}'(h), v\rangle\rangle = (h, \mathrm{Id}(v)), \quad \forall v \in V, h \in H,$$

where Id and Id′ denotes the continuous injections of V into H and of H into V', respectively.

We now consider the stationary problem (1.2). Following the exposition in [27] it is straight forward to show that (under suitable regularity assumptions on the data) it holds

$$(L^{\diamond}u, v) = \mathcal{A}(u, v) := \sum_{i,j=1}^{d} (a^{ij} \diamond D_j u, D_i v) + \sum_{i=1}^{d} (b^i \diamond D_i u, v) + (c \diamond u, v), \tag{3.1}$$

for any $u, v \in V$. From Proposition 2.11 the following result is evident.

Corollary 3.2. *Suppose that a^{ij}, b^i, c are in $\mathcal{F}_l(\mathcal{D})$ for l such that $k \leq 2l$ $(i, j = 1, \ldots, d)$. Then $\mathcal{A}$ defines a continuous bilinear form on V and it holds*

$$|\mathcal{A}(u, v)| \leq c\|u\|\|v\|, \quad (u, v \in V) \tag{3.2}$$

for some constant $c < \infty$.

Proof. This is just a term-wise application of relation (2.11). □

Definition 3.3. Assume given a right-hand side f in V' (for some parameter $k \in \mathbb{R}$). Then we define the variational formulation of (1.2) as

$$\text{Find } u \in V \text{ such that } \mathcal{A}(u, v) = (f, v), \qquad (v \in V). \tag{3.3}$$

Remark 3.4. The requirement $u \in V$ implies the boundary condition $u|_{\partial\mathcal{D}} = 0$. In the case of the more general boundary condition $u|_{\partial\mathcal{D}} = g$ with g in $(\mathcal{S})^{-1,k,1}$, we may rewrite the problem to the form in (3.3) by setting $\hat{u} = u - g$ and solving

$$\text{Find } \hat{u} \in V \text{ such that } \mathcal{A}(\hat{u}, v) = (f, v) - \mathcal{A}(g, v), \qquad (v \in V).$$

In the following we only consider the case $u|_{\partial\mathcal{D}} = 0$.

Before we give conditions securing existence of a unique solution to (3.3), we consider the bilinear form $\mathcal{A}$. Assume that a^{ij}, b^i, c are in $\mathcal{F}_l(\mathcal{D})$ ($i,j = 1,\ldots,d$, $k \le 2l$), and suppose that u is in $(\mathcal{S})^{-1,k,0}$. Then, if we set $v = wH_\gamma$ with $w \in H_0^1(\mathcal{D})$ and $\gamma \in \mathcal{I}$, we get from the definition of the inner product and the Wick-product that

$$\mathcal{A}(u, wH_\gamma) = \sum_{\alpha \preceq \gamma} \mathcal{B}_{\gamma-\alpha}(u_\alpha, w)(2\mathbb{N})^{k\gamma}, \tag{3.4}$$

where we have defined the bilinear operators

$$\mathcal{B}_\beta(h,w) := \sum_{i,j=1}^{d} (a^{ij}_\beta D_j h, D_i w)_0 + \sum_{i=1}^{d} (b^i_\beta D_i h, w)_0 + (c_\beta h, w)_0, \tag{3.5}$$

for each $\beta \in \mathcal{I}$, $h, w \in H_0^1(\mathcal{D})$. Recall that the bilinear form $\mathcal{B}_0$ is called coercive on $H_0^1(\mathcal{D})$ if there exits a constant $\theta_0 > 0$ such that

$$\mathcal{B}_0(h,h) \ge \theta_0 \|h\|_1^2,$$

for all h in $H_0^1(\mathcal{D})$. From [27] we have the following lemma:

Lemma 3.5. *If $\mathcal{B}_0$ is coercive on $H_0^1(D)$, then there exists constants, $k_0, \theta \in \mathbb{R}$, such that $k_0 \le 2l$, $\theta > 0$, and the bilinear form $\mathcal{A}$ satisfies*

$$\mathcal{A}(v,v) \ge \theta \|v\|^2, \tag{3.6}$$

for all v in V and all $k < k_0$ (k denotes the parameter used in the definition of the space V).

The next result is a consequence of Lemma 3.5 and the Lax-Milgram Theorem. A detailed proof can be found in [27].

Theorem 3.6. *Suppose that a^{ij}, b^i, c are in $\mathcal{F}_l(D)$ for some l such that $k \le 2l$ ($i,j = 1,\ldots,d$), and let f be in V'. Moreover, suppose that $\mathcal{B}_0$ is coercive on $H_0^1(D)$ and that the parameter k is small enough. Then the variational problem* (3.3) *has a unique solution.*

The relation (3.4) has an interesting consequence for the variational problem (3.3). If we choose $v = wH_\gamma(2\mathbb{N})^{-k\gamma}$ ($\gamma \in \mathcal{I}, w \in H_0^1(\mathcal{D})$) in (3.3), then it is clear from (3.4) that the chaos coefficients of u satisfies the following system of variational problems

For each $\gamma \in \mathcal{I}$ find $u_\gamma \in H_0^1(D)$ such that

$$\mathcal{B}_0(u_\gamma, w) = (f_\gamma, w)_0 - \sum_{\alpha \prec \gamma} \mathcal{B}_{\gamma-\alpha}(u_\alpha, w), \qquad (w \in H_0^1(D)). \tag{3.7}$$

Theorem 3.7. *Suppose that $a_\gamma^{ij}, b_\gamma^i, c_\gamma$ are in $L^\infty(D)$ and let f_γ be in $H^{-1}(\mathcal{D}) = (H_0^1(D))'$ $(i,j = 1,\ldots,d, \gamma \in \mathcal{I})$. Moreover, suppose that the bilinear form $\mathcal{B}_0$ is coercive on $H_0^1(D)$. Then there exist a unique set of functions $\{u_\gamma \in H_0^1(D) : \gamma \in \mathcal{I}\}$ solving* (3.7).

Proof. The proof of this result follows the lines of the proof of Theorem 4.2 in [27]. The central observation is that we may solve (3.7) as a sequence of problems where the right-hand side is known when the γth equation is reached. The result then follows by induction on γ and an application of Lax-Milgram theorem. □

The assumptions in Theorem 3.7 are weaker than those in Theorem 3.6. Thus, we can find the chaos coefficients of the solution of (3.3) by solving the set of coupled variational problems in (3.7). Moreover, the coupling in (3.7) is relatively simple in the sense that with a suitable ordering of the multi-indices $\mathcal{I}$ it is possible to solve (3.7) as a sequence of variational problems over $H_0^1(D)$, each problem depending only on a subset of the previously solved problems. The crucial point in such an ordering is to make sure that the set of chaos coefficients $\{u_\beta : \beta \prec \gamma\}$ is known when we reach the γth equation.

Remark 3.8. The reader should at this point note that the numerical solution of variational problems over $H_0^1(D)$ is well-known [4, 6, 26], thus, with an ordering of $\mathcal{I}$ as described above, we could create a numerical solution algorithm for (3.3) based on solving the system (3.7) using some deterministic solution solver, and in this way approximate the chaos expansion of the solution. This approach is the essence of the numerical method described in Section 5.

We now turn to the parabolic version of our problem, that is, we consider the time-dependent problem (1.1). We assume, for convenience, that the coefficients in the differential operator $L^\diamond$ are independent of $t \in [0,T]$. The problem of time-dependent coefficients is treated in [28]. Before we can give the variational formulation of this equation we recall some preliminary results.

Definition 3.9. Let X, Y be separable Hilbert spaces, and such that $Y \hookrightarrow X \hookrightarrow Y'$. Define

$$W(0,T;Y) := \{f \in L^2(0,T;Y) : D_t f \in L^2(0,T;Y')\}.$$

The following result is proved in [6].

Proposition 3.10. *The space $W(0,T;Y)$ forms a Hilbert space with respect to the inner product*

$$(f,g)_{W(0,T;Y)} := \int_0^T (f(s), g(s))_Y ds + \int_0^T (D_t f(s), D_t g(s))_{Y'} ds, \quad (3.8)$$

and

$$W(0,T;Y) \hookrightarrow C^0(0,T;X) \tag{3.9}$$

where $C^0([0,T];X)$ denotes the space of continuous functions from $[0,T]$ into X, equipped with the norm of uniform convergence.

Following the idea outlined in [28] we assume $f \in L^2(0,T;V')$ and $u \in W(0,T;V)$ and take the H-inner product of (1.1) with any $v \in V$. A partial integration (under suitable assumptions of regularity on the data) gives

$$(D_t u(t), v) + \mathcal{A}(u(t), v) = (f(t), v),$$

for almost every $t \in [0,T]$. Note that the embedding (3.9) implies that the trace $u(0)$ has sense, and also note that $(D_t u(\cdot), v) = D_t(u(\cdot), v)$ in distributive sense on $(0,T)$ [6].

Definition 3.11. Assume given a right-hand side f in $L^2(0,T;V')$ and an initial value u_0 in H (for some parameter $k \in \mathbb{R}$). Then we define the variational formulation of (1.1) as

$$\begin{aligned} &\text{Find } u \in W(0,T;V) \text{ such that } u(0) = u_0 \text{ and} \\ &\qquad D_t(u(\cdot), v) + \mathcal{A}(u(\cdot), v) = (f(\cdot), v), \qquad (v \in V). \end{aligned} \tag{3.10}$$

This equation is understood in a distributive sense on $(0,T)$, and the equality $u(0) = u_0$ in the sense of H.

From [28] we have the following theorem.

Theorem 3.12. *Suppose that a^{ij}, b^i, c are in $\mathcal{F}_l(D)$ for some l such that $k \le 2l$ $(i,j = 1,\ldots,d)$, let f be in $f \in L^2(0,T;V')$, and assume given an initial value u_0 in H. Moreover, suppose that $\mathcal{B}_0$ is coercive on $H_0^1(D)$ and that the parameter k is small enough. Then there exists a unique u in $W(0,T;V)$ satisfying* (3.10)*. Moreover, we have the energy inequality*

$$\max_{t\in[0,T]} |u(t)| + \int_0^T \|u(t)\| dt \le c\,\left(|u_0|^2 + \int_0^T \|f(t)\|_*^2 dt\right). \tag{3.11}$$

for some constant $c < \infty$.

As in the stationary case, the use of the Wick-product in the differential operator $L^\diamond$ has some interesting implications for the problem (3.10). More specificly, if we set $v = wH_\gamma(2\mathbb{N})^{-k\gamma}$ $(\gamma \in \mathcal{I}, w \in H_0^1(\mathcal{D}))$ in (3.10), then it is clear from (3.4) and the definition of the inner product in H, that the chaos coefficients of the solution u satisfies

$$\begin{aligned}&\text{For each } \gamma \in \mathcal{I}, \text{ find } u_\gamma \in W(0,T;H_0^1(\mathcal{D}))\\&\text{such that } u_\gamma(0) = u_{0,\gamma} \text{ and, for } w \in H_0^1(\mathcal{D}),\\&D_t(u_\gamma(\cdot),w)_0 + \mathcal{B}_0(u_\gamma(\cdot),w) = (f_\gamma(\cdot),w)_0 - \sum_{\alpha\prec\gamma} \mathcal{B}_{\gamma-\alpha}(u_\alpha(\cdot),w).\end{aligned} \tag{3.12}$$

This equation is understood in distributive sense on $(0,T)$, and the equality $u_\gamma(0) = u_{0,\gamma}$ in the sense of $L^2(\mathcal{D})$.

Theorem 3.13. *Let $a_\gamma^{ij}, b_\gamma^i, c_\gamma$ be in $L^\infty(D)$, let f_γ be in $L^2(0,T;H^{-1}(\mathcal{D}))$, assume that $u_{0,\gamma}$ is in $L^2(\mathcal{D})$ $(i,j = 1,\dots,d,\alpha \in \mathcal{I})$, and suppose that $\mathcal{B}_0$ is coercive on $H_0^1(D)$. Then there exists a unique set of functions $\{u_\gamma \in W(0,T;H_0^1(\mathcal{D})) : \gamma \in \mathcal{I}\}$ solving* (3.12).

A proof of this theorem can be found in [28]. Note that analogous to the stationary case, with a suitable ordering of the multi-indices $\mathcal{I}$ we can solve (3.12) as a sequence of time-dependent variational problems over $H_0^1(\mathcal{D})$, and this gives the chaos coefficients of the solution to (3.10). This idea is the essence of the numerical approach described in Section 6.

4. A Galerkin approximation of V

The numerical approximation of the variational problems (3.3) and (3.10) presented in this paper, are based on a Galerkin approximation of the space V. In this section we show briefly how to construct such a Galerkin approximation and in the following two sections we use this approximation to construct approximations of (3.3) and (3.10), respectively.

The Galerkin approximation presented here is based on the finite element method and appeared first (as far as we know) in [1]. See also [27] where a similar approximation is treated. Before we proceed, recall from [6] that a Galerkin approximation of a Hilbert space U is a family of finite-dimensional vector spaces $\{U_m : m \in \mathbb{N}\}$ satisfying: (1) $U_m \subset U$, and (2) $U_m \to U$ when $m \to \infty$ in the following sense: There exists a set $\mathcal{U}$ such that $\mathcal{U} \hookrightarrow U$, and such that for all $v \in \mathcal{U}$, we can find a sequence $\{v_m : m \in \mathbb{N}\}$ with the property: For all $m \in \mathbb{N}$ we have $v_m \in U_m$ and $v_m \to v$ in U as $m \to \infty$.

There is a natural approach to the construction of a Galerkin approximation for $V := (\mathcal{S})_0^{-1,k,1}$ $(k \in \mathbb{R})$. Each $f \in V$ has a chaos expansion representation

$$f = \sum_{\alpha\in\mathcal{I}} f_\alpha H_\alpha, \quad (f_\alpha \in H_0^1(\mathcal{D}), \alpha \in \mathcal{I}).$$

Thus, we may approximate f by including only a finite number of multi-indices in the chaos expansion and approximate each remaining chaos coefficients in a finite element space (defined below). In the rest of this section we will discuss details in this construction. We assume spatial dimension $d = 1, 2$ or 3 and for convenience we also assume that $\mathcal{D}$ is a polyhedral domain in $\mathbb{R}^d$. The last assumption could be relaxed since interpolation results exists for finite element spaces over domains with for example Lipschitz continuous boundary [4].

For $N, K \in \mathbb{N}$ define the cutting $\mathcal{I}_{N,K} \subset \mathcal{I}$ by

$$\mathcal{I}_{N,K} := \{0\} \cup \bigcup_{n=1}^{N} \bigcup_{k=1}^{K} \{\alpha \in \mathbb{N}_0^k \mid |\alpha| = n \text{ and } \alpha_k \neq 0\}. \tag{4.1}$$

This set can be shown to contain $(N + K)!/(N!K!)$ different multi-indices. The resulting space

$$(\mathcal{S}_{N,K})_0^{-1,k,1} := \{ f = \sum_{\alpha \in \mathcal{I}_{N,K}} f_\alpha H_\alpha : f_\alpha \in H_0^1(\mathcal{D}), \|f\|_{-1,k,1} < \infty\}, \tag{4.2}$$

is clearly a subspace of V for any choice of N, K.

The problem of finite element approximations of spaces like $H_0^1(\mathcal{D})$ has been studied extensively, and here we follow the approach described in [4, 26]. We start by giving some notation and definitions.

A triangulation of a polyhedral domain $\mathcal{D}$ is a finite collection of open triangles (or tetrahedra) $\{T_i : i = 1, \dots, l\}$ such that $T_i \cap T_j = \emptyset$ if $i \neq j$, $\bigcup_{i=1}^{l} \overline{T_i} = \overline{\mathcal{D}}$, and no vertex of any tetrahedron lies in the interior of an edge (or face) of any other tetrahedron. A family $\{\mathcal{T}_h : h \in (0, 1]\}$ of triangulations of the domain $\mathcal{D}$ is said to be non-degenerate if

$$\max\{\operatorname{diam} T : T \in \mathcal{T}_h\} \leq h \ \operatorname{diam} \mathcal{D},$$

and there exists a $\rho > 0$ such that for all $T \in \mathcal{T}_h$ and all $h \in (0, 1]$ it holds

$$\operatorname{diam} B_T \geq \rho \ \operatorname{diam} T,$$

where B_T is the greatest ball contained in the tetrahedron T. We call h the grid-size of a given triangulation $\mathcal{T}_h$.

Let $\{\mathcal{T}_h \mid h \in (0, 1]\}$ denote a non-degenerate family of triangulations of the domain $\mathcal{D} \subset \mathbb{R}^d$. For a given domain $T \subset \mathbb{R}^d$ we let $\mathbb{P}_n(T)$ denote the family of polynomials of degree less or equal to n on T. That is,

$$\mathbb{P}_n(T) := \{ \sum_{\beta \in \mathbb{N}^d, |\beta| \leq n} a_\beta x^\beta : x \in T\}.$$

For each $h \in (0,1]$, $n \in \mathbb{N}$ we define the finite element space

$$X_h^n(\mathcal{D}) := \{v \in C^0(\overline{\mathcal{D}}) : v|_{\partial\mathcal{D}} = 0, v \in \mathbb{P}_n(T) \text{ for each } T \in \mathcal{T}_h\}. \quad (4.3)$$

Note that a simple application of Green's formula shows $X_h^n(\mathcal{D}) \subset H_0^1(\mathcal{D})$.

Moreover, any function in $X_h^n(\mathcal{D})$ may be defined by prescribing the function on each tetrahedron $T \in \mathcal{T}_h$, and imposing the conditions $v \in C^0(\overline{\mathcal{D}})$ and $v|_{\partial\mathcal{D}} = 0$. In order to identify the polynomial $v|_T \in \mathbb{P}_n(T)$ it suffices to prescribe its value in $(d+n)!/(d!n!)$ distinct points (or nodes) in $\overline{T}$ (that is, we use Lagrange elements [4]). It is essential that these points are chosen such that elements sharing a common edge (face) use the same evaluation points on that edge (face) [28].

A basis for $X_h^n(\mathcal{D})$ is now easily constructed. In particular, let a_i $(i = 1, \dots, M)$ denote the global set of evaluation points (or nodes) in $\overline{\mathcal{D}}$. Then it suffices to choose functions $\psi_i \in X_h^n(\mathcal{D})$ such that $\psi_i(a_j) = \delta_{ij}$ $(i, j = 1, \dots, M)$. These functions are called the shape functions. With this basis, we define the interpolant $I^h : C^0(\overline{\mathcal{D}}) \to X_h^n(\mathcal{D})$ as the map

$$I^h(f) := \sum_{i=1}^{M} f(a_i)\psi_i, \qquad (f \in C^0(\overline{\mathcal{D}})).$$

For convenience, we will sometimes use the notation f^M to denote the interpolant $I^h(f)$.

Remark 4.1. It is important to notice that the support of each shape function is "small". It consists of only a few elements from the triangulation. This is one of the good properties of the finite element method, and it has the advantage that the resulting linear systems (see Section 5 and Section 6) become sparse.

We define our approximating spaces

Definition 4.2.

$$(\mathcal{S}_{N,K,M})_0^{-1,k,1,\mathcal{D}} := \{f \in (\mathcal{S})_0^{-1,k,1,\mathcal{D}} : f = \sum_{\alpha \in \mathcal{I}_{N,K}} f_\alpha^M H_\alpha, f_\alpha^M \in X_h^n(\mathcal{D})\}.$$

We let $V_m^{K,M,N}(\mathcal{D})$ $(m := \min(K, M, N))$ denote the approximating space $(\mathcal{S}_{N,K,M})_0^{-1,k,1,\mathcal{D}}$ and we omit the triple K, M, N and the domain $\mathcal{D}$ from the notation whenever it is clear from the context. Moreover, we assume that $k \le 0$ and write k as $-|k|$ whenever the sign is important for the interpretation of the text. This assumption is not restrictive, because from Theorem 3.6 and Theorem 3.12 and the inclusion in Lemma 2.3 it is clear that we may always choose $k \le 0$.

From [28, Corollary 5,4] we have

Corollary 4.3. *For $k \leq 0$ the family of spaces $\{V_m^{K,M,N} : M, N, K \in \mathbb{N}\}$ constitutes a Galerkin approximation of $V = (\mathcal{S})_0^{-1,-|k|,1}$. Furthermore,*

$$\dim\,(V_m^{K,M,N}) = \binom{N+K}{K} M.$$

5. A numerical method for the stationary case

In this section we describe a numerical method for the stationary problem (3.3). The work presented here can be found in more detail in [27].

The (finite-dimensional) Galerkin approximation of the variational problem (3.3) is given by

$$\text{Find } u_m \in V_m \text{ such that } \mathcal{A}(u_m, v) = (f, v), \qquad (v \in V_m). \tag{5.1}$$

It is a consequence of Theorem 3.6 and the relation $V_m \subset V$, that under the same assumptions as in Theorem 3.6, there exists a unique solution of (5.1) for each $m \in \mathbb{N}$. This leads to a sequence of approximations $\{u_m : m \in \mathbb{N}\}$ for the solution of the variational problem (3.3).

We are interested in how well u_m approximates the solution u. From [27] we have the following result.

Theorem 5.1. *If in addition to the assumptions in Theorem 3.6 we suppose the following: Let $k \leq 0$, $|k| = q + r$, with $q \geq 0$, $r > r^*$, where r^* solves*

$$\frac{r^*}{2^{r^*}(r^*-1)} = 1, \qquad (r^* \approx 1.54). \tag{5.2}$$

Moreover, let u denote the solution of (3.3), and assume that u is in $(\mathcal{S})_0^{-1,-|k|,1} \cap (\mathcal{S})^{-1,-|k|,2}$. Then

$$\|u - u_m\|^2_{-1,-|k|,0} \leq c B_{N,K}(r)^2 \|u\|^2_{-1,-q,1} + ch^2 \|u\|^2_{-1,-|k|,2}, \tag{5.3}$$

where c is some positive constant independent of m and where

$$\begin{aligned} B_{K,N}(r) &= \sqrt{C_1(r)K^{1-r} + C_2(r)\left(\frac{r}{2^r(r-1)}\right)^{N+1}}, \\ C_1(r) &= \frac{1}{2^r(r-1)-r}, \qquad C_2(r) = 2^r(r-1)C_1(r). \end{aligned} \tag{5.4}$$

This establishes the convergence of the approximation, and gives an estimate for the rate of convergence. The rate is weaker than known results for classical partial differential equations [6, 26]. It follows from the expression for $B_{N,K}(r)$ that the convergence in the stochastic dimension, that is in the cutting of the chaos expansion, may be rather slow.

The system (5.1) consists of $(N+K)!/(N!K!)M$ unknowns. Thus, a priori, it seems impractical to solve this problem for all but a few choices of $N, K, M \in \mathbb{N}$. Now, since its solution u_m is in the approximating space V_m, we may write

$$u_m(x) = \sum_{\alpha \in \mathcal{I}_{N,K}} u_{m,\alpha}(x) H_\alpha, \text{ where} \tag{5.5}$$

$$u_{m,\alpha}(x) := \sum_{j=1}^{M} c_{m,\alpha,j} \psi_j(x) \in X_h^n(\mathcal{D}),$$

with real constants $c_{m,\alpha,j}$ and where ψ_j denotes the jth shape function described in Section 4 $(x \in \mathcal{D}, \alpha \in \mathcal{I}_{N,K}, j = 1, \ldots, M)$. Let $<$ denote some ordering of the multi-indices $\mathcal{I}$ with the property

$$\{\alpha \in \mathcal{I} \,:\, \alpha \prec \gamma\} \subset \{\alpha \in \mathcal{I} \,:\, \alpha < \gamma\}. \tag{5.6}$$

The corresponding ascending ordering of the set $\mathcal{I}_{N,K}$ is denoted $\mathcal{O}_{N,K}$ whenever we want to emphasize the ordering of the set. If we for a given multi-index γ in $\mathcal{O}_{N,K}$ choose $v = \psi_i H_\gamma (2\mathbb{N})^{-k\gamma}$, it follows from relation (3.4) that the vector $c_{m,\gamma} = (c_{m,\gamma,j})$ $(j = 1, \ldots, M)$ solves the M-dimensional problem

$$B_0 c_{m,\gamma} = b_{m,\gamma} - \sum_{\alpha \prec \gamma} B_{\gamma-\alpha} c_{m,\alpha}, \tag{5.7}$$

where the vector $b_{m,\gamma}$ and matrices B_β are given by

$$[b_{m,\gamma}]_i := (f_\gamma, \psi_i)_{0,\mathcal{D}}, \quad [B_\beta]_{ij} := \mathcal{B}_\beta(\psi_j, \psi_i), \tag{5.8}$$
$$(i, j = 1, \ldots, M, \ \beta \in \mathcal{I}_{N,K}).$$

The reader should observe that if we solve (5.7) choosing the multi-indices γ according to the ordering of the set $\mathcal{O}_{N,K}$, then the right-hand side in (5.7) is known when we reach the γth equation (cf. relation (5.6)). Thus, we can solve (5.1) as a sequence of $(N+K)/(N!K!)$ problems, each having M unknown variables. This approach leads to a considerable reduction in the required work-load compared with solving the full system (5.1) directly.

Remark 5.2. There is another point of view to the above discussion: If we choose an appropriate ordering σ of the unknown variables $c_{m,\alpha,j}$ $(\alpha \in \mathcal{I}_{N,K}, j = 1, \ldots, M)$. Then (5.7) shows that (5.1) can be written on the form $Ac_m = b$ where $c_m = (c_{m,\sigma(\alpha,j)})$ and the matrix A is on a lower $M \times M$-block-triangular form with the constant matrix B_0 along the diagonal [28].

Remark 5.3. There is a clear relation between the sequence of equations in (5.7) and the set of (deterministic) variational problems in (3.7): A finite element approximation of (3.7), based on the space $X_h^n(\mathcal{D})$ defined in Section 4, leads to (5.7).

Once (5.1) has been solved, we are interested in the stochastic simulation of $u_m(x)$. This simulation is based on the following idea: Recall from Section 2 that

$$H_\alpha(\omega) = \prod_{j=1}^{l(\alpha)} h_{\alpha_j}(\langle \omega, \eta_j \rangle),$$

where the stochastic variable $\omega \mapsto (\langle \omega, \eta_1 \rangle, \ldots, \langle \omega, \eta_{l(\alpha)} \rangle)$ is standard Gaussian distributed. Thus, once the solution of (5.7) is known for all $\gamma \in \mathcal{I}_{N,K}$, we may do stochastic simulation of u_m by first generating K independent standard Gaussian variables $X(\omega) = (X_i(\omega))$ $(i = 1, \ldots, K)$ using some random number generator, and then forming the sum

$$\begin{aligned} u_m(x, \omega) &= \sum_{\alpha \in \mathcal{I}_{N,K}} u_{m,\alpha}(x) H_\alpha(X(\omega)) \\ &= \sum_{\alpha \in \mathcal{I}_{N,K}} u_{m,\alpha}(x) \prod_{j=1}^{K} h_{\alpha_j}(X_j(\omega)), \qquad (x \in \mathcal{D}). \end{aligned} \tag{5.9}$$

The advantage of this approach is that it enables us to generate random samples easy and fast. For example, in situations where one is interested in repeated simulations of solutions to some particular equation, one may compute the chaos coefficients in advance, store them, and produce the simulations whenever they are needed.

Remark 5.4. In general the solution $u(x)$ of (3.3) is a stochastic distribution. Thus, the stochastic simulation (i.e. point-wise evaluation in $\mathcal{S}'$) of $u(x)$ does not make sense. We can, however, take the point of view advocated in [28]: The solution $u_m(x)$ of (5.1) gives a regular approximation of the singular object $u(x)$. Moreover, $u_m(x)$ also represents an approximation of

$$u_{N,K}(x) := \sum_{\alpha \in \mathcal{I}_{N,K}} u_\alpha(x) H_\alpha,$$

which is the best approximation of $u(x)$ we can achieve by including only the finite set of chaos-coefficients $\mathcal{I}_{N,K}$. Stochastic simulation of $u_m(x)$, therefore, approximates the stochastic behavior of $u_{N,K}(x)$, which again can be viewed as an approximation to the stochastic behavior of $u(x)$. Of course, if the solution $u(x)$ is a regular stochastic process, then this simulation approach makes perfect sense.

Summarizing; the above ideas give the following algorithm for the numerical solution of (3.3). Here we assume that an appropriate triangulation $\mathcal{T}_h$ has been chosen, with a corresponding finite element space $X_h^n(\mathcal{D})$ and set of shape functions $\{\psi_j\}$ $(j = 1, \dots, M)$. Furthermore, we assume that there has been chosen $N, K \in \mathbb{N}$ to get a cutting as described in (4.1), we let $R \in \mathbb{N}_0$ denote the number of simulations, and we assume that the set of variational equations (3.7) has been formed.

Algorithm 5.5.

(1) Form the ordered set $\mathcal{O}_{N,K}$ and set $\gamma = (0, \cdots, 0)$

(2) While $\gamma \in \mathcal{O}_{N,K}$ do

- *(2.1) Calculate $b_{m,\gamma} = [(f_\gamma, \psi_j)_{0,D}]$*
- *(2.2) Find the set $\mathcal{L}_\gamma = \{\alpha \in \mathcal{I}_{N,K} : \alpha \prec \gamma\}$*
- *(2.3) For each $\alpha \in \mathcal{L}_\gamma$*
  - *(2.3.1) Calculate the matrices $B_{\gamma-\alpha} = [\mathcal{B}_{\gamma-\alpha}(\psi_j, \psi_i)]$*
  - *(2.3.2) Update the right hand side*
    $b_{m,\gamma} := b_{m,\gamma} - B_{\gamma-\alpha} c_{m,\alpha}$
- *(2.4) Solve the problem $B_0 c_{m,\gamma} = b_{m,\gamma}$*

(3) Find the next multi-index γ (using ordering (5.10)) and go to 2

(4) Create a sequence of RK independent Gaussian variables $\{X_i, i = 1, \dots, RK\}$

(5) For each $r = 1, \dots, R$ do

- *(5.1) Set $X^{(r)} := [X_{(r-1)K+j}]$ $(j = 1, \dots, K)$*
- *(5.2) Form simulation $u_m^{(r)}(x) = \sum_{\alpha \in \mathcal{I}_{N,K}} u_{m,\alpha}(x) H_\alpha(X^{(r)})$*

(6) Print and visualize the results.

Before the application of the algorithm can start, it is necessary to choose a specific ordering of the multi-indices. In the examples given later, we used the following ordering

$$\alpha < \beta \text{ if } \begin{cases} |\alpha| < |\beta| & \text{or} \\ |\alpha| = |\beta|, & \text{bin}(\alpha) < \text{bin}(\beta), \end{cases} \tag{5.10}$$

where $\text{bin}(\alpha)$ denotes the binary number given by

$$\text{bin}(\alpha) := \underbrace{11 \cdots 1}_{\alpha_K} 0 \underbrace{11 \cdots 1}_{\alpha_{K-1}} 0 \cdots 0 \underbrace{11 \cdots 1}_{\alpha_1}. \tag{5.11}$$

That is, the binary number generated by piecing together strings of 1's (where the ith string has length α_i) separated by 0's. This ordering satisfies the order property (5.6). More details on the implementation of Algorithm 5.5 can be found in [27] (see also [28]). Here we only want to point out a couple of important points not mentioned in the above reference.

First, during the stochastic simulation in Step 5.2 we need to do $\binom{N+K}{K}RK$ evaluations of Hermite polynomials of order ranging from 0 to N. We apply a type of Horner's algorithm [13] for these evaluations. In particular, it is well-known that the Hermite polynomials satisfy the recurrence relation

$$\begin{aligned} &h_0(x) = 1, \quad h_1(x) = 1, \\ &h_n(x) = xh_{n-1}(x) - (n-1)h_{n-2}(x), \quad (n = 2, 3, \ldots). \end{aligned} \tag{5.12}$$

A straight forward induction argument using (5.12) shows that

$$\begin{aligned} &h_n(x) = x^{n \mathrm{mod} 2}(x^2(x^2(\cdots(x^2 + a^{(n)}_{\lfloor n/2 \rfloor -1})\cdots) + a^{(n)}_1) + a^{(n)}_0), \\ &(n = 2, 3, \ldots), \end{aligned} \tag{5.13}$$

for suitable constants $a^{(n)}_j$ $(j = 0, \ldots, \lfloor n/2 \rfloor - 1)$. Thus, once these constants are known, we can calculate $h := h_n(x)$ stable and efficient with the representation in (5.13). For more information on generation of Hermite polynomials see, for example, [5], or [11] and the references therein.

Second, it is of interest to have a stable method for the evaluation of Hermite functions (2.2). The reader should note that the formula given in (2.2) is not good for numerical evaluation due to the behavior of the Hermite polynomials, leading to loss of accuracy for large $|x|$ and n. From the definition (2.2) and using the relation (5.12) we find

$$\begin{aligned} &\xi_1(x) = \pi^{-1/4} e^{-x^2/2}, \qquad \xi_2(x) = \sqrt{2}\, \xi_1(x), \\ &\xi_n(x) = \sqrt{2/(n-1)}\, x\, \xi_{n-1}(x) - \sqrt{(n-2)/(n-1)}\, \xi_{n-2}(x), \\ &(n = 3, 4, \ldots). \end{aligned} \tag{5.14}$$

This relation allows for a recursive computation of Hermite functions $\xi := \xi_n(x)$ which is fast and stable.

6. The numerical method for the time-dependent case

In this section we describe a numerical method for the time-dependent variational problem (3.10). The work presented here can be found in more detail in [28].

Let $\{V_m : m \in \mathbb{N}\}$ denote the Galerkin approximation of V introduced in Section 4. Let $\Delta t := T/N_T$ ($N_T \in \mathbb{N}$) be a given fixed time-step, set $t_n := n \cdot \Delta t$ ($n = 0, 1, \ldots, N_T$) and let U_m^n denote the approximation of $u(t_n, x)$ in V_m. Then our fully-discrete implicit Galerkin approximation scheme for (3.10) is given by

$$\begin{aligned}
&\text{Find } U_m^n \in V_m (n = 1, \ldots, N_T) \\
&\text{such that for some initial value } U_m^0 \text{ in } V_m \text{ it holds} \\
&\Big(\frac{1}{\Delta t}(U_m^n - U_m^{n-1}), v\Big) + \mathcal{A}(U_m^n, v) = (f(t_n), v), \quad (v \in V_m). \qquad (6.1)
\end{aligned}$$

We have the following theorem from [28].

Theorem 6.1. *Given the same assumptions as in Theorem 3.12 and in addition assume that f is in $C^0(0, T; H)$. Then the problem* (6.1) *has a unique solution $\{U_m^n : n = 1, \ldots, N_T\}$ and it holds*

$$\max_{n \in \{1,\ldots,N_T\}} |U_m^n|^2 + \Delta t \sum_{k=0}^{N_T} \|U_m^n\|^2 \le |U_m^0|^2 + c\Delta t \sum_{k=0}^{N_T} |f(t_n)|^2, \quad (6.2)$$

for some positive constant c independent of m and Δt.

Remark 6.2. The estimate (6.2) is a discrete version of (3.11) and implies that the approximation is stable as $N_T \to \infty$.

The elliptic projection $R : V \to V_m$ is defined as the mapping $V \ni u \mapsto Ru \in V_m$ where Ru is the unique element in V_m such that

$$\mathcal{A}(Ru, v) = \mathcal{A}(u, v), \qquad (v \in V_m). \qquad (6.3)$$

It is a consequence of the Lax-Milgram Theorem that this projection is well-defined. A proof of the following convergence result can be found in [28].

Theorem 6.3. *In addition to the assumptions in Theorem 3.12 we assume the following: We let $k \le 0$, $|k| = q + r$, with $q \ge 0$, $r > r^*$ (with r^* given by* (5.2)*), we assume $U_m^0 = Ru_0$ and $f \in W(0, T; V)$, and we assume that the solution of* (3.10) *satisfies*

$$\begin{aligned}
&u \in C^1([0, T]; V), && (6.4)\\
&D_t u(t) \in \mathcal{V} := (\mathcal{S})_0^{-1,-|k|,1} \cap (\mathcal{S})^{-1,-|k|,2}, \quad (t \in [0, T]). && (6.5)
\end{aligned}$$

Then it holds

$$\begin{aligned}
&\|u(t_n) - U_m^n\|_{-1,-|k|,0}^2 \\
&\leq cB_{N,K}(r)^2(\|u(t_n)\|_{-1,-q,1}^2 + \int_0^T \|D_s u(s)\|_{-1,-q,1}^2 ds) \\
&+ ch^2(\|u(t_n)\|_{-1,-|k|,2}^2 + \int_0^T \|D_s u(s)\|_{-1,-|k|,2}^2 ds) \\
&+ c\Delta t^2(\int_0^{t_n} (\|D_s u(s)\|_{-1,-|k|,1}^2 + \|D_s f(s)\|_{1,|k|,-1}^2) ds),
\end{aligned} \tag{6.6}$$

for some positive constant $c < \infty$ independent of m and Δt.

Theorem 6.3 establishes the convergence of the method: The method converges linearly both in time (Δt) and space (h). This rate is weaker than known results for ordinary partial differential equations (see [6, 26]) where the typical result achieves quadratic convergence in space. Again, due to the form of $B_{N,K}(r)$ in (5.4) it is clear that the convergence in the stochastic dimension, that is in the cutting of the chaos expansion, may be rather slow.

Remark 6.4. The assumption on the starting value U_m^0 could be relaxed, but this would introduce the additional term $c\|U_m^0 - Ru_0\|_{-1,-|k|,0}$ on the right-hand side of the estimate (6.6) [28].

The solution of (6.1) amounts to solving N_T linear systems, each with $(N+K)!/(N!K!)M$ unknown. In order to reduce the work-load, we apply essentially the same approach as in the stationary case. Since U_m^n is in V_m, we may write

$$U_m^n = \sum_{\alpha \in \mathcal{I}_{N,K}} u_{m,\alpha}(t_n, x) H_\alpha, \text{ where} \tag{6.7}$$

$$u_{m,\alpha}(t_n, x) := \sum_{j=1}^M C_{m,\alpha,j}^n \psi_j(x) \in X_h^n(\mathcal{D}),$$

with real constants $C_{m,\alpha,j}^n$ and shape functions ψ_j ($x \in \mathcal{D}, \alpha \in \mathcal{I}_{N,K}, j = 1, \ldots, M, n = 0, \ldots, N_T$). Choosing γ from the ordered set $\mathcal{O}_{N,K}$, setting $v = \psi_i H_\gamma (2\mathbb{N})^{-k\gamma}$, and using (3.4) we find that the vector $C_{m,\gamma}^n = (C_{m,\gamma,j}^n)$ ($j = 1, \ldots, M$) solves the M-dimensional problem

$$(H_0 + B_0) C_{m,\gamma}^n = H_0 C_{m,\gamma}^{n-1} + b_{m,\gamma}^n - \sum_{\alpha \prec \gamma} B_{\gamma - \alpha} C_{m,\alpha}^n, \tag{6.8}$$

with the matrices B_β given as in (5.8), and where

$$\begin{aligned}
&[b_{m,\gamma}^n]_i := (f_\gamma(t_n), \psi_i)_0, \\
&[H_0]_{ij} = \frac{1}{\Delta t}(\psi_j, \psi_i)_0, \quad (i, j = 1, \ldots, M, \gamma \in \mathcal{I}_{N,K}).
\end{aligned} \tag{6.9}$$

This enables us to solve the nth linear problem in (6.1) as a sequence of $(N+K)/(N!K!)$ linear problems (6.8), each having M unknown variables.

Remark 6.5. Note that when solving the systems (5.7) and (6.8) the matrix to be inverted is independent of the multi-index γ (and n). Thus, additional work-load improvements can be achieved if we, for example, do a LU-factorization when solving the first equation, and reuse this factorization when solving the subsequent equations.

The stochastic simulation of the solution $U_m^n(x)$ follows the idea outlined in Section 5. The only difference being that the coefficients in (6.7) now depend on time

Remark 6.6. Similar to the elliptic case there is a relation between the set of variational problems in (3.12) and the system (6.8). A finite element approximation of (3.12) based on the space $X_h^n(\mathcal{D})$ together with an implicit finite difference approximation in time, leads to (6.8).

We summarize our numerical method in algorithmic form. We assume given a non-degenerate triangulation $\mathcal{T}_h$ and a corresponding function space $X_h^n(\mathcal{D})$ with a set of basis-functions $\{\psi_i : i = 1, \ldots, M\}$. Also, we assume given positive integers N, K, N_T and R for the cutting $\mathcal{I}_{N,K}$, the time-discretization $\Delta t = T/N_T$, and the number of simulations we want of the solution, respectively. Moreover, we assume chosen an approximation U_m^0 in V_m for the initial value u_0.

Algorithm 6.7.

(1) Form the set ordered $\mathcal{O}_{N,K}$ and initialize $C_{m,\alpha}^0$ ($\alpha \in \mathcal{O}_{N,K}$)
(2) For each $n = 1, \ldots, N_T$
  (2.1) Set $t_n = n \cdot \Delta t$ and $\gamma = (0, \ldots, 0)$
  (2.2) While $\gamma \in \mathcal{O}_{N,K}$
    (2.2.1) Calculate $[b_\gamma^n]_i := (f_\gamma(t_n), \psi_i)_0$.
    (2.2.2) Find the set $\mathcal{L}_\gamma = \{\alpha \in \mathcal{I}_{N,K} : \alpha \prec \gamma\}$
    (2.2.3) For each $\alpha \in \mathcal{L}_\gamma$
      (2.2.3.1) Calculate the matrices $[B_{\gamma-\alpha}]_{ij} := \mathcal{B}_{\gamma-\alpha}(\psi_j, \psi_i)$
      (2.2.3.2) Update the right-hand side $b_\gamma^n := b_\gamma^n - B_{\gamma-\alpha} C_{m,\alpha}^n$
    (2.2.4) Solve the problem $(H_0 + B_0) C_{m,\gamma}^n = H_0 C_{m,\gamma}^{n-1} + b_\gamma^n$
  (2.3) Find the next multi-index γ (using ordering (5.10)) and go to 2.2.
(3) Create a sequence of RK independent Gaussian variables $\{X_i, i = 1, \ldots, RK\}$
(4) For each $r = 1, \ldots, R$
  (4.1) Set $X^{(r)} := [X_{(r-1)K+j}]$ ($j = 1, \ldots, K$)

(4.2) Form simulation $U_m^{n\ (r)} = \sum_{\alpha\in\mathcal{I}_{N,K}} U_{m,\alpha}^n H_\alpha(X^{(r)})$
(5) Print and visualize the results.

The comments given in Section 5 for Algorithm 5.5 apply also to Algorithm 6.7.

7. The stochastic Poisson equation

In this section we study the Poisson equation driven by smoothed and singular white noise using the method from Section 5. In particular, we investigate numerically the one-dimensional version of the problem

$$\begin{aligned} -\Delta u(x) &= f(x), \quad (x \in \mathcal{D}),\\ u(x) &= 0, \qquad\quad (x \in \partial\mathcal{D}), \end{aligned} \tag{7.1}$$

where the right-hand side equals singular or smoothed white noise (cf. Example 2.5). The stochastic Poisson equation has been studied by many authors, for example [22, 25, 33]. The approach found in [22] is relevant for our point of view. From [22] we have the following result.

Theorem 7.1. *The smoothed and singular Wick-stochastic Poisson equations have unique* $(\mathcal{S})^{-1}$*-valued distribution process solutions, denoted* $u_\phi(x)$ *and* $u(x)$*, respectively. Moreover,*

$$u_\phi(x) = \int_{\mathbb{R}^d} G(x,y)W_\phi(y)dy \ \in L^p(\mu), \quad (p < \infty, x \in \mathcal{D}), \tag{7.2}$$

and

$$u(x) = \sum_{i=1}^{\infty} \int_{\mathbb{R}^d} G(x,y)\eta_i(y)dy\ H_{\epsilon_i} \ \in (\mathcal{S})^*, \quad (x \in \mathcal{D}), \tag{7.3}$$

where G *denotes the classical Green's function of* $\mathcal{D}$ *and* $(\mathcal{S})^*$ *denotes the space of Hida distributions.*

Remark 7.2. Theorem 7.1 provides a simple formulae for the solution of both the smooth and singular white noise case of (7.1). Moreover, (7.2) implies that in the smoothed noise case we have a path-wise (up to sets of measure zero) defined solution for all dimensions $d \in \mathbb{N}$, and from (7.3) we deduce that in the singular case we get a path-wise defined solution for those dimensions where the Greens function $G(x,\cdot)$ is square integrable. It is well-known that $G(x,\cdot)$ is square integrable on the bounded domain $\mathcal{D} \subset \mathbb{R}^d$ for dimension $d \leq 3$.

The weak formulation of (7.1) becomes

$$\text{Find } u \in V \text{ such that } (\nabla u, \nabla v) = (f, v), \qquad (v \in V). \tag{7.4}$$

A term-wise application of Poincaré's inequality shows that the corresponding bilinear form $\mathcal{A}(u,v) := (\nabla u, \nabla v)$ is coercive on V for any parameter k. Thus, there exists a unique solution in V provided the parameter k is chosen small enough to make f an element in V'. It suffices to choose $k < -1$ for the singular white noise case, and $k \leq 0$ for the smoothed white noise case.

Remark 7.3. Note that in this example we get a decoupling of the corresponding sequence of problems in (3.7) and (5.7), which again leads to simplifications in Algorithm 5.5. The decoupling is due to the lack of stochastic behavior in the coefficients of the differential operator used in (7.1). That is, there are no Wick-products mixing the chaos coefficients.

Example 7.4. Our one-dimensional example for the approximation of Poisson equation (7.1) includes two specific cases, the smoothed and singular the white noise case. In both cases use the cutting $(N, K) = (1, 1000)$ and set $\mathcal{D} = [-5, 5]$. We triangulate $\mathcal{D}$ using 800 equal intervals and use Lagrange elements with linear functions. In the smoothed white noise case we use as smoothing function

$$\phi(x) := \frac{2}{\sqrt{\pi}} e^{-4x^2}, \qquad (x \in \mathcal{D}). \tag{7.5}$$

To illustrate the numerical results we use various types of plots. In order to investigate the convergence in the space dimension we plot the maximal absolute error in the chaos coefficients

$$\sup_{\alpha \in \mathcal{I}_{N,K}} \|u_\alpha - u_{m,\alpha}\|_\infty, \tag{7.6}$$

for different number of intervals in the triangulation of $\mathcal{D}$ (cf. Figure 1). To illustrate the convergence in the stochastic dimension we plot simulations of the solution after different number of terms has been included in the chaos expansion in (5.9). Here we use the same path (ω) in the plots, so that we can see the effect of including additional terms in the expansion.

The difference between the plots will give an indication of how fast our expansion converges (cf. Figures 2, 3, 5 and 6). We also make plots of the sup-norm of the chaos coefficients of the solution against the ordered set of multi-indices $\mathcal{O}_{N,K}$ (cf. Figure 7). This type of plots tells us something about the stochastic regularity of the solution. Finally, we try to illustrate the typical behavior of the chaos coefficients by plotting some of them (cf. Figure 4).

We have some comments on the results:

The singular white noise case: The plot in Figure 1 indicates quadratic convergence in space. This is a better result than what we are able to

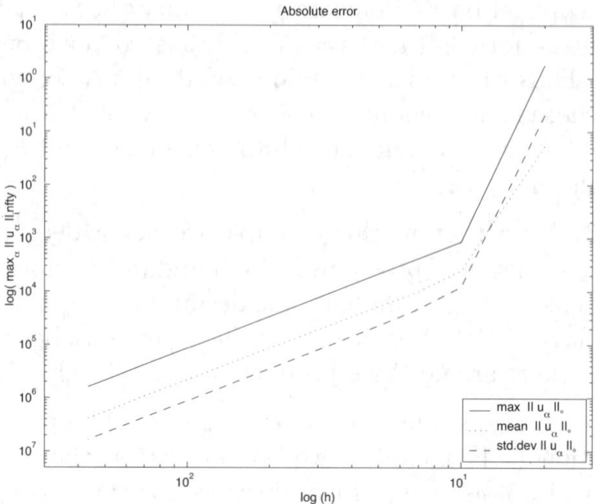

Figure 1. Example 7.4. The figure shows a logarithmic plot of the maximal absolute error in the approximation of the chaos coefficients for the singular white noise case, compared to the chosen grid $h = 10/(M+1)$. We have also included the mean absolute error and the standard deviation. The plot indicates quadratic convergence in space.

prove in general (cf. Theorem 5.1). Because of the decoupling of our problems, we basically get the same rate of convergence as for the deterministic case. In Figure 2 we plot simulations of the solution after 6, 35 and 1000 terms are included in the chaos expansion. Figure 3 shows the corresponding approximations of the right-hand side $W(x)$. The reader should note that in Figure 2 there is almost no observable difference in the plot of the simulation with 35 terms in the expansion and the simulation plot with 1000 terms. This behavior indicates a relatively fast convergence in the stochastic dimension, and it is caused by a decay in the numerical value of the chaos coefficients the solution. In Figure 7 we have plotted the sup-norm of each chaos coefficients and the decay is clearly visible. This decay can be explained as follows: From (7.3) we know that the nth chaos in the solution is given by a pairing with the classical Green's function for the Poisson equation and the nth Hermite function. Typically, for large n, the Hermite functions oscillate around zero, and because of the regular behavior of the Green's function, these oscillations cause cancellation in the integrals in (7.3). This gives the fast decay in the numerical value of the chaos coefficients. In Figure 4 we have plotted some typical chaos coefficients. The oscillations visible here are caused by the oscillations in ξ_n. For higher-dimensional problems, the Greens' function does not posses the regular behavior, and this

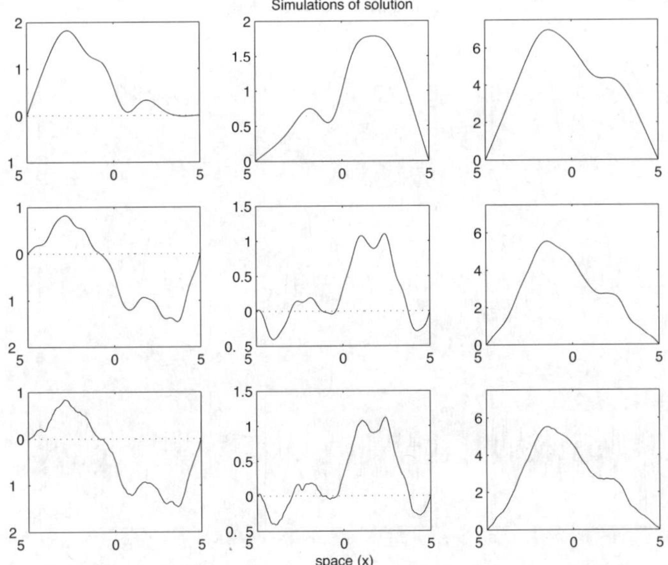

Figure 2. Example 7.4. The figure shows simulations of the solution for the singular white noise case. Each column corresponds to a different simulation, and each row corresponds to the case when 6, 35 and 1000 terms have been included in the chaos expansion, respectively. The dotted line is the averaged solution. Notice the small difference in the plot of the simulation with 35 terms in the expansion and the plot with 1000 terms. This indicates relatively fast convergence in the stochastic dimension. This figure should be compared with Figure 3 where we give the corresponding plots of the right-hand side.

reduces the stochastic regularity of the solution of (7.1) (cf. Theorem 7.1 and Remark 7.2).

The smoothed white noise case: In this case we expect a faster convergence in the stochastic dimension because of the higher stochastic regularity of the solution. This is also what our numerical results indicates. In Figure 5 we see a faster convergence in the stochastic dimension compared to the singular case. Figure 6 shows the corresponding simulations of right-hand side. In Figure 7 we have plotted the sup-norm of the chaos coefficients and we see that they decay much faster for the smooth noise case than for the singular noise case. After about 250 terms the sup-norms are in the order of machine precision. Thus, from our numerical experiments we would expect the solution to have quite regular stochastic behavior. This result is supported by Theorem 7.1.

Remark 7.5. The fast convergence in the stochastic dimension that we can observe in Figure 2 and Figure 5 is related to the fact that the chaos expansion of the solution converges in $L^2(\mu)$ (cf. Theorem 7.1).

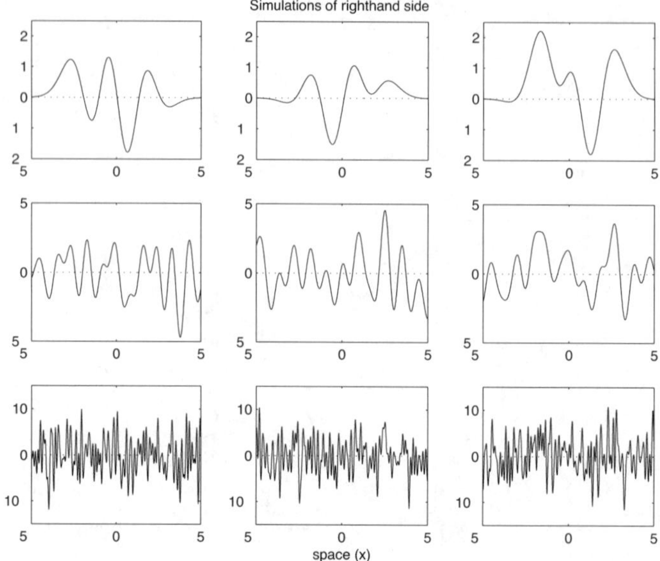

Figure 3. Example 7.4. The figure shows simulations of singular white noise. We have plotted the simulations of the right-hand side corresponding to the plots in Figure 2.

If this was not the case, we would expect quite irregular and diverging behavior of the approximations in this kind of plots. For example, we know that the singular white noise $W(x)$ is a proper element in the space of Hida distributions $(\mathcal{S})^*$. In Figure 3 we have plotted singular white noise $W(x)$ for 6, 35 and 1000 terms included in the chaos expansion (5.9). The plots show dramatic changes in the behavior as we increase the number of terms.

8. The Wick-stochastic pressure equation

In this section we study numerically the following parabolic boundary value problem

$$\begin{aligned} &D_t u(x,t) - \nabla(K(x) \diamond \nabla u(x,t)) = f(x,t), \quad (x \in \mathcal{D}, t \in [0,T]), \\ &u(x,t) = 0, \qquad (x \in \partial\mathcal{D}, t \in [0,T]), \\ &u(x,0) = u_0, \qquad (x \in \mathcal{D}), \end{aligned} \tag{8.1}$$

using the method from Section 6. This equation was introduced in [21] as a stochastic model for the pressure in a fluid flowing through a porous medium, for example, through a sand stone in an oil reservoir. In the model, u denotes the pressure in the fluid, K denotes the permeability

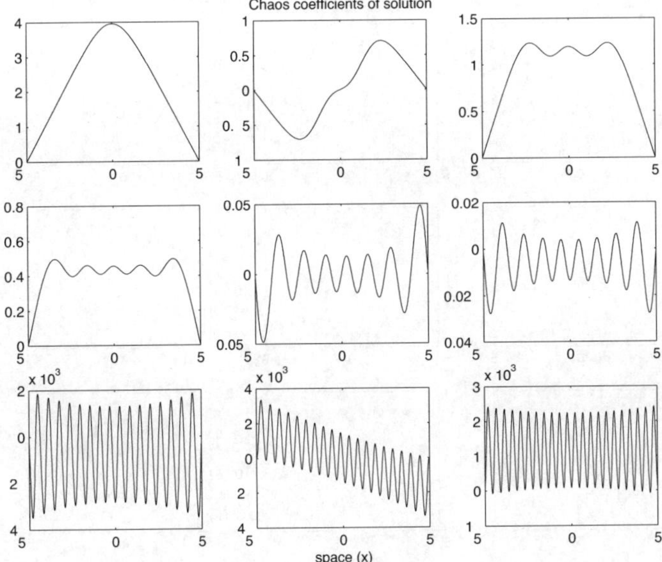

Figure 4. Example 7.4. The figure shows some chaos coefficients for the solution in the singular white noise case. In particular, the plots show the 2, 5, 6, 10, 15, 20, 60, 81 and 100th chaos coefficient of the solution (counting row-wise from left to right). The oscillations are caused by the oscillations in the corresponding Hermite functions (cf. equation (7.3)).

of the porous medium, and f denotes the source rate of the fluid. The stochasticity is usually a result of letting the permeability be some given (positive) random field. This stochastic pressure equation (in particular the stationary version) has been studied extensively, see for example [1, 3, 18, 21, 27, 28, 32, 31].

We let the permeability K be given as the Wick exponential of space white noise

$$K(x) := \exp^{\diamond}(W(x)) = \sum_{\alpha \in \mathcal{I}} \frac{\eta^{\alpha}(x)}{\alpha!} H_{\alpha}, \qquad (x \in \mathcal{D}), \tag{8.2}$$

and investigate the one-dimensional case numerically for different choices of the right hand side f and initial value u_0. Equation (8.1) with permeability given as in (8.2) was studied in [21], where the authors prove existence of a strong solution in $(\mathcal{S})^{-1}$.

The variational formulation of (8.1) with the above choice for the permeability (8.2) becomes

$$\begin{aligned}&\text{Find } u \in W(0,T;V) \text{ such that } u(0) = u_0 \text{ and}\\ &D_t(u(\cdot), v) + (K \diamond D_x u(\cdot), D_x v) = (f(\cdot), v), \qquad (v \in V).\end{aligned} \tag{8.3}$$

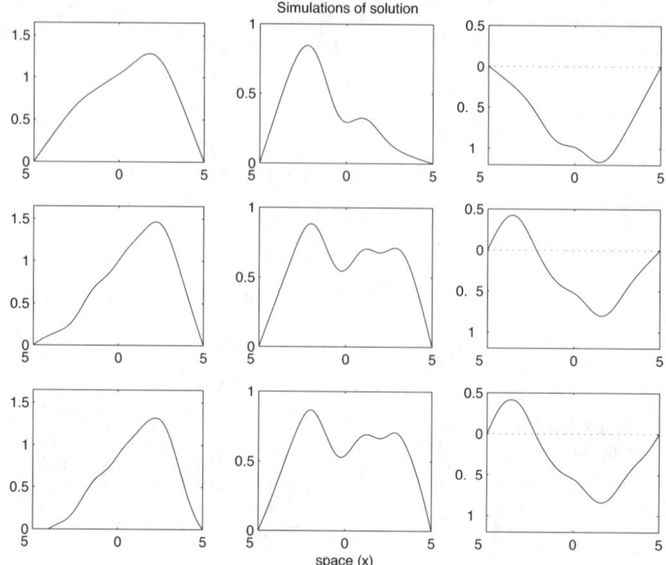

Figure 5. Example 7.4. The figure shows simulations of the solution for the smoothed white noise case. This plot is of the same type as in Figure 2. Each column corresponds to a different simulation of the solution, and each row corresponds to the case when 6, 10 and 1000 terms have been included in the chaos expansion (5.9), respectively. The dotted line is the averaged solution.

Case	**A**	**B**	**C**
Right hand side (f)	1	$\sum_\alpha 1 \cdot H_\alpha$	1
Initial pressure (u_0)	0	0	$\sum_\alpha 1 \cdot H_\alpha$

Table 1. Data used in the numerical simulations in Example 8.1.

Note that $\exp^\diamond(W(x))$ is in $\mathcal{F}_l(\mathcal{D})$ for $l < -1$ (cf. Example 2.12) and since $E[\exp^\diamond(W(x))] = 1$ it follows from Poincaré's inequality that the corresponding bilinear form $\mathcal{B}_0$ from (3.5) is coercive on $H_0^1(\mathcal{D})$. Thus, existence of a unique solution of (8.3) in V follows from Theorem 3.12 provided we choose the parameter k small enough.

Example 8.1. Our numerical example for the approximation of the Wick-stochastic pressure equation (8.1) include three specific cases: Case A, B, and C, corresponding to the different choices for f and u_0 given in Table 1. In all three cases we use $\Delta t = 0.2$, $(N, K, M) = (3, 15, 99)$ and solve on the domain $\mathcal{D} = [-5, 5]$ using a uniform triangulation with Lagrange interval elements.

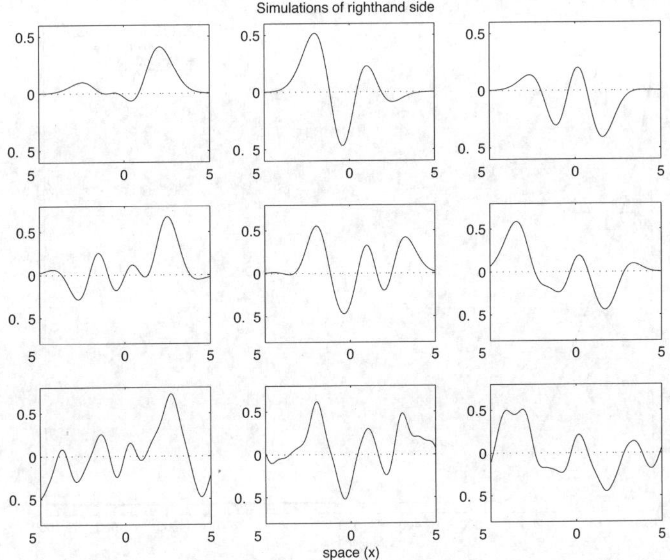

Figure 6. Example 7.4. The figure shows simulations of smoothed white noise. Here we have plotted the simulations of the right-hand side, $W_\phi(x)$, corresponding to the simulations in Figure 5.

We have some comments on the results:

Case A: Here we set $f = 1$ and $u_0 = 0$. Thus, the only source of random behavior is through the permeability. In each simulation the solution from Algorithm 6.7 approaches the corresponding (pathwise) stationary approximation smoothly and relatively fast, and even though we start with non-random initial data the solution will soon contain a random part. This is of course a result of the random permeability. In Figure 8 we se how the sup-norm of the chaos choefficients increase from zero (for all non-zero multi-indices) to some positive value. Also note the stationary nature of the solution, with almost no noticable difference between the third and fourth sub-plot. In Figure 9 we show some simulations of the solution at different times.

Case B: Here we set $f = \sum_{\alpha \in \mathcal{I}} H_\alpha$ and $u_0 = 0$. That is, the right-hand side is singular, and the source of random behavior is through both the right-hand side and the permeability. Again the simulations approach the corresponding (pathwise) stationary approximation smoothly and relatively fast, but now the sup-norm of the chaos choefficients (Figure 10) have greater numerical value and not the decay for higher order multi-indices we see in Figure 8. This indicates an increased singular behavior of the pressure compared to Case A. In Figure 11 we show some

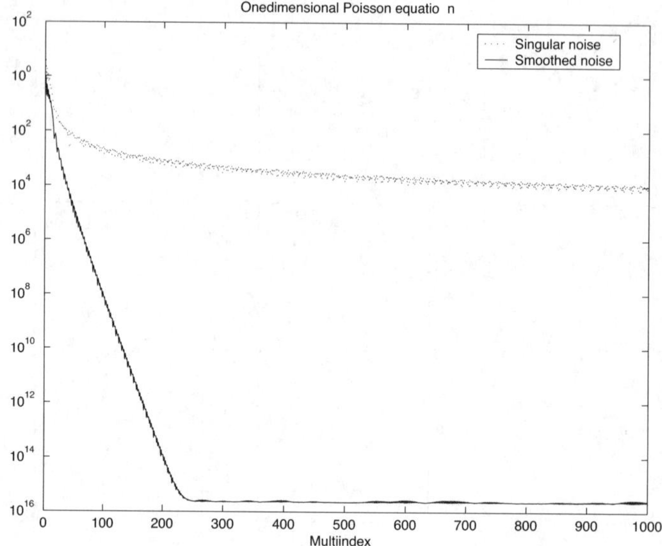

Figure 7. Example 7.4. This plot shows the sup-norm of the chaos coefficients plotted against the order of the multi-indices for both cases. Notice how fast the sup-norm decay for the smoothed white noise case compared with the singular white noise case. This indicates better stochastic regularity for the solution in the smoothed case, compared to the singular case.

simulations of the solution at different times. Also here the increased singular behaviour is evident from the variation in these simulations.

Case C: In this case we set $f = 1$ and $u_0 = \sum_{\alpha \in \mathcal{I}} H_\alpha$. That is, the right-hand side is again non-random but the initial value is singular, and the source of random behavior is through both the initial value and the permeability. If we compare this case to Case A we see from Figure 12 and Figure 13 that the effect of the change in u_0 is a slight increased singular behavior of the pressure. But much less than what we saw in case B. This indicates a smoothing (in the stochastic dimension) by the solution operator; a singular starting value becomes less singular after a short time.

9. The viscous Wick-stochastic Burgers equation

In this section we attempt to apply our numerical method to a non-linear Wick-stochastic partial differential equation. In particular, we consider the one-dimensional viscous Wick-stochastic Burgers' equation

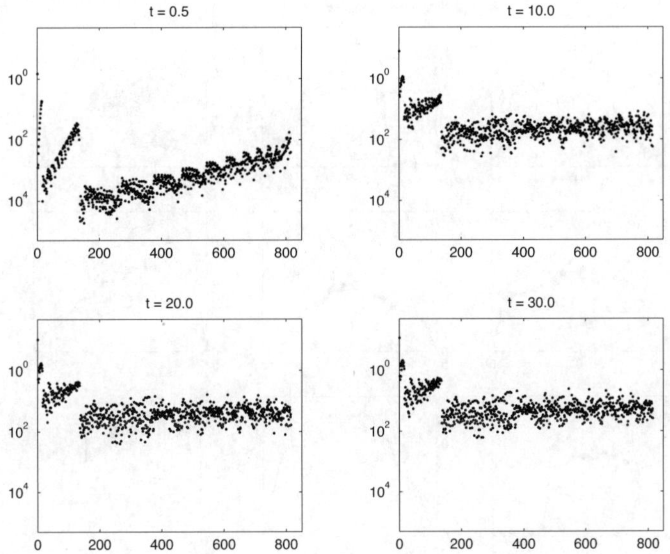

Figure 8. Example 8.1, Case A: These four plots show the sup-norm of the chaos-coefficients plotted against our ordering of the multi-indices for different times. Note how the value for higher order multi-indices increase from zero to a stationary value.

with stochastic source:

$$
\begin{aligned}
&D_t u(x,t) + \lambda u(x,t) \diamond D_x u(x,t) - \nu D_{xx} u(x,t) = f(x,t), \\
&u(x,0) = h(x), \quad (x \in \mathbb{R},\, t \in (0,T]),
\end{aligned} \tag{9.1}
$$

where λ and ν are positive constants and where we allow f and h to be generalized stochastic processes.

The deterministic form of equation (9.1) has been used as a prototype for modeling various non-linear phenomena, and different stochastic versions Burgers' equation has been studied, see for example [15, 20, 19, 22]. A multi-dimensional version of (9.1) was introduced in [20] as a stochastic model for growth on interfaces of solids. It has been proven (see, for example, Theorem 4.5.4 in [22]) that under certain assumptions on the data there exists a unique strong $(\mathcal{S})^{-1}$-valued solution of (9.1). That is, there exists a unique solution in $C^{1,2}((0,T] \times \mathbb{R}; (\mathcal{S})^{-1})$. Moreover, this existence result holds also in the multi-dimensional setting, and the uniqueness holds among all solutions of gradient form (that is, as the gradient of some process in $C^{1,3}((0,T] \times \mathbb{R}; (\mathcal{S})^{-1})$). The proof of these results utilizes a Wick Cole-Hopf transform, which transforms (9.1) into a stochastic heat equation with a random potential. We recommend [20, 22] for more details.

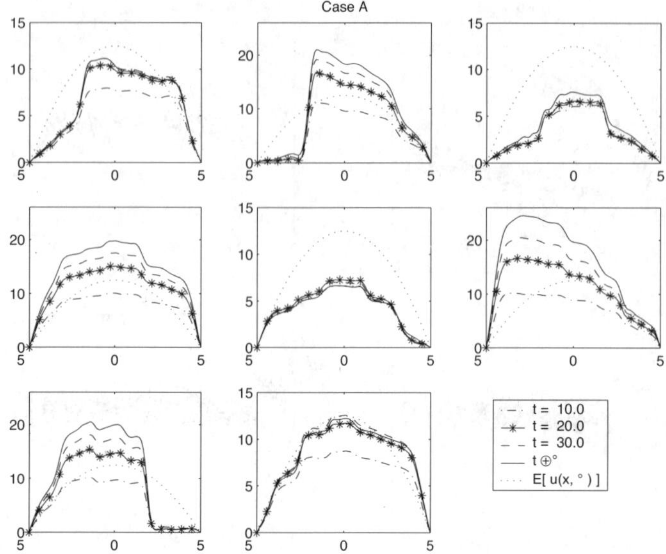

Figure 9. Example 8.1, Case A: The figure displays eight different simulations of the pressure. Each of the sub-plots shows one of the simulations evaluated at the times $t = 10$, $t = 20$, and $t = 30$. In each sub-plot we also plot the stationary solution for this case and the averaged stationary solution.

We want to solve (9.1) numerically. To this end, we formulate the equation in weak form and show that the chaos coefficients of the solution satisfies a sequence of variational problems. We give sufficient conditions for this sequence of problems to have a unique solution. Next, we apply the usual Galerkin procedure and solve numerically, thus obtaining a sequence of approximations for the solution. Here it should be noted that the Galerkin approach has been successfully applied to the deterministic version of Burgers' equation, see for example [24], and because of the close relationship between the stochastic equation and a sequence of deterministic equations, we believe that our numerical method do converge also in this non-linear case, although we have at the present not been able to prove this.

We proceed as in Section 3 with a formal justification of the variational formulation. Let $\mathcal{D}$ denote some bounded open interval in $\mathbb{R}$, and let (as usual) V and H denote the spaces $(\mathcal{S})_0^{-1,k,1,\mathcal{D}}$ and $(\mathcal{S})^{-1,k,0,\mathcal{D}}$, respectively. Assume that the right-hand side f is in $L^2(0,T;H)$ and that the initial value h is in H. Moreover, let u denote a solution of (9.1) and assume that u is in $W(0,T;(\mathcal{S})_0^{-1,q,1,\mathcal{D}})$ with $k+1 < q$. Note that this assumption on u implies that $u(t) \diamond u(t)$ is in H for all t in $[0,T]$

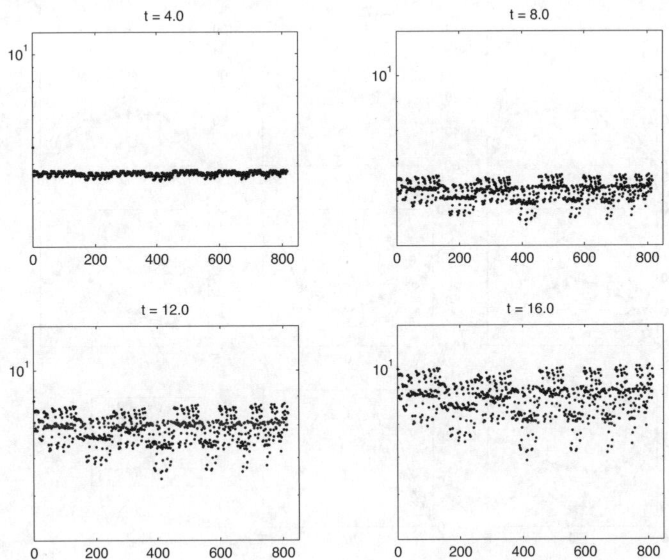

Figure 10. Example 8.1, Case B: The four plots show the sup-norm of the chaos-coefficients plotted against our ordering of the multi-indices for different times. Note how the value for higher order multi-indices increase from zero to a positive stationary value. Comparing with Figure 10 this indicates an increased irregularity if the solution, an effect caused by the added stochastic behavior of the right-hand side.

(cf. Corollary 2.14). Taking the H-inner product of (9.1) with some v in V, and a partial integration, leads to the (non-linear) variational formulation ($v \in V$):

$$\text{Find } u \in W(0,T;(\mathcal{S})_0^{-1,q,1,\mathcal{D}}) \text{ such that } u(0) = h \text{ and} \tag{9.2}$$

$$D_t(u(t),v) - \frac{\lambda}{2}(u(t) \diamond u(t), D_x v) + \nu(D_x u(t), D_x v) = (f(t),v).$$

Remark 9.1. If the right-hand side f is in $C^0([0,T] \times \mathcal{D}; (\mathcal{S})^{-1,k,\mathbb{R}})$, a subspace of $L^2(0,T;H)$, and if the initial-value h is in $C^0(\mathcal{D}; (\mathcal{S})^{-1,k,\mathbb{R}})$, a subspace of H, then the conditions of Theorem 4.5.4 in [22] are satisfied, securing the existence of a unique $(\mathcal{S})^{-1}$-valued strong solution of (9.1).

In the variational equation (9.2) we may choose $v = wH_\gamma(2\mathbb{N})^{-k\gamma}$ with γ from the ordered set $\mathcal{O}$. Then by the definition of the Wick-product it follows that the chaos coefficients of u must satisfy the following sequence

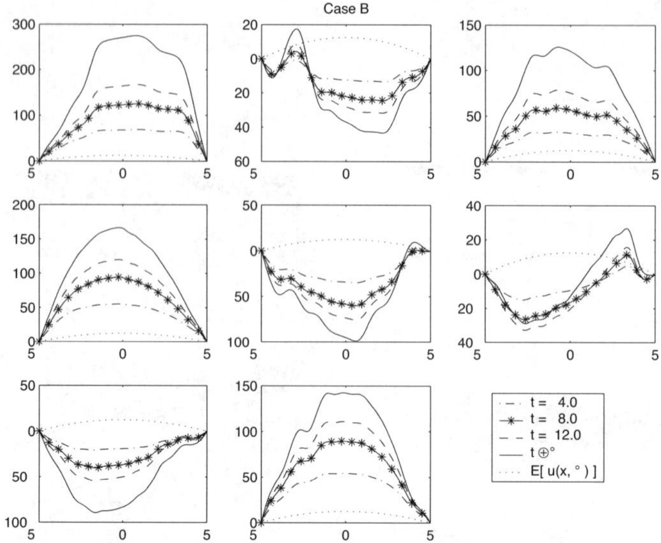

Figure 11. Example 8.1, Case B: The figure displays eight different simulations of the pressure. Each of the sub-plots shows one of the simulations evaluated at the times $t = 4.0$, $t = 8.0$, and $t = 12.0$. In each sub-plot we also plot the stationary solution for this case and the averaged stationary solution.

of variational problems

$$\begin{aligned}
&\text{Find } u_0 \text{ in } W(0,T;H_0^1(\mathcal{D})) \text{ such that} \\
&u_0(0) = h_0 \text{ and } (w \in H_0^1(\mathcal{D})) \qquad (9.3)\\
&(D_t u_0(t), w)_0 - \frac{\lambda}{2}(u_0^2(t), D_x w)_0 + \nu(D_x u_0(t), D_x w)_0 = (f_0(t), w)_0,
\end{aligned}$$

$$\begin{aligned}
&\text{For each } \gamma \text{ in } \mathcal{O}\backslash\{0\} \text{ find } u_\gamma \text{ in } W(0,T;H_0^1(\mathcal{D})) \text{ such that} \\
&u_\gamma(0) = h_\gamma \text{ and } (w \in H_0^1(\mathcal{D})) \\
&(D_t u_\gamma(t), w)_0 - \lambda(u_0(t)u_\gamma(t), D_x w)_0 + \nu(D_x u_\gamma(t), D_x w)_0 \\
&= (f_\gamma(t), w)_0 + \frac{\lambda}{2}\sum_{0\prec\beta\prec\gamma}(u_{\gamma-\beta}(t)u_\beta(t), D_x w)_0. \qquad (9.4)
\end{aligned}$$

Theorem 9.2. *Suppose that f_γ is in $L^2([0,T]\times\mathcal{D})$ and that h_γ is in $L^2(\mathcal{D})$ $(\gamma\in\mathcal{I})$. Moreover, assume that* (9.3) *has a unique bounded solution u_0 in $W(0,T;H_0^1(\mathcal{D}))$. Then there exists a unique set of functions $\{u_\gamma\in W(0,T;H_0^1(\mathcal{D})) : \gamma\in\mathcal{I}\}$ solving* (9.3)–(9.4).

Remark 9.3. Equation (9.3) represents a weak formulation over $H_0^1(\mathcal{D})$ of the deterministic viscous Burgers' equation. This equation is well-

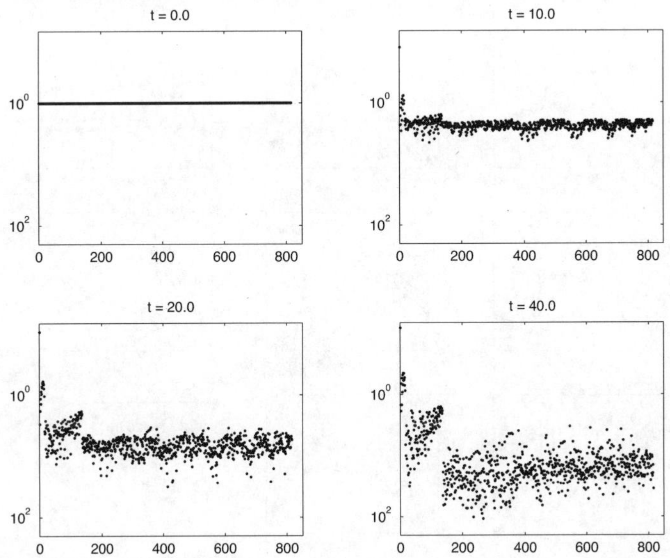

Figure 12. Example 8.1, Case C: The four plots show the sup-norm of the chaos-coefficients plotted against our ordering of the multi-indices for different times. Note how the value for higher order multi-indices decrease from 1 to a smaller stationary value. This indicates a smoothing effect for the solution operator.

studied and conditions securing existence of a unique bounded solution can be found, for example, in the book by Whitham [34].

Proof of Theorem 9.2. The result follows by an induction argument on γ. Define the bilinear form

$$\mathcal{B}_0(t; g, h) := -\lambda(u_0(t)g, D_x h)_0 + \nu(D_x g, D_x h)_0, \qquad (g, h \in H_0^1(\mathcal{D})).$$

Then by our assumptions on u_0 it is easy to show that $\mathcal{B}_0(t; \cdot, \cdot)$ is bounded on $H_0^1(\mathcal{D}))$ (uniformly in $t \in [0, T]$), and from Poincaré's inequality we have $(t \in [0, T], g \in H_0^1(\mathcal{D}))$

$$\mathcal{B}_0(t; g, g) \geq \|D_x g\|_0^2 - \lambda\|u_0(t)\|_\infty \|g\|_0 \|D_x g\|_0 \geq c_1\|g\|_1^2 - c_2\|g\|_0^2,$$

for suitably chosen positive constants $c_1, c_2 < \infty$. Consider the γth equation in (9.4). Our ordering of the set of multi-indices $\mathcal{I}$ implies that the right-hand side in (9.4) is known and defines a continuous linear form on $H_0^1(\mathcal{D})$. Thus, from Theorem XVIII.3.3.2 in [6] there exists a unique solution u_γ in $W(0, T; H_0^1(\mathcal{D}))$. □

Remark 9.4. Theorem 9.2 gives sufficient conditions for the existence of a unique candidate for the solution of (9.2). Moreover, it provides a

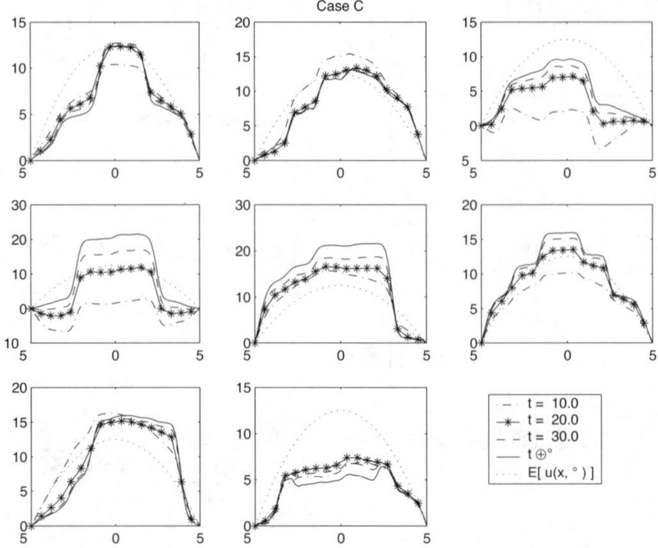

Figure 13. Example 8.1, Case C: The figure displays eight different simulations of the pressure. Each of the sub-plots shows one of the simulations evaluated at the times $t = 10$, $t = 20$, and $t = 30$. In each sub-plot we also plot the stationary solution for this case and the averaged stationary solution.

way of constructing this candidate by solving a sequence of variational problems over $H_0^1(\mathcal{D})$, each giving one chaos coefficient in the solution. In order to establish the existence of a solution of (9.2) it only remains to localize this candidate in a suitable space $W(0, T; (\mathcal{S})_0^{-1,q,1,\mathcal{D}})$ $(k+1 < q)$. In the case when we have a unique $(\mathcal{S})^{-1}$-valued strong solution (cf. Remark 9.1) such a localization is clearly possible.

The usual Galerkin approximation of V together with a finite-difference approximation in time, leads to a fully-discrete implicit Galerkin approximation scheme for (9.2).

$$
\begin{aligned}
&\text{Find } U_m^n \in V_m \ (n = 1, 2, \dots, N_T) \text{ such that} \\
&\text{for some initial value } U_m^0 \text{ in } V_m \text{ it holds } (v \in V_m) \\
&\Big(\frac{1}{\Delta t}(U_m^n - U_m^{n-1}), v\Big)_0 - \frac{\lambda}{2}\Big(U_m^n \diamond U_m^n, D_x v)_0 + \nu(D_x U_m^n, D_x v\Big)_0 \\
&= (f(t_n), v)_0.
\end{aligned} \tag{9.5}
$$

Using the notation from (6.7) and (6.9) we can rewrite (9.5) as the following sequence of M-dimensional algebraic equations. For each $\gamma \in$

$\mathcal{O}_{N,K}$ find $C^n_{m,\gamma}$ $(n = 1, \ldots, N_T)$ such that it holds

$$(H_0 + A_0 + H(C^n_{m,0}))C^n_{m,0} = H_0 C^{n-1}_{m,0} + b^n_{m,0}, \qquad (\gamma = 0), \tag{9.6}$$

$$\begin{aligned}&(H_0 + A_0 + 2H(C^n_{m,0}))C^n_{m,\gamma}\\ &= H_0 C^{n-1}_{m,\gamma} + b^n_{m,\gamma} - \sum_{0\prec\alpha\prec\gamma} H(C^n_{m,\gamma-\alpha})C^n_{m,\alpha}, \qquad (\gamma \neq 0).\end{aligned} \tag{9.7}$$

Here we have defined matrices A_0 and $H(\cdot)$ by

$$[A_0]_{ij} = (D_x\psi_j, D_x\psi_i)_0,$$

$$[H(v)]_{ij} := -\frac{\lambda}{2}\sum_{l=1}^{M}(\psi_j\psi_l, D_x\psi_i)_0 v_l, \quad (i, j = 1, \ldots, M, v \in \mathbb{R}^M),$$

where ψ_j denotes the jth shape function described in Section 4. The system (9.6)–(9.7) corresponds to a classical Galerkin approximation and implicit time-discretization of the system (9.3)–(9.4), and it may be solved numerically following the lines of Algorithm 6.7. The main difference being that the first equation at each time step (that is, the equation for $\gamma = 0$) is non-linear. This first equation can be solved, for example, by applying Newton's method.

Example 9.5. Our numerical example for the approximation of the viscous Burgers' equation (9.1) assumes $\mathcal{D} = [-5, 10]$, $T = 1.75$, and the right-hand side f given by $f(x) = \phi(x)W(x)$, where $W(x)$ denotes singular white noise (cf. Example (2.5)), and where ϕ denotes a smooth function with compact support on $\mathcal{D}$ and with $\phi(x) = 0.1$ for $x \in [-4.9, 9.9]$. Moreover, we set $\lambda = 2$ and assume the initial value h given by

$$h(x) = \begin{cases} 0 & \text{if } x < 0 \text{ or } 7/2 \leq x, \\ 1 & \text{if } 0 \leq x < 2, \\ (7-2x)/3 & \text{if } 2 \leq x < 7/2. \end{cases} \tag{9.8}$$

We investigate two different choices for the diffusion parameter ν. That is, we set $\nu = 0.05$ and $\nu = 1.0$, respectively. In both cases we use the cutting $(N, K) = (1, 400)$, we triangulate $\mathcal{D}$ using 600 equal intervals with Lagrange elements with linear functions (cf. Section 4), and we use the time-step $\Delta t = 0.0125$.

We have some comments on the numerical results:

First, we consider the case $\nu = 0.05$. In Figure 14 we plot three different simulations of the solution, evaluated at the times $t = 0.25$, 1.0 and 1.75. We can make several observations: There is clearly an increase over time of the irregularity of the simulation. We see a waveform moving towards the right, and this wave follows the form of the averaged solution u_0 (marked with a dotted line in the plots). There seems to happen something along the wave-front of u_0 which leads to the spike we see in the first line of plots in the figure. We have done more simulations than what we show here, and in some of them we see a similar extreme value develop along the wave-front. In Figure 15 we plot the 3rd, 15th and 50th chaos coefficients of the solution. For the 3rd order chaos coefficient, we see a spike develop where the wave-front of the averaged solution is located. Moreover, we observe an increase in the numeric value of the chaos coefficients as time advances, and if we study the plots carefully, we see that some of the increase in the numeric value seems linear, while the spike on the wave-front increase faster than linear. The plot of the sup-norms in Figure 16 also indicate an increase in the numerical values of the chaos coefficients as time advances. Note the particular increase for some of the lower order chaos coefficients. This increase in the sup-norm is due to a development of spikes like the one we see in Figure 15. The shape of the chaos coefficients explains the behavior of the simulations in Figure 14.

We will now give a formal explanation for the behavior of the chaos coefficients. First note that the averaged solution u_0 is a weak solution of the viscous Burgers' equation without a source term and with initial value $u_0(0, x) = h(x)$ $(x \in \mathcal{D})$. For $\gamma = \epsilon_k$ $(k = 1, \ldots, K)$ it follows from (9.4) that u_{ϵ_k} is a weak solution of

$$D_t u_{\epsilon_k} + g D_x u_{\epsilon_k} = \nu D_{xx} u + f_{\epsilon_k} - (D_x g) u_{\epsilon_k} \tag{9.9}$$

with initial value $u_{\epsilon_k}(0, x) = 0$ and where $g(x) = \lambda u_0(x)/2$, and $f_{\epsilon_k}(x) = \phi(x)\eta_k(x)$ $(x \in \mathcal{D})$. We see from (9.9) that several mechanisms are causing the observed behavior of the chaos coefficients. First, for those $x \in \mathcal{D}$ where $u_0(x) \approx 0$ the advection term $g D_x u$ has little effect on the solution, while for those $x \in \mathcal{D}$ where $u_0(x) \approx 1$ we get a drift in the positive x-direction in the solution. Such a drift can be observed in Figure 15 for $x \in (0.0, 3.5)$. Moreover, the source term f_{ϵ_k} gives rise to linear growth for those $x \in \mathcal{D}$ where $\eta_k(x) > 0$, and to linear decline where $\eta_k(x) < 0$. The oscillating behavior of the Hermite functions thus explains the oscillations we observe in Figure 15 and also some of the increased irregular behavior of the simulations in Figure 14. The diffusive term $\nu D_{xx} u_{\epsilon_k}$ causes a smearing of the solution, similar to what we observed for the Poisson equation in Section 7, and we believe a higher stochastic regularity of the solution, because more of the irregularities imposed by the source term are smoothed out. Finally, the last term on the right-hand side of (9.9) explains the spikes we observe in the lower order chaos coefficients of u_0. On the wave-front the derivative $D_x u_0$ is negative and has high numeric value. Thus, the term $(D_x g)u_{\epsilon_k}$ causes exponential growth in the solution along the characteristics in this region.

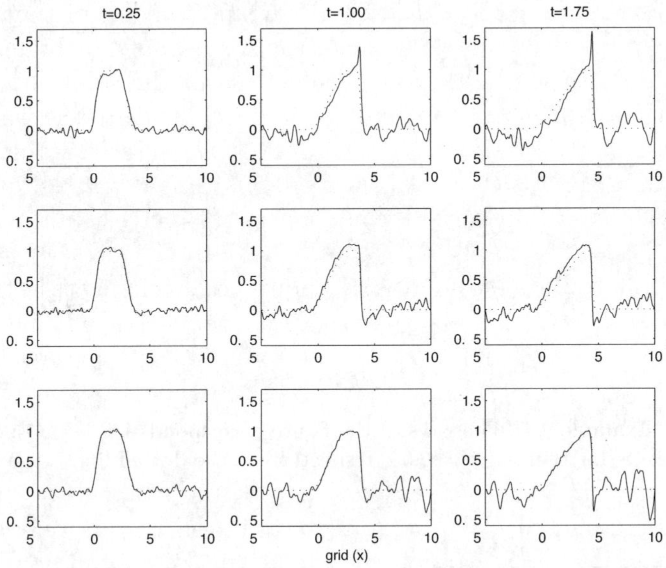

Figure 14. Example 9.5. Each row in the figure corresponds to a simulation of the solution for the case $\nu = 0.05$. The dotted line is the averaged solution.

We now turn to the case $\nu = 1.0$. We have in the Figures 17, 18, and 19 made the same type of plots as for the case $\nu = 0.05$. We see that the simulations in Figure 17 behave much more regularly than what we saw in Figure 14 for the case $\nu = 0.05$. Again, we observe some increase in the irregularity of the simulation as time advances, but this increase is slower than for the case $\nu = 0.05$. Moreover, we do not see the behavior along the wave-front we saw in Figures 14 and 15. The chaos coefficients do not display the spikes along the wave front we saw for the case $\nu = 0.05$ (cf. Figure 18). In the sup-norm plot in Figure 19 we also see less increase in the numerical value of the chaos coefficients, and there are fewer chaos coefficients with higher values for the sup-norm. This behavior can be explained using the same formal argument as above, but now, since $\nu = 1.0$, the diffusive term has more influence on the behavior of the solution. This leads to more smearing of the chaos coefficients and a higher stochastic regularity of the solution. Note that derivative $D_x u_0$ causing the exponential growth on the wave-

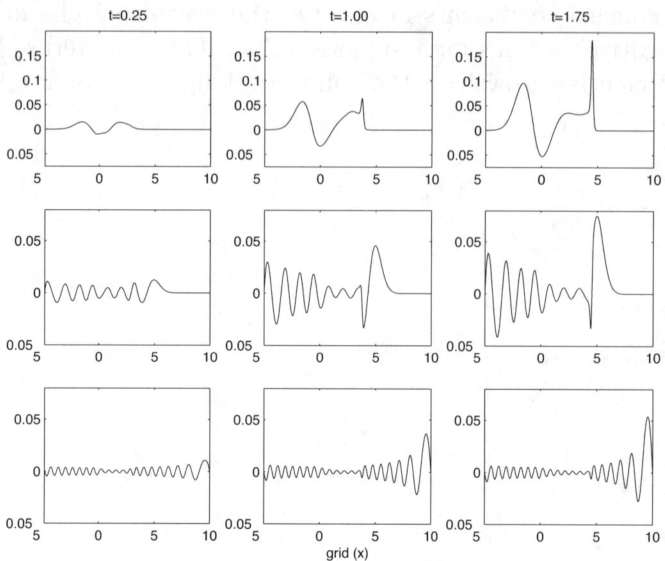

Figure 15. Example 9.5. The rows in the figure corresponds to the 3, 15 and 50th chaos coefficients for the case $\nu = 0.05$, respectively. The dotted line is the averaged solution.

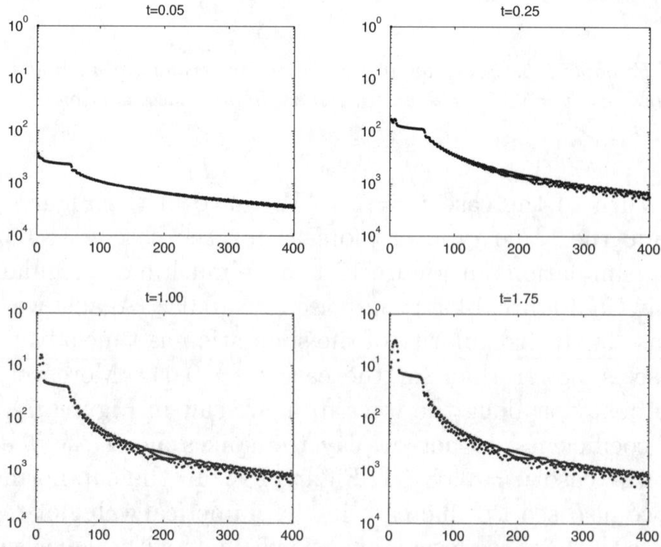

Figure 16. Example 9.5. These four plots show the sup-norm of the chaos coefficients for the case $\nu = 0.05$, plotted against our ordering of the multi-indices. Note how the values increase as time advances, and compare with Figure 19.

front, has much lower numeric value because of smearing of u_0 caused by the diffusion.

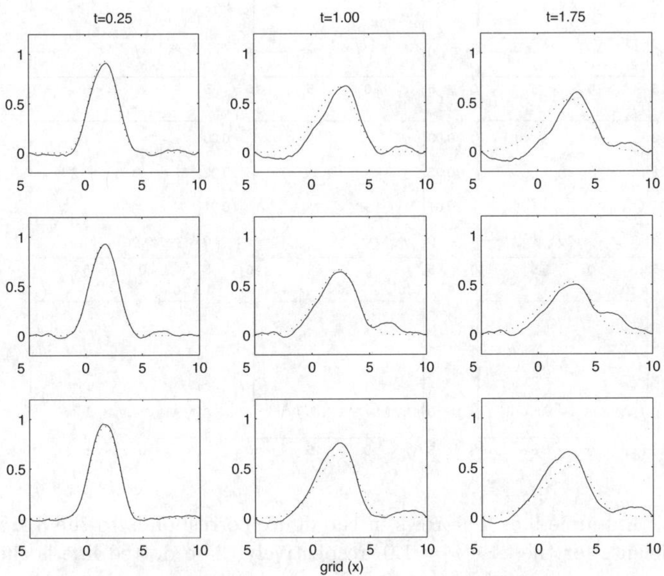

Figure 17. Example 9.5. Each row in the figure corresponds to a simulation of the solution for the case $\nu = 1.0$. The dotted line is the averaged solution.

Remark 9.6. Finally, we would like to point out that it could be interesting to study the numerical solution these type of non-linear stochastic equations further. The understanding of multiplication using the Wick product, makes the numerical approximation of non-linear stochastic PDEs somewhat easier to handle. We typically get the situation we have seen in this example. We can solve a non-linear stochastic problem as a sequence of deterministic problem, with the first equation being a non-linear equation while the rest of the equations in the sequence are linear. In conclusion, the numerical solution of non-linear stochastic PDEs of Wick type is essentially not more difficult than solving the first non-linear deterministic equation.

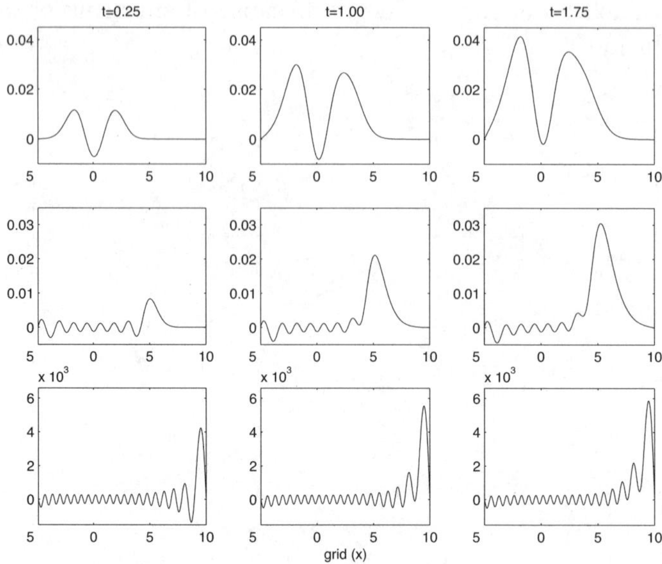

Figure 18. Example 9.5. The rows in the figure corresponds to the 3, 15 and 50th chaos coefficients for the case $\nu = 1.0$, respectively. The dotted line is the averaged solution.

References

[1] Benth, F.E., Gjerde, J.: Numerical Solution of the Pressure Equation for Fluid Flow in a Stochastic Medium, in "Stochastic analysis and related topics, VI (Geilo, 1996)", 175–186, Progr. Probab. **42**, Birkhäuser Verlag, 1998.

[2] Benth, F.E., Gjerde, J.: Convergence Rates for Finite Element Approximations of Stochastic Partial Differential Equations, *Stochastics Stochastics Rep.* **63** (1998) 313–326.

[3] Benth, F.E., Theting, T.G.: Some regularity results for the Stochastic Pressure Equation of Wick-type, Manuscript, University of Mannheim, Mannheim 2000.

[4] Brenner, S.C., Scott, L.R.: The Mathematical Theory of Finite Element Methods, TAMS, **15**, Springer-Verlag, Berlin 1994.

[5] Chorin, A.J.: Hermite expansions in Mmonte Carlo computation, *J. Computational Phys.* **8** (1971) 472–482.

[6] Dautray, R., Lions, J.-L.: *Mathematical Analysis and Numerical Methods for Science and Technology*, Springer-Verlag, Berlin 1988–1993.

[7] Davie, A.M., Gaines, J.G.: Convergence of numerical schemes for the solution of parabolic stochastic partial differential equations, *Math. Comp.* **70** (2001) 121–134.

[8] Filinkov, A., Sorensen, J.: Differential Equations in Spaces of Abstract Stochastic Distributions, *Stochastics Stochastics Rep.* **72** (2002) 129–173.

[9] Filipova, O., Haenel, D.: Lattice Boltzmann Methods - A new tool in CDF, in "Computational Fluid Dynamics for the 21st Century" Eds: Hafez, M. et al.,

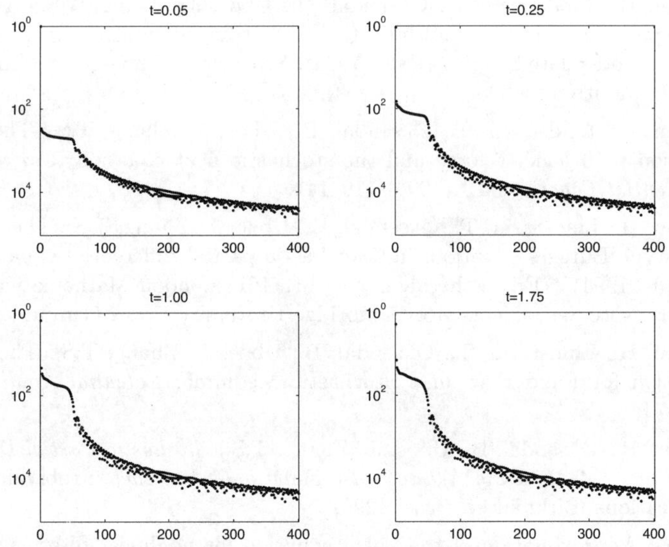

Figure 19. Example 9.5. These four plots show the sup-norm of the chaos coefficients for the case $\nu = 1.0$, plotted against our ordering of the multi-indices. Note how the values increase as time advances and compare with Figure 16 for the case $\nu = 0.05$.

117, Springer Notes on Numerical Fluid Mechanics, Vol. 78, Springer-Verlag, Berlin 2001.

[10] Germani, A., Piccioni, M.: Semidiscretization of stochastic partial differential equations on $\mathbb{R}^d$ by a finite-element technique, *Stochastics* **23** (1998) 131–148.

[11] Ghanem, R.G., Spanos, P.D.: *Stochastic Finite Elements: A Spectral Approach*, Springer-Verlag, Berlin 1991.

[12] Gjerde, J., Holden, H., Øksendal, B., Ubøe, J., Zhang, T.S.: An equation modelling transport of a substance in a stochastic medium, in "Seminar on Stochastic Analysis, Random Fields and Applications (Ascona, 1993)", 123–134, Progr. Probab. **36**, Birkhäuser, Basel 1995.

[13] Golub, G.H., Van Loan, C.F.: *Matrix Computations*, 3rd ed., The Johns Hopkins University Press, Baltimore and London 1996.

[14] Gyöngy, I., Nualart, D.: Implicit scheme for stochastic parabolic partial differential equations driven by space-time white noise, *Potential Anal.* **7** (1997) 725–757.

[15] Gyöngy, I., Nualart, D.: On the stochastic Burgers' equation in the real line, *Ann. Probab.* **27** (1999) 782–802.

[16] Hida, T., Kuo, H., Potthoff, J., Streit, L.: *White Noise, An Infinite Dimensional Calculus*, Mathematics and its Applications, Vol. 253, Kluwer Acad. Pub., Dordrecht, 1993.

[17] Holden, H., Hu, Y.: Finite Difference Approximation of The Pressure Equation for Fluid Flow in a Stochastic Medium – A Probabilistic Approach, *Comm. Partial Differential Equations* **21** (1996) 1367–1388.

[18] Holden, H., Lindstrøm, T., Øksendal, B., Ubøe, J., Zhang, T.S.: A comparison experiment for Wick multiplication and ordinary multiplication, Stochastic analysis and related topics (Oslo, 1992). Stochastics Monogr., **8**, Gordon and Breach, Montreux, 1993.

[19] Holden, H., Lindstrøm, T., Øksendal, B., Ubøe, J., Zhang, T.S.: The Burgers equation with a noisy force and the stochastic heat equation, *Comm. Partial Differential Equations* **19** (1994) 119–141.

[20] Holden, H., Lindstrøm, T., Øksendal, B., Ubøe, J., Zhang, T.S.: The stochastic Wick-type Burgers equation, in "Stochastic partial differential equations (Edinburgh, 1994)" (Ed.: Etheridge, A.), 141–161, London Mathematical Society Lecture Note Series, Vol. 216, Cambridge University Press, Cambridge 1995.

[21] Holden, H., Lindstrøm, T., Øksendal, B., Ubøe, J., Zhang, T.S.: The Pressure Equation for Fluid Flow in a Stochastic Medium", *Potential Anal.* **4** (1995) 655–674.

[22] Holden, H., Øksendal, B., Ubøe, J., Zhang, T.S.: *Stochastic Partial Differential Equations – A Modeling, White Noise Functional Approach*, Probability and its Applications, Birkhäuser, Basel 1996.

[23] Itô, K.: Approximation of the Zakai equation for nonlinear filtering, *SIAM J. Control Optimization* **34** (1996) 620–634.

[24] Johnson, C., Szepessy, A.: On the convergence of a finite element method for a nonlinear hyperbolic conservation law, *Math. Comp.* **49** (1987) 427–444.

[25] Lindstrøm, T., Øksendal, B., Ubøe, J., Zhang, T.S.: Stability properties of stochastic partial differential equations, *Stochastic Anal. Appl.* **13** (1995) 177–204.

[26] Quarteroni, A., Valli, A.: *Numerical Approximation of Partial Differential Equations*, Springer Series in Computational Mathematics, Vol. 23, Springer-Verlag, Berlin 1994.

[27] Theting, T.G.: Solving Wick-stochastic boundary value problems using a finite element method, *Stochastics Stochastics Rep.* **70** (2000) 241–270.

[28] Theting, T.G.: Solving Parabolic Wick-stochastic Boundary Value Problems Using a Finite Element Method, Preprint, Norwegian University of Technology and Science, 2001.

[29] Våge, G.: A General existence and uniqueness theorem for Wick-sdes in $\mathcal{S}^n_{-1,k}$, *Stochastics Stochastics Rep.* **58** (1996) 259–284.

[30] Våge, G.: Stochastic differential equations and Kondratiev Spaces, PhD thesis, The Norwegian Institute of Technology, 1995.

[31] Våge, G.: Hilbert space methods applied to elliptic stochastic partial differential equations, in "Stochastic analysis and related topics, V (Silviri, 1994)", Progr. Probab. **32**, 281–294, Birkhäuser Verlag, Basel 1998.

[32] Våge, G.: Variational methods for PDEs applied to stochastic partial differential equations, *Math. Scand.* **82** (1998) 113–137.

[33] Walsh, J.B.: An introduction to stochastic partial differential equations, in "Ècole d'été de probabilités de Saint-Flour, XIV—1984", 265–439, Lecture Notes in Mathematics **1180**, Springer-Verlag, Berlin 1986.

[34] Whitham, G.B.: *Linear and nonlinear waves*, Wiley, New York 1974.

[35] Yoo, H.: Semi-discretization of stochastic partial differential equations on $\mathbb{R}^1$ by a finite-difference method, *Math. Comp.* **69** (2000) 653–666.

[36] Yosida, K.: *Functional Analysis*, Classics in Mathematics, Springer-Verlag, Berlin 1995.